STUDIES IN INTERFACE SCIENCE

Organized Monolayers and Assemblies: Structure, Processes and Function

STUDIES IN INTERFACE SCIENCE

SERIES EDITORS
D. Möbius and R. Miller

Organized Monolayers and Assemblies: Structure, Processes and Function

Edited by

D. Möbius

Max-Planck-Institut für Biophysikalische Chemie,
P.O. Box 2841, Göttingen, Germany

R. Miller

Max-Planck-Institut für Kolloid- und Grenzflächenforschung,
Am Mühlenberg 1, Potsdam/Golm, Germany

2002
ELSEVIER
Amsterdam – Boston – London – New York – Oxford – Paris – Tokyo – San Diego
San Francisco – Singapore – Sydney

ELSEVIER SCIENCE B.V.
Sara Burgerhartstraat 25
P.O. Box 211, 1000 AE Amsterdam, The Netherlands

First edition 2001

Library of Congress Cataloging in Publication Data
A catalog record from the Library of Congress has been applied for.

ISBN: 0 444 51256 X
ISSN: 1383 7303

♾ The paper used in this publication meets the requirements of ANSI/NISO Z39.48-1992 (Permanence of Paper).
Printed in The Netherlands.

PREFACE

The gas-water interface provides excellent possibilities to organise appropriate molecules in ultrathin films that are called monomolecular layers (monolayers or Langmuir and Gibbs films, respectively), since generally these films are only one molecule thick. Different to ultrathin films formed by surface chemical reaction on solid surfaces (self-assembly), the intermolecular interactions and resulting order may be controlled externally in the monolayers by variation of relevant parameters like the area per molecule or the lateral pressure. Therefore, these ultrathin films are particularly suited as model systems for the investigation of interfacial phenomena as well as membrane processes.

The various contributions in this monograph range from reviews of fundamental studies of long range order e.g. of the hydrocarbon chains of simple insoluble amphiphilic molecules to reports and discussion of potential applications in sensors or opto-electronic devices. Monolayers of long chain fatty acids or alcohols exhibit a rich phase diagram as detected by optical techniques and X-ray diffraction. The incorporation of functional molecules like dyes or components of molecular recognition systems has opened a wide field of constructing molecular machines, i.e. devices on the nanometer scale. A large variety of analytical tools including different spectroscopic methods has been used to collect sufficient information for the evaluation of molecular orientation and association and ordering of large aggregates as well as understanding the underlying processes. Examples are the interplay between insoluble and soluble amphiphiles at the air-water interface, the influence of monolayer rheology on aggregate flow orientation, or the effects of trigger molecules and subtle variations of the molecular structure on orientation and association.

The monolayers organised at the air-water interface may be transferred to solid substrates, and molecular architectures may be constructed by sequential transfer. For possible applications, however, the ultrathin organic layers formed by chemisorption and/or by molecular recognition with strong interaction of the components as in the system biotin/streptavidin may be more stable and easier to build. In the last two chapters of this monograph the potential of such ultrathin organic films for practical application is discussed providing a projection of the field to the near future.

July 2002 D. Möbius and R. Miller

Contents

CHAPTER 5
ENZYMATIC REACTIONS AT INTERFACES **207**

Marie Hélèn Ropert, Gerald Brezesinski and Helmuth Möhwald

CHAPTER 6
ELECTRON SPIN RESONANCE SPECTROSCOPY OF LANGMUIR-BLODGETT FILMS CONTAINING FUNCTIONAL MOLECULES **247**

Shin-ichi Kuroda

CHAPTER 1
ORDER IN LANGMUIR MONOLAYERS AND IN THE AQUEOUS SUBPHASE

Pulak Dutta

Dept. of Physics & Astronomy, Northwestern University, Evanston, IL 60208, USA

A variety of recent experiments have revealed a wide range of phases and phase transitions in simple amphiphilic monolayers on the surface of water. This paper describes some of the ordering phenomena revealed by X-ray diffraction. Recent experiments probing order in the underlying subphase are also described.

1. INTRODUCTION

Floating monolayers of insoluble amphiphilic molecules---Langmuir films---are deceptively easy to make and to manipulate. A grainy back-and-white 1940's newsreel from General Electric shows Irving Langmuir casually putting these monolayers through their paces using tools available in every kitchen. (The monolayers were made 'visible' by the use of a macroscopic floating film of oil that is displaced by the amphiphiles.) It can be a shock to realize that, in this movie, Langmuir is playing with materials at the molecular scale---he is manipulating molecules.

In other words, it's not necessarily difficult or expensive to do nanoscience. Until recently, however, a rather basic question about Langmuir monolayers remained: exactly what are the molecules doing when we manipulate the monolayer (e.g. change its pressure, density or temperature)? Hypotheses had been made and some were generally accepted in the field, but they had not been proved. It is reasonable that at the highest possible areal densities, the molecules are oriented so as to occupy the smallest area (simple linear-chain molecules would be oriented vertically). But how are they packed within the monolayer plane, and what happens as the temperature and pressure change? These monolayers are named after Langmuir because he was the first to see isotherm discontinuities indicating phase transitions, but pressure and density are macroscopic averages and do not tell us what is happening to the molecules at these phase transitions. This was not known until new experimental techniques (including X-ray diffraction and a variety of microscopies) were applied to this system starting in the 1980's and especially in the 1990's.

It is probably safe to say that, of the various techniques, synchrotron X–ray diffraction has had the most impact in this area. In general one should not overestimate the power of X-ray scattering: it does not generate real-space pictures, and it is not useful in disordered systems such as monolayers on very rough surfaces or monolayers with no lateral order or tilt order. But in the case of Langmuir monolayers of simple, rod-like molecules (e.g. saturated fatty acids and alcohols), X-ray scattering has provided important and vivid new information that has explicitly contradicted the conventional wisdom. Many phases of these monolayers turned out not to be liquids ('liquid expanded', 'liquid condensed', 'superliquid' etc); rather, they are mesophases with some degree of positional order and are therefore characterizable by X-ray diffraction.

A 1999 review [1] describes the general features of the fatty acid/alcohol Langmuir monolayer phase diagram in some detail, and contains much more information than can be included here. The following section contains a brief overview of X-ray results. In subsequent sections some newer information is surveyed that was not available at the time Ref. 1 was published, regarding the ordering of the molecular backbones and positional order in the underlying aqueous subphase.

2. THE PHASE DIAGRAM OF FATTY ACID AND ALCOHOL LANGMUIR MONOLAYERS

In 1945, Ställberg-Stenhagen and Stenhagen [2] published an unexpectedly complex pressure-temperature phase diagram for fatty acid monolayers, containing a variety of phase boundaries and not explainable by the popular solid/ liquid expanded/ liquid condensed hierarchy of phases. These results were considered dubious and widely ignored; most studies continued to be performed only at room temperature. The 1966 book by Gaines [3], which was the bible of this field through the 1980's, says that "In view of the experimental difficulties attendant on the demonstration of these effects, it seems that further study is needed before any of them can be considered well understood. No detailed discussion of them is given here..."

When such further studies were finally conducted [4], it was found that the phase diagram of ref. 2 was largely accurate. A 'modern' phase diagram is shown in Fig.1; the labels used for the phases have historical origins and don't really mean anything. All phase boundaries except the Ov-L_2 boundary can be seen as discontinuities in isotherms. The phase diagram is applicable to all insoluble fatty acids irrespective of chain length if the temperature range is appropriately shifted; the pressures change slightly as well, but the topology is not changed. The L_2 phase is not seen in saturated fatty alcohols.

X-ray diffraction can also be used to locate the phase boundaries, but it is much easier to do this with isotherms, Brewster angle microscopy, etc. Once the boundaries are located, and the next step is to identify what these phases and phase transitions are, that is when X-rays become invaluable. An explanation of how this technique is used to determine not just the lattice but also the tilt magnitude and direction is given in Ref. 1. The schematic diagrams in Fig. 1 show the structures thus determined. There are

(a) two phases with distorted hexagonal (DH) i.e. centered rectangular lattices and vertical molecules: CS and S;

(b) one phase with a hexagonal lattice and vertical molecules (LS);

(c) two phases with DH lattices and chains tilted towards a near neighbor (L_2, L_2");

(d) two phases with DH lattices and tilts towards a next-nearest neighbor (L_2', Ov).

All these phases have at least medium-range order. The high-pressure low-temperature CS phase appears to be long-range-ordered. The L_2 phase is anisotropically ordered (better-ordered in one direction than in the direction normal to it). At higher temperatures there is finally a liquid phase (not shown). Another phase not shown in Fig. 1 is a thin sliver of the 'I' phase between the L_2 and L_2' phases, in which the tilt direction is intermediate between the near-neighbor and next-nearest-neighhbor directions. A simple Landau theory [5] perfectly reproduces the details of the phase transition between the L_2 and I phases (first order) and between the I and L_2' phases (continuous).

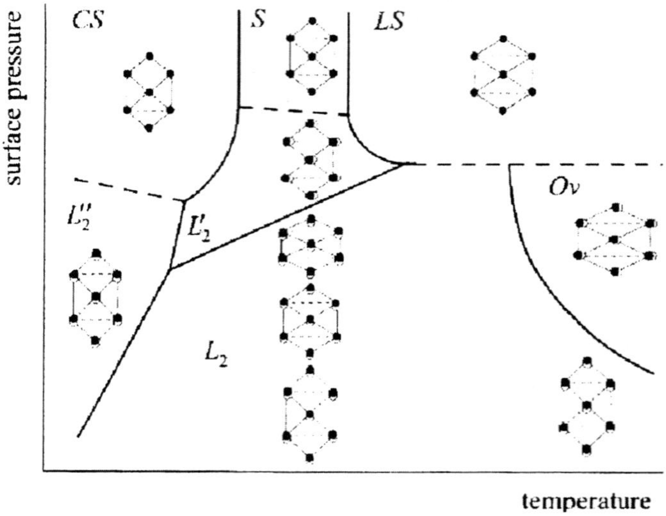

Fig.1 Saturated fatty acid monolayer phase diagram, schematically showing the lattice distortions and tilt directions determined by X-ray diffraction data. [From Ref. 1.]

The story of the recently identified Ov phase illustrates the strengths and weaknesses of the various experimental techniques and the advantages of bringing multiple techniques to bear on this problem. The L_2 /Ov phase boundary is not seen in isotherms; in other words, there is no discontinuity in the density and in that narrow sense there is no phase transition at all. This phase boundary was located by Overbeck and Möbius [6] using Brewster Angle Microscopy: they saw changes in texture when crossing specific points in the isotherm. X-ray scattering had never been used in this region to look for phase boundaries, since this is time-consuming and there was no reason to explore this area. Once the mysterious phase boundary was detected, however, we rather quickly established [7] using X-rays that there was a distinct change in the diffraction pattern upon crossing the phase boundary (Fig. 2). The tilt direction changes, from towards a nearest-neighbor (L_2 phase) below the phase boundary to towards a next-nearest-neighbor (Ov phase) above the phase boundary. We also discovered why isotherm studies had seen nothing: the tilt angle and the packing of the tilted molecules remain unchanged even though the tilt direction (relative to the lattice) changes. Thus there is no change in the areal density, but there is a change of symmetry. The phase transition is first-order even though there is no flat coexistence region in the isotherm.

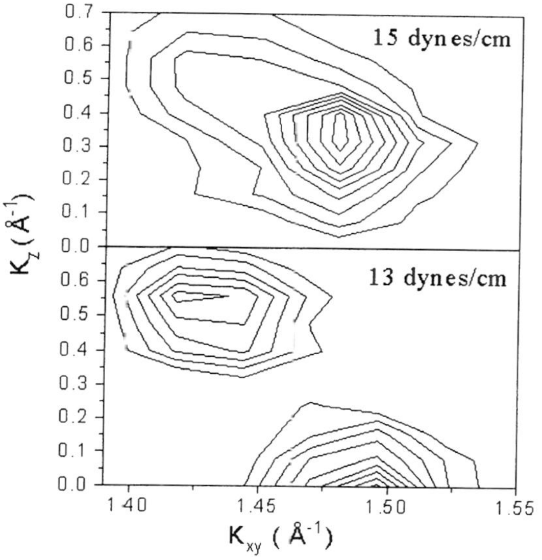

Fig. 2: Sample X-ray diffraction data from C_{19} acid monolayers at 30°C, just below the transition to the Ov phase (13 mN/m) and just above (15 mN/m). Data are plotted as K_p/K_z intensity contours; each contour represents an equal intensity step (arbitrary units) above the contour immediately outside it. [From Ref. 7]

3. BACKBONE ORDERING IN FATTY ACID MONOLAYERS

The various phases listed above differ in the lattice symmetry; tilt or lack thereof; tilt direction; and range of positional order. However, there is one more parameter. The alkane chains are not cylinders with circular cross-sections; to fully specify how the molecules are arranged, one needs to also specify the orientation of the molecule about its long axis. Since the carbon atoms in the alkane tail define a plane, specifying the orientation of this backbone plane is a convenient way to specify the orientation of the entire molecule.

In single crystals where a large number of diffraction peaks can be observed, it is possible to deduce, from the peak intensities, the molecular form factor and thus the orientation relative to the lattice. In Langmuir monolayers, even the second-order diffraction peaks are sometimes hard to see, and most structures are derived (or rather surmised) from the first-order peak positions. Recently [8], however, we obtained direct X-ray evidence of backbone order. Fig. 2 shows diffraction peaks for the CS phase indexed in terms of a rectangular non-primitive unit cell. We see not only the known first-order (11) and (20) peaks [9], and the known second-order (02) and (31) peaks [10], but a (21) peak. Now the rectangular unit cell contains two molecules, and if these molecules are identical, the (21) peak is forbidden. This immediately tells us that there are two inequivalent sites. Any differences between inequivalent sites will cause the (21) peak to be visible, but they will in general also cause the (01) peak to be visible; we do not see the (01) peak. This argues strongly for a herringbone arrangement of molecular cross-sections, just like some phases of paraffins. As can be seen from the inset to Fig. 2, when there is herringbone order the backbones at the two inequivalent sites make equal (although opposite) angles relative to the (01) direction; therefore the molecules at the two sites have identical structure factors for the (01) peak and this peak remains forbidden. However, relative to the (21) direction, the backbones at the two sites are oriented very differently (again, see inset); thus the two molecular structure factors are not the same and the (21) peak is not forbidden.

Similar results are obtained for the tilted L_2'' phase; these data are not shown here but can be found in ref. 8. We do not see 'forbidden' peaks in any of the other phases, which means they can have at most short-range backbone order.

4. ORDER IN THE AQUEOUS SUBPHASE

When polyvalent metal ions are added to the subphase, the monolayer generally becomes much better ordered [11], changes its viscoelastic response [12], and transfers more easily to a solid substrate [13]. As always, a change in properties indicates a change in structure. As we shall see below, what changes is that the underlying aqueous solution becomes ordered. It is tempting to think of this as an ordering or the metal ions, since their presence is necessary to see this effect, but that may not be true.

Some years ago, X-ray diffraction was used by another group to detect the formation of a monolayer superlattice in the subphase when Cd^{2+} ions were present [14]. The unit cell was mysteriously large (a 2x3 supercell of the Langmuir monolayer unit cell). We have confirmed the observation in our own experiments, and we have also observed superlattices in the presence of Mn^{2+}, Mg^{2+} and Pb^{2+}. The presence of each of these ions results in a different subphase lattice.

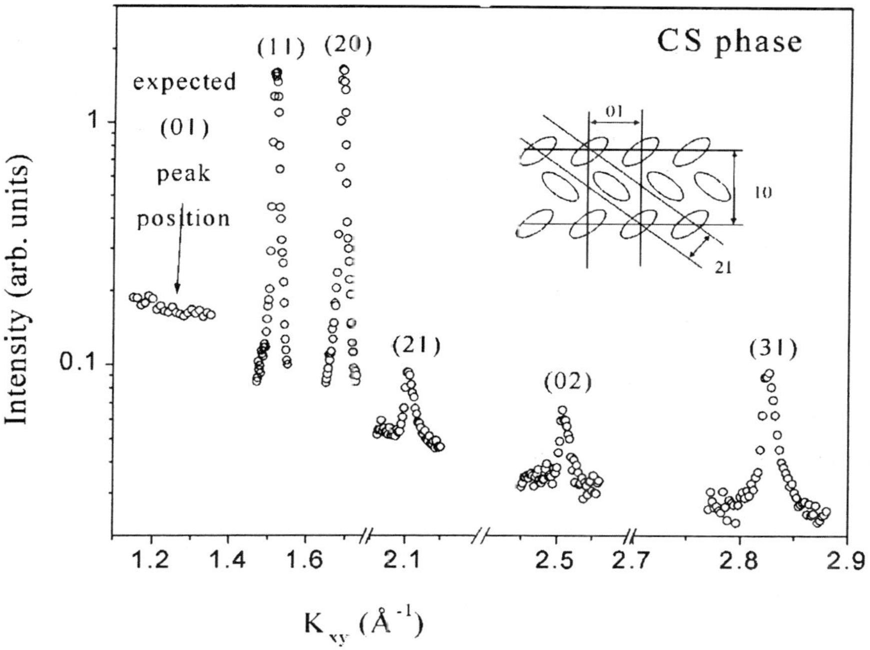

Fig. 3 Sample x-ray data (in the horizontal plane, i.e. at $K_z \approx 0$) from a heneicosanoic (C_{21}) acid monolayer in the CS phase (T=7°C, π=30 mN/m). Peaks are indexed according to a centered orthorhombic lattice with two molecules per unit cell. The inset shows a centered orthorhombic lattice with a herringbone arrangement of the molecular backbones. [From Ref. 8]

With Mn^{2+} we see a 1x2 superlattice of the organic structure; with Mg we see a 2x2 superlattice [15]. Perhaps the strangest subphase structure is seen in the presence of Pb^{2+}, and we will therefore show the Pb^{2+} data [16] here as an example. Fig. 4 shows the large number of peaks seen. These can all be indexed, and the resulting unit cell is rotated with respect to the organic lattice and is very large (bottom of Fig. 4). Its area is 14 times the organic unit cell, i.e. nearly 280Å^2. While the diffraction data are unambiguous, they make it entirely impossible to believe that what we are seeing are lattices of isolated ions.

We therefore sought to test for the presence of lead within the superlattice. If the superlattice peaks were solely due to lead ions, the contribution to the intensity of each weak peak should be proportional to the square of the lead form factor. Tuning the x-ray energy from 12.885KeV (below the L_3 absorption edge of the lead atom) to 13.100KeV (slightly above the edge) should decrease the absolute intensity of a diffraction peak due to scattering from lead atoms by ~20%.

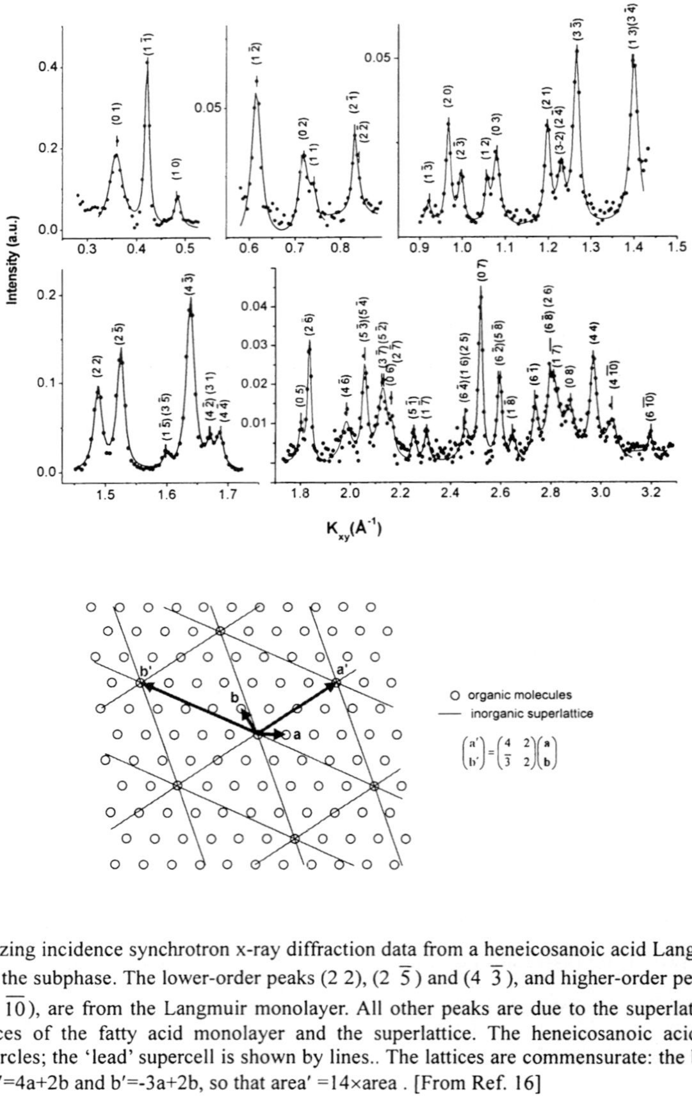

Fig. 4: TOP: Grazing incidence synchrotron x-ray diffraction data from a heneicosanoic acid Langmuir monolayer with lead ions in the subphase. The lower-order peaks (2 2), (2 $\bar{5}$) and (4 $\bar{3}$), and higher-order peaks (0 7), (6 $\bar{1}$), (6 $\bar{8}$), (4, 4), (4 $\overline{10}$), are from the Langmuir monolayer. All other peaks are due to the superlattice. BOTTOM: Real space lattices of the fatty acid monolayer and the superlattice. The heneicosanoic acid molecules are represented by circles; the 'lead' supercell is shown by lines.. The lattices are commensurate: the basis vectors are related through a′=4a+2b and b′=-3a+2b, so that area′ =14×area . [From Ref. 16]

We have measured the intensities of 13 weak peaks with K_{xy} between 0.6 Å^{-1} and 1.4 Å^{-1} at these two energies (data for some representative superlattice peaks, and also some organic peaks, are shown in Fig. 5). Within an experimental error of ~5%, we observed no change for any of these peaks.

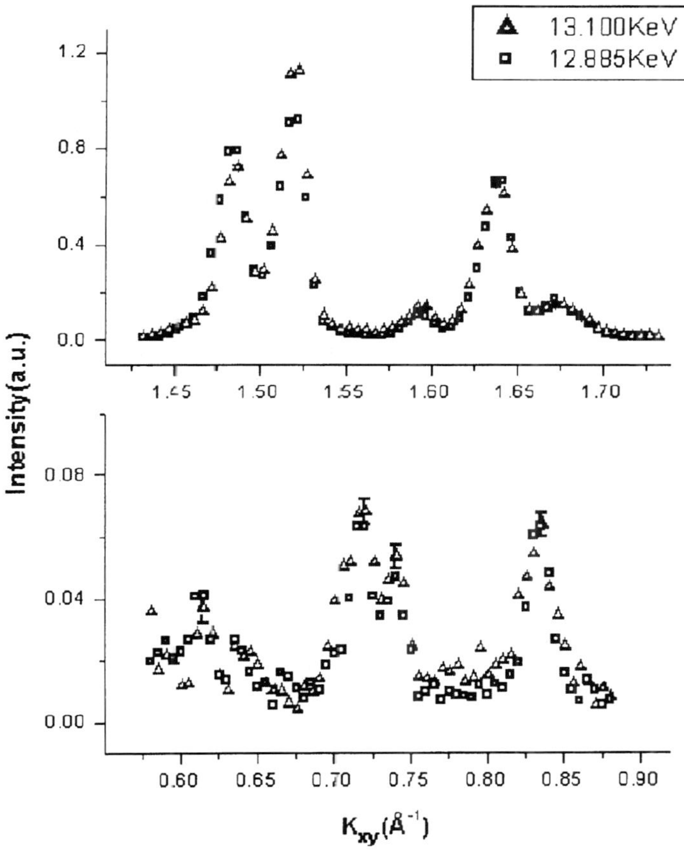

Fig. 5 In-plane diffraction scan with x-ray energy below and slightly above the L_3 absorption edge of lead. (a) Three peaks from the organic monolayer plus weak superlattice peaks. (b) Four representative superlattice peaks. We also looked at 9 other superlattice peaks not shown here, with the same results. If the superlattice consisted of lead atoms only, a 20% change in the peak intensities would be expected. There was no observable change in intensity for any of the peaks studied. [From ref. 16]

If lead does not contribute measurably to the diffraction peak intensity, scattering from the superstructure must be dominated by other atoms. In principle, superlattice reflections can arise from a periodic superstructure within the organic monolayer, but the Bragg rod widths we observe are not consistent with horizontal or vertical density modulations in the entire organic monolayer. The only reasonable possibility is that the superlattice peaks are due to a thin layer in the aqueous subphase. Since no superlattice is seen when no multivalent metal ions are added to the subphase (even at very high pH), and different lattices are seen when different ions are added, the lattice cannot be due to impurities in the water. However, it is well established [17] that metal ions undergo hydrolysis and hydration in aqueous solution, and a wide range of complex polynuclear oxo(hydroxo)-bridged structures can be formed. The observation of changes in the organic monolayer structure as a function of the subphase pH and subphase ion concentration [18] is further evidence that pH- and concentration-sensitive hydrolysis products, rather than isolated ions, are present at the interface.

Thus, we expect that hydrolysis products and water molecules arrange themselves under the Langmuir monolayer, interacting with the carboxyl head groups in such a way as to form a large commensurate superlattice. The ratio of metal atoms to other atoms (hydroxyl ions, water) can be small enough that their presence is undetectable in our anomalous scattering studies.

5. ONGOING AND FUTURE STUDIES

Langmuir monolayers have been used by others [19, 20] as templates to grow inorganic crystals. This is a somewhat simplified experimental model of the real-world process of biomineralization. More generally, biomineralization is just one example of the possible use of an organic template to control the nucleation of an inorganic mineral, and one can ask how and when such template-directed nucleation occurs and whether there are any advantages to using controllable organic lattices such as Langmuir monolayers as templates.

The monolayer structures mentioned in the previous section can be considered precursors to bulk crystal growth. We had originally expected that unusual interfacial structures would serve to mediate between otherwise incompatible monolayer and mineral structures. Unfortunately, it is not easy to nucleate bulk Mn^{2+}, Mg^{2+} and Pb^{2+} crystals at the interface because of the difficulty of making stable supersaturated solutions. It is much easier to make supersaturated $CaCO_3$, which is of great interest because it is a biological mineral. Others have grown such crystals at Langmuir monolayers, and reported oriented growth on the basis of selected crystals lifted off the surface [19]. However, in our in situ studies we see that calcium carbonate does grow preferentially at the interface but is misoriented - a powder. There is also no relationship between the calcite (or vaterite) structure and the monolayer structure observed in situ.

Another crystal that grows well under a Langmuir monolayer is BaF_2 [20]. Our in situ studies confirm that in this case, perfectly aligned crystals grow under a Langmuir monolayer. Unexpectedly, the BaF_2 lattice spacings vary with the solution concentration: at high concentrations they are exactly the same as for bulk BaF_2, but at the lowest concentrations (smallest nuclei) the lattice is commensurate with the Langmuir monolayer lattice. Strained epitaxy is well known in inorganic thin film deposition under high vacuum, but if confirmed, this will be the first report of strained epitaxial growth during nucleation from solution. Further studies are in progress.

6. ACKNOWLEDGEMENT

The work of our group was supported by the US Department of Energy under grant no. DE-FG02-ER45125.

7. REFERENCES

1. V.M. Kaganer, H. Möhwald and P. Dutta, Revs. Mod. Phys.,**71**, 779 (1999)
2. S. Ställberg-Stenhagen and E. Stenhagen, Nature 156, 239 (1945)
3. G.L. Gaines, "Insoluble monolayers at liquid-gas interfaces", Interscience Publishers, New York, 1966, pg. `184
4. A.M. Bibo and I.R. Peterson, *Adv. Mater.* **2**, 309 (1990).
5. M.K. Durbin, A. Malik, A.G. Richter, R. Ghaskadvi, T. Gog and P. Dutta, J. Chem. Phys. **106**, 8216 (1997)
6. G.A. Overbeck, and D. Möbius, D. *J. Phys. Chem.* **97**, 7999 (1993)
7. M.K. Durbin, A. Malik, R. Ghaskadvi, M.C. Shih, P. Zschack and P. Dutta, J. Phys. Chem. **98**, 1753 (1994)

8. M.K. Durbin, A. Richter, C-J. Yu, J. Kmetko, J.M. Bai and P. Dutta, Phys. Rev. **E 58**, 7686 (1998)

9. B. Lin, M.C. Shih, T.M. Bohanon, G.E. Ice, and P. Dutta, *Phys. Rev. Lett.* **65**, 191 (1990)

10 T. Bohanon, B. Lin, M. Shih, G. Ice and P. Dutta, *Phys. Rev. B (Rapid Comm.)* **41**, 4846 (1990)

11. D.K. Schwartz, *Surf. Sci. Rep.* **27**, 241 (1997)

12. R.S. Ghaskadvi, , S. Carr and M. Dennin, *J. Chem. Phys.* **111**, 3675 (1999)

13. J.A. Zasadzinski, R. Viswanathan, L.Madsen, J. Garnaes, and D.K. Schwartz, *Science* **263,** 1726 (1994)

14. F. Leveiller, D. Jacquemain, M. Lahav, L.Leiserowitz, M.Deutsch, K. Kjaer and J. Als-Nielsen, , *Science* **252**, 1532 (1991)

15. J. Kmetko, A. Datta, G. Evmenenko, and P. Dutta, *J. Phys. Chem.B* **105,** 10818 (2001)

16. J. Kmetko, A. Datta, G. Evmenenko,M.K. Durbin, A.G. Richter, P. Dutta,*Langmuir* **17**, 4697 (2001)

77. D.T. Richens, *The chemistry of aqua ions*: Wiley, New York, 1997; Chapter 2.5.4

18. A. Datta, J. Kmetko, C.J.Yu, A,G. Richter, K.S.Chung, J.M. Bai, and P. Dutta, *J. Phys. Chem. B 104*, 5797 (2000)

19. S. Mann, in *Inorganic materials* D. W. Bruce, D. O'Hare, Eds. (Wiley, Chichester, England, 1992) pp. 238.; S. Mann, *Nature* **365**, 499-505 (1993); S. Mann, et al., *Science* **261**, 1286-1292 (1993); B. R. Heywood, S. Mann, *Adv. Mater.* **4**, 278-282 (1992); B. R. Heywood, S. Mann, *Langmuir* **8**, 1492-1498 (1992).; B. Heywood, S. Mann, *J. Am. Chem. Soc.* **114**, 4681-4686 (1992).

20. B. Li, et al., *Langmuir* **15**, 4837-4841 (1999).; L. Lu, H. Cui, W. Li, H. Zhang, S. Xi, *J. Mater. Res.* **16**, 2415-2420 (2001).

CHAPTER 2

ANALYTIC MODEL OF FLOW ORIENTATION IN LANGMUIR-BLODGETT FILMS

Michio Sugi[1], Yuka Tabe[2] and Keiichi Ikegami[2]

[1] Faculty of Engineering, Toin University of Yokohama, Aoba-ku, Yokohama (Japan)

[2] Nanotechnology Research Institute, National Institute of Advanced Industrial Science and Technology, Tsukuba, Ibaraki (Japan)

1. INTRODUCTION

The Langmuir-Blodgett (LB) technique, devised in the middle of 1930's, consists of the manipulation of a monolayer condensed at an air-water interface and the deposition of it by dipping and raising through the interface onto a flat substrate which is kept upright [1]. This technique is often referred to as the vertical dipping method. Three different types are recognized in the manner of deposition: Y-type which picks up monolayers onto the substrate for both dipping and raising, X-type picking only for the dipping, and Z-type picking for the raising [1].

The products, called Langmuir-Blodgett films or simply LB films, have been attracting continued interest in various fields of science and engineering. The origin of the present interest can be traced back to the early 1960's to the pioneer work of H. Kuhn and his collaborators, who utilized the LB technique as a means of arranging various molecules in the form of ultrathin ordered films as a new class of assembly systems [2].

Characterization of the precursors of LB films, i.e., monolayers at the air-water interface, has been fairly advanced as well as that of the end products, i.e., LB films. Now there are indeed many guidelines for "good LB films" referring to some of the characteristic aspects associated with the LB deposition process [3]. Most of them are, however, empirical laws to give qualitative guidelines, and only a few analytic approaches are found to be effective in deriving quantitative guidelines. The in-plane anisotropy in LB film system, for instance, is reported in the early stages [4], and its occurrence has been known to depend on the deposition conditions. Until the late 1980's, however, the origin of the in-plane anisotropy had been far from fully understood.

Let us consider now the vertical dipping method. During the deposition, the substrate moves either upward or downward between the upper and the lower dead centers. The direction of the movement, however, does not matter very much if the consideration is confined to the motion of the monolayer on the water surface: the monolayer is drawn closer to the upright substrate and fades away from the water surface as the deposition proceeds, i.e., the deposition process is inevitably associated with the flow and the deformation of monolayer.

The LB deposition process might be compared to a two-dimensional cosmos with black holes distributed on a finite line with a uniform density. The former is the water surface with the monolayer being the interstellar matter and the latter is the substrate moving upward or downward. The monolayer is, however, different from the dispersed interstellar matter, a condensed film like a two-dimensional solid [1]. These considerations suggest that the flow and the deformation of the solid state should be appropriately taken into account for the better understanding of the LB deposition process.

The continuous deformation of a solid by an external force is referred to as the plasticity [5]. The standard mechanical processing techniques as forging, rolling, squeezing and plucking of metals, for instance, are nothing other than the utilization of the plasticity. A tiny pellet of a few grams of gold is plastically extended to cover some ten square meters, and copper lead wires, thick or thin, are produced using the plucking process. It is well known that these plastic deformations are often associated with the changes in the microscopic texture such as size, shape and orientation of the crystallites, resulting in the changes in the physical properties such as mechanical, magnetic and electric ones. What would then happen in the case of two-dimensional monolayers during the LB deposition process, and how can we describe it quantitatively in relation to the in-plane anisotropy observed in the resultant LB films?

In this respect, endeavors have been made to describe the in-plane anisotropy observed in LB films employing the morocyanine - fatty acid mixed films as a model system. Their chemical structures are shown in Fig. 1, where DX with X=O, S or Se and C_n with n=16 - 22 refer to the merocyanine and the fatty acid, respectively.

(a)

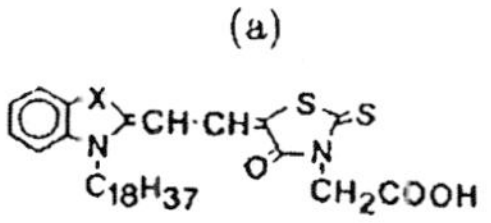

DX
(X=O, S or Se)

(b)

$CH_3(CH_2)_{n-2}COOH$ C_{20}

Fig. 1 Chemical structures of (a) surface-active merocyanine dye (DX with X=O, S or Se) and (b) straight-chain fatty acid (C_n with n=16 – 22).

The merocyanine dye family is generally characterized as a conjugate π-electron system that has a donor nucleus at one end and an accepter at the other. Another feature of this family is the resonance between two extreme forms, non-polarized and polarized, caused by the intramolecular charge transfer between the donor and the acceptor. The non-polarized form of DX is referred to in Fig. 1. The resonance structure is represented by >N–C=C–C=C–C=O ⇌ >N$^+$=C–C=C – C=C–O$^-$. The donor nucleus, shown at the left-hand side, is either of benzoxazole, benzothiazole and benzoselenazole nuclei for X=O, S and Se, respectively, each substituted with a straight-chain long alkyl, while the acceptor is a rhodanine acetic acid radical in each case.

On the basis of the experimental remarks on the in-plane anisotropy in the DX-C_n mixed system, we have derived the "analytical model of flow orientation", which expresses the orientation effect due to the plastic flow of monolayers by six parameters as representing the LB vertical dipping process. This approach, assuming a local thermal equilibrium, has been further extended to lead to new methods to control the in-plane anisotropy, clarifying the limitations, at the same time, to be amended for further development in this approach.

This article deals with the present status of the flow orientation model together with its problems to be solved. The qualitative features of the in-plane anisotropy in the LB films are exemplified in §2, referring to the optical absorption and the ESR measurements of the DS-C_{20} and other mixed films, which have led to the quantitative model of flow orientation. The framework of the model is summarized in §3 together with a brief description of the experimental results, referring to the simplest case of the LB deposition using a single substrate and a sufficiently large trough. §4 deals with the attempts of extending the model to the more complicated cases such as batch production of LB films, deposition using a trough of finite dimensions, transfer processes with peeling-off of the monolayer, slipping, stretching, anisotropy due to the compression and other anomalies, and also the problems associated with them. In §5, another approach, describing the LB deposition as a Markov process, is reviewed together with the results of numerical calculation. The thermal equilibrium assumption adopted in the analytic approach is critically examined referring to the Markovian approach, and the in-plane anisotropy due to the compression is also touched upon. §6 describes the advanced methods to control the in-plane anisotropy in LB films by regulating the plastic flow in a monolayer by the use of rotating disks, reviewing the quantitative model together with the experimental results.

2. IN-PLANE ANISOTROPY IN LB FILMS

It is well known that the DX-C_{20} mixture form stable monolayers at the air-water interface when it is dissolved in chloroform and spread on an aqueous subphase containing Cd^{2+}, and that, for the molar ratio up to [DX] : [C_{20}]=1 : 2, these monolayers are easily deposited to form Y-type LB films [6]-[17] . The as-deposited LB films of DS-C_{20} and DSe-C_{20} as well, are blue in color and associated with a characteristic red-shifted band (J-band [18]-[20]) in the optical absorption spectrum.

It is indicated that an ordered arrangement of the dye molecules, referred to as the J-aggregate, is formed in these LB films, and that they are associated with stable spins of π-electron character in the dark state. Both optical and ESR spectra have been found to exhibit the in-plane anisotropy that has opened the way to the analytic model of flow orientation in the present form. The results from the DS-C_{20} films are outlined in the following.

2.1. Optical Anisotropy of J-Aggregate

Figure 2 shows an example of anisotropic optical absorbance observed in a mixed LB film of

[DS] : [C_{20}]=1 : 2 with a J-band centered around 590 nm, where the solid and the dashed lines refer to the absorbances $A_{\parallel}$ and $A_{\perp}$ measured using the incidents normal to the layer plane with the electric vector parallel and perpendicular to the dipping direction, respectively [21], [22].

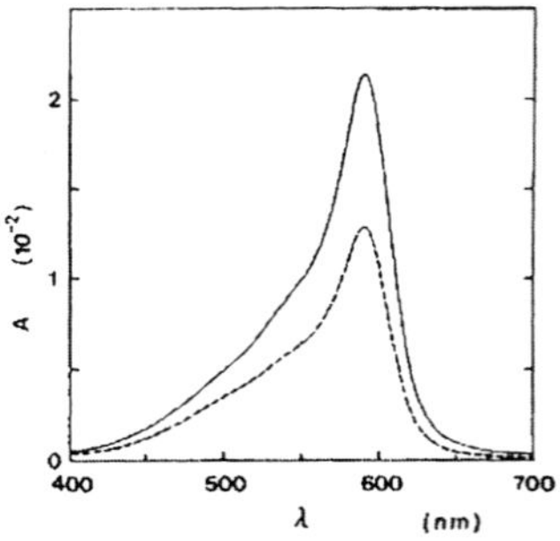

Fig. 2 Anisotropic optical absorption spectrum in a DS-C_{20} 1 : 2 mixed LB film for a linearly polarized light with an incident angle of 90° to the substrate. The absorbance per monolayer A is plotted against the wavelength λ. The solid and the dashed lines refer to $A_{\parallel}$ and $A_{\perp}$ for the electric vector parallel and perpendicular to the dipping direction in the film plane, respectively. (Reproduced from [22] by permission of the Physical Society of Japan.)

The dichroic ratio defined by $R=A_{\parallel}/A_{\perp}$ is conventionally used to represent the in-plane anisotropy observed in the optical absorbance: the transition dipole moment of the dye chromophore is preferentially oriented in the direction of dipping and raising the substrate if $R>1$; it prefers the direction perpendicular to the dipping direction if $R<1$, and it may be isotropically distributed if $R=1$, while the angle of 45° between the dipole moment and the dipping direction will also result in $R=1$.

The larger magnitude for $A_{\parallel}$ seen in Fig. 2 indicates that $R>1$, i.e., the transition dipole moment is preferentially oriented in the dipping direction. The value of R has been found to increase with the deposition velocity or the speed of dipping and raising the substrate v_d, but to remain in any case appreciably greater than unity at the J-band peak: $R\geq 1.3$, typically [22].

It has been found that the J-band can be dissociated and restored by the secondary treatments using heat, acid and base [13], [23]. When an as-deposited film, blue in color, is kept at 90°C for 30 min, for instance, the J-band and its in-plane anisotropy disappear and the film turns red with the absorption maximum shifted to 540 nm. If the film is further treated by an exposure to the vapor from ammonia water, then it turns again blue, with the J-band and its anisotropy restored. These facts indicate that the DS chromophores in the J-aggregate are bound by such relatively weak forces as the dipole-dipole interactions. Here, it is noted that the spectral shape of the DS-C_{20} mixed film is found to be insensitive or essentially invariant to the molar mixing ratio down to [DS] : [C_{20}]=1 : 20 [9] It is suggested that DS and C_{20} are immiscible with each other forming a phase-separated monolayer consisting of the DS aggregates embedded in a C_{20}-rich matrix, and that the 540-nm isotropic band in the heat-treated film reflects the disorders introduced in the aggregates.

2.2 Anisotropy in ESR Line Shape

Electron spin resonance (ESR) studies by Kuroda et al. have shown that the DS-C_{20} mixed LB film is associated with stable spins of π-electron character with a concentration as high as one spin per several hundred DS molecules in the dark state [10], [15], [24]-[27]. The spins have been found to involve two different species, A and B, of which the former is anisotropic and the latter is featureless in ESR line shape as shown in Fig. 3 [24].

The concentrations of A and B are essentially equal to each other as seen in the figure, suggesting that A and B form pairs in occurrence.

If we assume that a spin A and a spin B are associated with each other in an aggregate of several hundred DS molecules, the aggregate will be more or less one-dimensional in shape so that the spin pair is stabilized at its ends: a situation reminiscent of a cation or an anion radical stabilized at each end of a polymer chain [24]. This picture is quite compatible with the fact that both A and B disappear when the film is heat-treated at 90°C for 30 min as is the case with the J-band and its anisotropy mentioned in §2.1.

The species A was later successfully assigned to the donor radical by the ESR studies using the ^{15}N-substituted donor and acceptor [28]-[30]. It is therefore indicated that the intermolecular distance in the J-aggregate is small enough to allow the intermolecular charge transfer, and that A and B originate from the cation and the anion radicals, respectively, located at the opposite ends of the one-dimensionally developed J-aggregate.

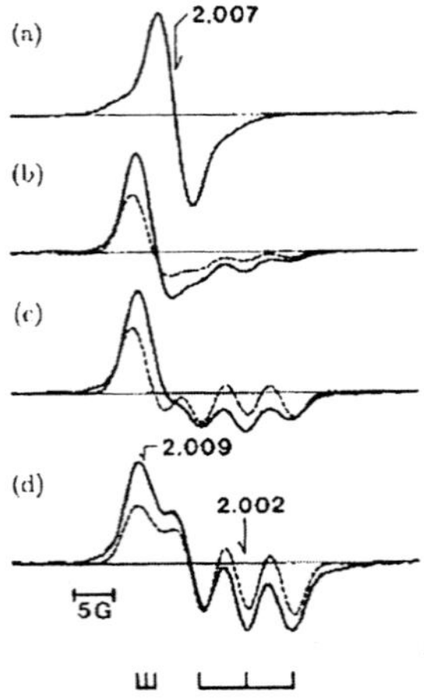

Fig. 3 First-derivative ESR spectra of a 1000-layered DS-C_{20} 1 : 2 mixed LB films at room temperature (solid curves). The external magnetic field is normal to the film plane (a). It makes the angle of 0°, 45° and 90° with the dipping direction of the substrate from (b) to (d), respectively. Numbers show *g*-values of the fields as indicated. Dashed lines show the simulated curves of the spin species A. Stick lines at the bottom of (d) show the resonance fields corresponding to principal components of *g* and hyperfine tensors. (Reproduced from Ref. [24] by permission of the Physical Society of Japan.)

Analysis of the anisotropic line shape of A has revealed that the $p\pi$ orbital axis of DS chromophore lies in the layer plane with its distribution function simulated by the superposition of two components having maxima at about ±60° with respect to the dipping direction [15].

Figure 4 shows the model of the in-plane alignment of the DS chromophores deduced from the analysis. This is compatible with the dichroic ratio R>1 of the optical absorbance if the aggregate of DS in the ESR study is identified with the optical J-aggregate, since the $p\pi$ orbital axis and the transition dipole moment are approximately orthogonal to each other, suggesting the deviation angle of $\alpha \approx \pm 30°$ which indicates $R>1$.

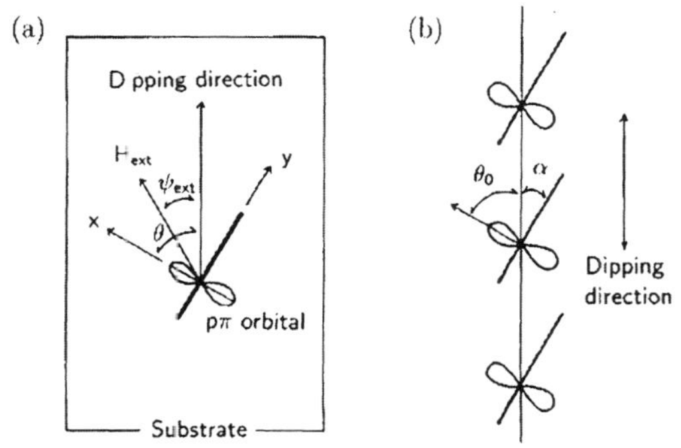

Fig. 4 Model of the in-plane alignment of DS chromophores. (a) The external magnetic field and the $p\pi$ orbital axis make angles ψ_{ext} and $\theta_{p\pi}$ with the dipping direction, respectively. The long axis of the chromophore is schematically expressed by a thick line. (b) The long axis makes an angle α with the direction connecting the centers of adjacent molecules. $<\theta_{p\pi}>$ denotes the average of $\theta_{p\pi}$. (Reproduced from Ref. [24] by permission of the Physical Society of Japan.)

Models of the J-aggregate structure have been proposed from various viewpoints including the remarks from the optical and the ESR measurements. A comparable value $\alpha \approx \pm 30°$ has been estimated, e.g., by Nakahara et al. [16] by analyzing the optical anisotropy based on an extended dipole model, although there still remain controversies to be solved among the models [16], [20], [24], [30]-[34]. The detailed discussion is, however, beyond the scope of the present article.

2.3. Possible Origin of In-plane Anisotropy

Let us imagine a stream in autumn, flowing gently, with its surface covered with a lot of leaves fallen from the trees on the banks. Each of them is drifting down the stream as if it were destined to trace its own path. Looking down from a height, however, one would recognize a steady pattern like a bundle of stripes woven from numberless leaves drifting on the water surface. If they are elongated in shape like those of willow, one would find that they are aligned one after another in each stripe, forming head-to-tail, side-by-side or randomly stacked chains.

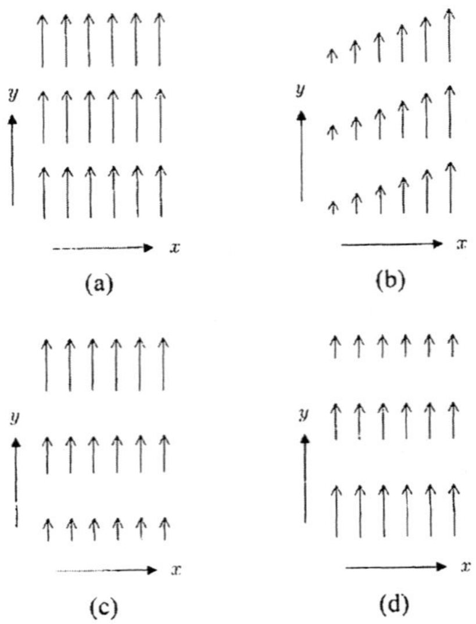

Fig. 5 Examples of simple flows and types of stacking. In each case, all the streamlines are assumed to be parallel to the y-axis, i.e., v_x=0. (a) $v_y=v_0$, random; (b) $v_y=v_0+gx$, head-to-tail; (c) $v_y=v_0+gy$ with $g>0$, head-to-tail; (d) $v_y=v_0-gy$ with $-g<0$, side-by-side.

Such a pattern reflects the local variation of the velocity field on the water surface. The center of the leaf at a given place moves in accordance with the local velocity vector, as a matter of course. Then, why a willow leaf sometimes prefers orienting its long axis towards a specific direction? The clue for answering the question may be found in the velocity gradient that exerts a torque on the leaf to orient to a specific direction [36], [37].

Let us consider four extreme cases (a) - (d) as shown in Fig. 5. In each case, all the streamlines are assumed to be parallel to the y-direction, i.e., $v_x(x,y)=0$ for simplicity.

Case (a) refers to the uniform velocity $v_y(x,y)=v_0=\text{const}$. All the elements of the velocity gradient tensor **G** are then equal to zero. As a result, the orientation of each leaf remains unchanged, and the leaves may be randomly stacked.

Case (b) exemplifies the x-dependent velocity with a function $v_y(x,y)=v_0+gx$, which results in a non-zero element in **G**, i.e., $G_{yx}=\partial v_y/\partial x=g$, orienting the long axis to the y-direction. A head-to-tail stacking of the leaves will be realized as a consequence.

Case (c) represents the increasing function of y by $v_y(x,y)=v_0+gy$ with $g>0$, resulting in a non-zero element $G_{yy}=\partial v_y/\partial y=g>0$ to orient the axis to the y-direction. The resultant stacking will also be head-to-tail.

Case (d) refers to a decreasing function $v_y(x,y)=v_0-gy$ with $G_{yy}=-g<0$, which orients the axis to the x-direction. In this case, therefore, a side-by-side stacking will be realized, in contrast to Cases (b) and (c).

There arises a question as to what extent this naive picture can be effective as the starting point for elaborating a quantitative model to explain the in-plane anisotropy actually observed in the LB films of DS-C_{20} and others. The endeavor of answering this question has led to the analytic model of flow orientation [21], [22], [38]-[40], of which the basic framework will be reviewed in the next section.

3. BASIC FRAMEWORK OF THE MODEL

In the analytic model of flow orientation for the conventional LB dipping process, the monolayer on the water surface is assumed to be a continuum,[1] and the model consists of two stages: the evaluation of the velocity potential on the water surface and the description of the rotatory motion of a particle in this potential field.

3.1. Velocity Potential of the Monolayer Flow

In evaluating the flow of monolayer at the air-water interface, we have assumed:

(1) the monolayer is a two-dimensional ideal fluid;

(2) the substrate, i.e., a flat plate on which the monolayer is deposited, is a two-dimensional line sink;

(3) the monolayer is infinite in its dimensions [21], [22].

3.1.1. Ideal Fluid Approximation

The assumption (1) involves the incompressible and irrotational conditions. The former and the latter enables us to introduce the velocity potential $U(x,y)$ and the stream function $V(x,y)$ to describe the flow in an x-y plane which represents the water surface.

Here, the velocity vector $\boldsymbol{v}=(v_x,v_y)^T$ around a point (x,y) is given by,

$$v_x = \frac{\partial U}{\partial x} \equiv U_x, \quad v_y = \frac{\partial U}{\partial y} \equiv U_y, \tag{1}$$

and also by,

---

[1] The continuum approximation is implausible to very brittle monolayers, for which case Matsumoto et al. have proposed the "channel-and-iceberg" model in Refs. [41] and [42].

$$v_x = \frac{\partial V}{\partial y} \equiv V_y, \quad v_y = -\frac{\partial V}{\partial x} \equiv -V_x. \tag{2}$$

Equations (1) and (2) are nothing but the well-known Cauchy-Riemann relation, $U_x = V_y$ and $U_y = -V_x$, which guarantees the existence of the complex velocity potential $W(z)=U+iV$ in the Gaussian plane, where $z=x+iy$ [35], [36]. The velocity vector v is then written using the first derivative dW/dz as,

$$v_x = \mathrm{Re}\left(\frac{dW}{dz}\right), \quad v_y = -\mathrm{Im}\left(\frac{dW}{dz}\right). \tag{3}$$

The deformation rates of an area element are given by the velocity gradient tensor **G** written as,

$$\mathbf{G} = \begin{pmatrix} G_{xx} & G_{xy} \\ G_{yx} & G_{yy} \end{pmatrix}, \tag{4}$$

where $G_{xx}=\partial v_x/\partial x$, $G_{xy}=\partial v_x/\partial y$, $G_{yx}=\partial v_y/\partial x$ and $G_{yy}=\partial v_y/\partial y$. In Eq. (4), the diagonal elements, G_{xx} and G_{yy}, denote the expansion-compression rates, and the off-diagonal elements, G_{xy} and G_{yx}, are the shear rates. Using the second derivative d^2W/dx^2, each element in Eq. (4) is written as,

$$G_{xx} = -G_{yy} = \mathrm{Re}\left(\frac{d^2W}{dz^2}\right), \tag{5}$$

and,

$$G_{xy} = G_{yx} = -\mathrm{Im}\left(\frac{d^2W}{dz^2}\right). \tag{6}$$

The incompressible condition is referred to by Eq. (5), i.e., the expansion or compression of the area element in the x-direction is canceled by that in the y-direction. Equation (6) is related with the irrotational condition: **G** is a symmetric tensor associated with a pair of principal axes, orthogonal to each other, at which directions the flow torque vanishes.

3.1.2 Normal Deposition with a Single Substrate

The assumptions (2) and (3) refer to the neglect of the thickness of the substrate in comparison with its width, and also the complicated boundary conditions due to the shape and dimensions of the trough and the barriers.

Let us consider the normal case that the monolayer is symmetrically deposited onto both surfaces of one single substrate, which is represented by a line sink. If the substrate occupies a range $-a/2 \leq x \leq a/2$ on the real axis, the complex potential $W=U+iV$ is written as [21],

$$W(z) = -\frac{v_d}{\pi} \int_{-a/2}^{a/2} \log(z - x')\, dx', \tag{7}$$

where v_d is the deposition velocity, i.e., the speed of dipping and raising the substrate, and we assume that the transfer ratio of $r_d = 1$, indicating that the monolayer is transferred from the water surface onto the substrate without compression or expansion. Eq. (7) is regular except for $|x| \leq a/2$ on the x-axis, and the factor $-v_d/\pi$ at the right-hand side is so chosen that the condition $v_y \to \mp v_d$ for $|x| < a/2$ is satisfied when $y \to \pm 0$.

Equation (7) is easily integrated as,

$$W(z) = \frac{v_d}{\pi} \left\{ \left(z - \frac{a}{2} \right) \log\left(z - \frac{a}{2} \right) - \left(z + \frac{a}{2} \right) \log\left(z + \frac{a}{2} \right) + a \right\}, \tag{8}$$

with the real and the imaginary parts,

$$U(x,y) = \frac{v_d}{2\pi} \left[\left(x - \frac{a}{2} \right) \log\left\{ \left(x - \frac{a}{2} \right)^2 + y^2 \right\} - \left(x + \frac{a}{2} \right) \log\left\{ \left(x + \frac{a}{2} \right)^2 + y^2 \right\} \right.$$

$$\left. + 2y \left\{ \arctan\left(\frac{x - a/2}{|y|} \right) - \arctan\left(\frac{x + a/2}{|y|} \right) \right\} + a \right], \tag{9}$$

and,

$$V(x,y)=\frac{v_d}{2\pi}\left[2\left(x-\frac{a}{2}\right)\arctan\left(\frac{y}{x-a/2}\right)-2\left(x+\frac{a}{2}\right)\arctan\left(\frac{y}{x+a/2}\right)\right.$$

$$\left.+y\log\left\{\left(x-\frac{a}{2}\right)^2+y^2\right\}-y\log\left\{\left(x+\frac{a}{2}\right)^2+y^2\right\}\right]. \quad (10)$$

respectively.

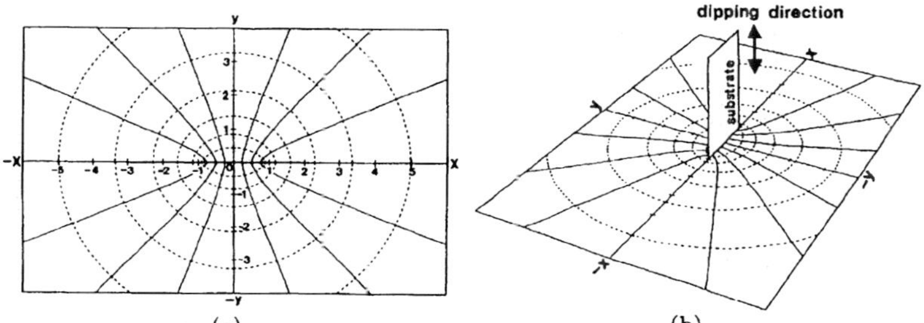

(a) (b)

Fig. 6 (a) Streamlines and equipotential lines shown by the solid and the dashed lines, respectively for a two-dimensional ideal fluid with a line-sink on the x-axis represented by the thick rod. (b) Schematic representation of the normal LB deposition process. (Reproduced from Ref. [22] by permission of the Physical Society of Japan.)

Figure 6 shows the equipotential and the streamlines, $U(x,y)$=const and $V(x,y)$=const, respectively, for the single-substrate case [22]. The first derivative of $W(z)$ is given as,

$$\frac{dW}{dz}=\frac{v_d}{\pi}\left\{\log\left(z-\frac{a}{2}\right)-\log\left(z+\frac{a}{2}\right)\right\}, \quad (11)$$

with the velocity vector expressed by,

$$v_x(x,y)=\mathrm{Re}\left(\frac{dW}{dz}\right)=\frac{v_d}{2\pi}\log\left\{\frac{(x-a/2)^2+y^2}{(x+a/2)^2+y^2}\right\},$$

$$v_y(x,y)=-\mathrm{Im}\left(\frac{dW}{dz}\right)=-\frac{v_d}{\pi}\left\{\arctan\left(\frac{y}{x-a/2}\right)-\arctan\left(\frac{y}{x+a/2}\right)\right\}, \quad (12)$$

the latter of which leads to,

$$\lim_{y \to \pm 0} v_y(x, y) = \mp v_d, \quad \text{for } |x| < \frac{a}{2},$$

$$= 0, \text{ for } |x| > \frac{a}{2}, \tag{13}$$

where the double signs are taken in the same order.

The second derivative d^2W / dx^2 is given by,

$$\frac{d^2W}{dz^2} = \frac{v_d}{\pi}\left(\frac{1}{z - a/2} - \frac{1}{z + a/2}\right), \tag{14}$$

and the elements of **G** are derived as,

$$G_{xx} = -G_{yy} = \frac{v_d}{\pi}\left\{\frac{x - a/2}{(x - a/2)^2 + y^2} - \frac{x + a/2}{(x + a/2)^2 + y^2}\right\},$$

$$G_{xy} = G_{yx} = \frac{v_d}{\pi} y\left\{\frac{1}{(x - a/2)^2 + y^2} - \frac{1}{(x + a/2)^2 + y^2}\right\}, \tag{15}$$

using Eqs. (5) and (6).

3.2. Rotatory Motion of Particles

Let us examine the motion of a particle, elongated in shape, put in the velocity potential field. This particle may be a rod-like crystallite in the actual case. In this stage, the ideal fluid approximation adopted for evaluating the potential functions, *W*, etc., ceases to work, and the friction between the particle and the surrounding medium should now be explicitly taken into account.

3.2.1. Equation of Angular Distribution Function

The center of the particle moves along a streamline with a velocity vector v (e.g., cf. Eq. (1)). The orientation of the particle is governed by the velocity gradient tensor **G** (e.g., cf. Eq. (4))

and the rotatory Brownian motion, while the inertia term is assumed to be neglected. Let (ξ, η) be a coordinate with the ξ- and η-axes parallel to the x- and y-axes, respectively, and its origin is fixed at the center of particle (x, y), as shown in Fig. 7. The relative velocity ($\dot{\xi}, \dot{\eta}$) of point (ξ, η) with respect to the origin is expressed by,

$$\begin{pmatrix} \dot{\xi} \\ \dot{\eta} \end{pmatrix} = \begin{pmatrix} G_{xx} & G_{xy} \\ G_{yx} & G_{yy} \end{pmatrix} \begin{pmatrix} \xi \\ \eta \end{pmatrix}, \tag{16}$$

using Eq. (4).

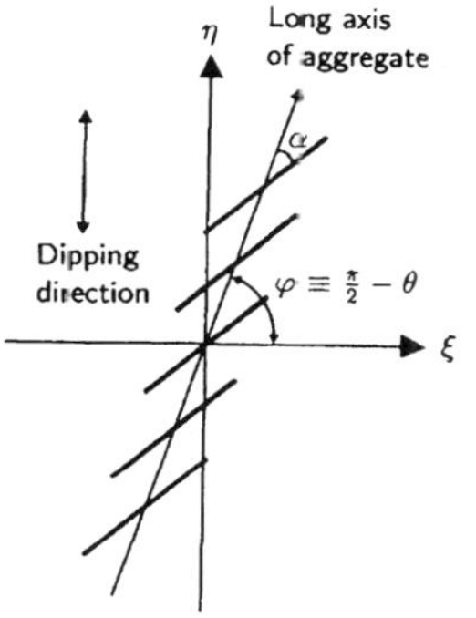

Fig. 7 Definition of ξ, η-coordinate with the origin fixed at the center of particle. The long axis of the aggregate forms angles φ and θ with the ξ- and η-axes, respectively. α refers to the angle between the long axis of the aggregate and the transition dipole moment. ϕ is the angle between the electric vector of the incident and the η-axis. (Reproduced from Ref. [22] with modification by permission of the Physical Society of Japan.)

Let φ be the angle between the position vector and the ξ-axis as shown in Fig. 7, i.e., $\xi = r\cos\varphi$ and $\eta = r\sin\varphi$ with $r = \sqrt{\xi^2 + \eta^2}$. The angular velocity is written as,

$$\dot{\varphi} \equiv \frac{d\varphi}{dt} = \frac{G_{yy} - G_{xx}}{2} \sin 2\varphi + G_{yx} \cos^2 \varphi - G_{xy} \sin^2 \varphi, \tag{17}$$

or,

$$\dot{\theta} \equiv \frac{d\theta}{dt} = \frac{G_{xx} - G_{yy}}{2} \sin 2\theta + G_{yx} \cos 2\theta + (G_{xy} - G_{yx}) \cos^2 \theta, \tag{18}$$

if we use the angle $\theta \equiv \pi/2 - \varphi$ instead of φ.

In the incompressible, irrotational case, we derive,

$$\dot{\theta} = G_{xx}\sin 2\theta + G_{yx}\cos 2\theta, \tag{19}$$

referring to Eqs. (5) and (6). Substituting $\dot{\theta} = 0$ for Eq. (19), we obtain the expressions for the principal axes as,

$$\theta_{\pm} = -\frac{1}{2}\arctan\left(\frac{G_{yx}}{G_{xx}}\right), \text{ with, } \pm\left(\frac{d\dot{\theta}}{d\theta}\right)_{\theta=\theta_{\pm}} < 0, \tag{20}$$

and the associated eigenvalues are,

$$G_{\pm} = \pm\sqrt{G_{xx}^2 + G_{xy}^2}, \tag{21}$$

where the double signs are taken in the same order. Here, the positive and negative eigenvalues, G_+ and G_-, correspond to the maximum expansion and compression rates, respectively, and the mechanically stable point is identical with θ_+. The extreme values of the shear rate and their positions are easily derived in a similar manner using the conditions with $d\dot{\theta}/d\theta = 0$ and $\pm d^2\theta/d\theta^2 < 0$ as,

$$\dot{\theta}_{\max} \equiv \left(\frac{d\theta}{dt}\right)_{\max} = G_+, \text{ and, } \dot{\theta}_{\min} \equiv \left(\frac{d\theta}{dt}\right)_{\min} = -G_+, \tag{22}$$

which are located at $\theta = \theta_{\pm} - \pi/4$, respectively. Thus, we have reconfirmed the naive view in §2.3, i.e., the preferred direction of the long axis is not identified with the tangential of a streamline, but determined by the velocity gradient.

The angular distribution function P can be derived starting from an equation of continuity [43],

$$\frac{\partial P}{\partial t} = -\frac{\partial S}{\partial \theta}, \tag{23}$$

where $P(\theta)d\theta$ gives the probability for θ to lie in a range θ – $\theta+d\theta$, and S is the flow of probability passing through a point θ. Here, S is written as,

$$S = \dot{\theta}P - D\frac{\partial P}{\partial \theta}, \tag{24}$$

where D is the rotatory diffusion coefficient. In Eq. (24), the first term at the right-hand side refers to the drift flow to orient the long axis of a particle towards the preferred direction, and the second term corresponds to the diffusion flow to randomize the orientation.

Substituting Eq. (24) for Eq. (23), we obtain the diffusion equation,

$$\frac{\partial P}{\partial t} = -\frac{\partial(\dot{\theta}P)}{\partial \theta} + D\frac{\partial^2 P}{\partial \theta^2}, \tag{25}$$

which expresses the compromise between the flow due to $\dot{\theta}$ and the Brownian motion.

The rotatory diffusion coefficient D is related with the rotatory friction coefficient ζ between the particle and the medium through the Einstein relation,

$$D = \frac{k_B T}{\zeta}, \tag{26}$$

where k_B and T are the Boltzmann constant and the absolute temperature, respectively. It is noted that D is generally a function of t since ζ may change as the particle moves towards the substrate.

3.2.2. Local Thermal Equilibrium Approximation

If we assume that thermal equilibrium is quickly attained in comparison to the change in **G**, i.e., the local thermal equilibrium approximation, only the values of $\dot{\theta}$ and D near the substrate are referred to as determining the orientation [21], [22]. In this case, $\partial P / \partial t = 0$, and Eq. (25) is readily integrated as,

$$\dot{\theta}P - D\frac{\partial P}{\partial \theta} = S_0, \tag{27}$$

where S_0 is the integration constant. The left-hand side of Eq. (27) is identical with the flow of probability S given by Eq. (24). As a consequence, $S_0 \neq 0$ indicates that there exists a steady flow of probability. This does not contradict the conservation of the total probability since θ is a periodic variable, and it may well happen in a vortical velocity field.[2]

For the ideal fluid, however, $S_0=0$ due to the irrotationality. Thus, the expression for P is written as,

$$\dot{\theta}P - D\frac{\partial P}{\partial \theta} = 0. \tag{28}$$

For the normal deposition with a single substrate, we derive from Eq. (15),

$$G_{xx} = -G_{yy} = -G, \quad G_{xy} = G_{yx} = 0, \quad |x| \leq a/2 \text{ and } y \to \pm 0, \tag{29}$$

and,

$$G = \left(\frac{4v_d}{\pi a}\right)\left(1 - \frac{4x^2}{a^2}\right)^{-1}, \tag{30}$$

which corresponds to the positive eigenvalue G_+ in Eq. (21) for the normal deposition, as a matter of consequence. The preferred direction is obtained as $\theta_+=0$ using Eq. (20), i.e., the long axis tends to orient in parallel with the y-axis. Here, the area element is compressed in the x-direction and stretched in the y-direction. The y-direction is identical with the direction of dipping and raising, if the orientation of the particle is assumed to remain as it was immediately before the transfer onto the substrate.

As a consequence, Eq. (28) is rewritten as,

---

[2] A non-zero S_0 corresponds to a steady rotation of the particle. This is actually the case with the rotating-disk method described in §6.

$$(G\sin 2\theta)P + \left(\frac{k_B T}{\zeta}\right)\frac{\partial P}{\partial \theta} = 0, \tag{31}$$

using Eqs. (19), (26) and (29). We integrate Eq. (31) to derive the angular distribution function P(θ) of Boltzmann-type as,

$$P(\theta) \propto \exp\left\{-\left(\frac{\zeta G}{k_B T}\right)\sin^2\theta\right\}, \tag{32}$$

whose maximum is located at θ=0. Allowing for the normalization condition $\int_{-\pi/2}^{\pi/2} P(\theta)d\theta = 1$, Eq. (32) is rewritten as,

$$P(\theta) = \left(\frac{e^c}{\pi I_0(c)}\right)exp(-2c\,sin^2\theta), \tag{33}$$

where I_0 is the modified Bessel function of the zeroth order, and c is defined as,

$$c = \frac{\zeta G}{2k_B T}, \tag{34}$$

which is referred to as the flow orientation parameter or the c-parameter [22].

According to Eqs. (33) and (34), $P(\theta)$ is determined by the c-parameter alone. The width of $P(\theta)$, for instance, is evaluated with a quantity $<\sin^2\theta>$ given by,

$$<\sin^2\theta> = \int_{-\pi/2}^{\pi/2}\sin^2\theta P(\theta)d\theta = \frac{1}{2}\left(1 - \frac{I_1(c)}{I_0(c)}\right), \tag{35}$$

where I_1 is the modified Bessel function of the first order.

3.3. Dichroic Ratio

Experimentally, we obtain the dichroic ratio $R \equiv A(0)/A(\pi/2)$ as representing the orientation of the particle. Here, $A(\phi)$ is the absorbance of incident light polarized to an angle ϕ with respect to the direction of dipping and raising of the substrate during the deposition process.

For evaluating R, the particle is assumed to be an aggregate of molecules as shown in Fig. 7.

The transition dipole moment of each molecule forms an angle α with the long axis of the aggregate, which is at an angle θ from the dipping direction. The absorbance $A(\phi)$ is written as,

$$A(\phi) \propto \int_{-\pi/2}^{\pi/2} \left\{\cos^2(\theta+\alpha-\phi)+\cos^2(\theta-\alpha-\phi)\right\} P(\theta)d\theta, \tag{36}$$

allowing for the symmetry with respect to the dipping direction [22]. The dichroic ratio is therefore given by,

$$R \equiv \frac{A(0)}{A(\pi/2)} = \frac{\int_{-\pi/2}^{\pi/2} \left\{\cos^2(\theta+\pi)+\cos^2(\theta-\pi)\right\} P(\theta)d\theta}{\int_{-\pi/2}^{\pi/2} \left\{\sin^2(\theta+\pi)+\sin^2(\theta-\pi)\right\} P(\theta)d\theta}, \tag{37}$$

If we substitute Eq. (33) derived assuming the local thermal equilibrium for Eq.(37), we obtain the expression for the normal deposition with a single substrate as,

$$R = \frac{I_0(c)+I_1(c)\cos 2\alpha}{I_0(c)-I_1(c)\cos 2\alpha}, \tag{38}$$

using Eq. (35) and a relation $<\cos^2 2\theta> = 1 - <\sin^2 2\theta>$. Therefore, R is governed by c and α in this case: it tends to $R=1$ for $c \to 0$ and $R=\cot^2\alpha$ for $c \to \infty$.

3.4. Bingham Model of Plasticity

The rotatory friction coefficient ζ is determined by the size and shape of the particle and the viscosity of the surrounding medium. If the medium is a Newtonian fluid, ζ is a constant independent of the shear rate. It is however well known that condensed monolayers at the air-water interface often show plasticity [1], which is characterized with the shear-dependence of ζ.

Of many models of visco-elasticity available so far [5], [44], a Bingham fluid model [45] has been introduced to take the plastic behaviour into account. ζ is then a function of the maximum shear rate $\dot{\theta}_{max} = G_+ = G$ (cf. Eqs. (22) and (30)) and given by

$$\zeta = \zeta(G) = \zeta' + \frac{\tau_0}{G}, \tag{39}$$

where ζ' and τ_0 refer to the Newtonian viscosity and the Bingham yield value, respectively [22]. Accordingly, the expression of ζG is replaced by,

$$\zeta G = \zeta' G + \tau_0, \tag{40}$$

which indicates that the system remains an elastic body if the maximum torque ζG acting on the particle does not exceed τ_0, and that plastic deformation takes place for $\zeta G > \tau_0$ and a proportional relationship similar to the Newtonian case is held between the excessive torque and the shear rate. The flow orientation parameter for the Bingham case is therefore expressed as,

$$c = \frac{\zeta' G + \tau_0}{2k_B T}, \tag{41}$$

by substituting Eq. (40) for Eq. (34). It is noted that c $\rightarrow \tau_0/(2k_BT)>0$ even when $G \rightarrow 0$.

3.5. Predictions from the Model

According to Eqs. (30), (33) and (41), the distribution function $P(\theta)$ for the Bingham case is governed by six parameters involved in the c-parameter, which is explicitly written as,

$$c = \left(\frac{2}{\pi}\right)\left(\frac{\zeta' v_d}{ak_B T}\right)\left(1 - \frac{4x^2}{a^2}\right)^{-1} + \frac{\tau_0}{2k_B T}, \tag{42}$$

by substituting Eq. (30) for Eq. (41) [22].

The six parameters are the temperature T, the width of the substrate a, the deposition velocity v_d, the relative position on the substrate $|x|/a$, and the two parameters ζ' and τ_0 involved in the Bingham model of rotatory friction coefficient. According to Eq. (42), the dichroic ratio R is determined by seven parameters, i.e., the above six and the deviation angle of the transition dipole moment α. Among the seven, a, v_d and $|x|/a$ are characteristic of the LB deposition process, while ζ', τ_0 and α are the properties of the monolayer.

3.5.1 Law of Similarity

The c-parameter given by Eq. (42) involves three dimensionless terms. They are (1) $\zeta' v_d/(ak_BT)$, (2) $(1-4x^2/a^2)^{-1}$ and (3) $\tau_0/(2k_BT)$, each leading to a law of similarity which governs the degree of orientation.

(1) $\zeta' v_d/(ak_BT)$ involves four parameters, of which the deposition velocity v_d and the substrate width a are adjustable independently of any other parameter involved in c, while the Newtonian friction coefficient ζ', and the Bingham yield value τ_0 as well, are generally known to be temperature-dependent. v_d and a are competitive against each other in their contributions to the c-parameter: c increases with v_d, but decreases with a, and the increment in c for 2 v_d is exactly equal to that for $a/2$.

(2) $(1-4x^2/a^2)^{-1}$ represents the effect of the position on the substrate. It exhibits its minimum at the center $|x|/a=0$, and increases slowly when $|x|/a>>1/2$, but diverges to infinity for $|x|/a \rightarrow 1/2$. Accordingly, the c-parameter, with its minimum $c_{\min}$ at the center, may have any value up to $c \rightarrow \infty$. In other words, starting from the lowest degree of orientation at the center, any high degree up to the complete orientation is realized in one and the same substrate towards the edges.

(3) $\tau_0/(2k_BT)$ refers to the effect of Bingham plasticity. For $\tau_0>0$, the c-parameter remains non-zero even for $v_d \rightarrow 0$. In this case, c is determined by $\tau_0/(2k_BT)$ alone, and its dependence on the dimensionless terms (1) and (2) disappears, resulting in a constant c-value all over the substrate except for the singular points $|x|/a=1/2$ associated with the edges. As a consequence,

the dichroic ratio R is always $R>1$ for $\alpha<45°$ and $R<1$ for $\alpha>45°$.

3.5.2. Effect of the Length of Particles

Both ζ' and τ_0 depend on the geometry of the particle and the viscoelasticity of the monolayer as a two-dimensional system. In the case of rod-like particles, e.g., rigid polymers or needle-like crystallites, it is well known that,

$$\zeta' \approx \eta' b^2,$$

$$\tau_0 \approx f_0 b^2, \tag{43}$$

where b is the length of the particle, and η' and f_0 are the surface Newtonian viscosity and the corresponding Bingham yield value, respectively [37]. Here, the Bingham viscosity η at a shear rate g is written as,

$$\eta = \eta(g) = \eta' + \frac{f_0}{|g|}. \tag{44}$$

The sharpness of orientation increases therefore with the length b of the particle. Furthermore, η' and f_0 of the monolayer may also be increased with b, resulting in the enhanced flow orientation effect.

3.6. Comparison with Experimental Results

The analytic model summarized in §3.1 − §3.5 is compared with the dichroic behaviour observed in the absorption spectra of the mixed LB films of merocyanine DS and arachidic acid C_{20} as exemplified in Fig. 2 [22]. The chemical structures of DS and C_{20} are shown in Fig.1.

The dichroic data to be reviewed are from the 590-nm J-band seen in the Y-type LB films of a mixture [DS] : [C_{20}]=1 : 2, prepared using the standard vertical dipping method with one single substrate. The monolayers were formed at 20°C on an aqueous subphase containing Cd^{2+}

(pH=6.1 − 6.3) and transferred under the surface pressure of 25 mN/m. The glass substrates were a=6.5 mm − 52 mm wide and 1mm thick. The trough used was of two-side-compression type with an approximately rectangular water surface of a constant width of 230 mm and a variable length ranging from 435 mm down to four times the substrate width $4a$. The deposition was carried out around the center of the water surface where the point symmetry is approximately held as to the monolayer flow. The deposition velocity was kept constant for all the dipping and raising strokes for each sample in a range of v_d=10 mm/min − 500 mm/min, where the transfer ratio was close to unity. For further details, the readers are referred to the original report [22].

3.6.1. Actual Behaviour of Dichroic Ratio

The experiment for the finite monolayers is found to coincide well with the model for the infinite case reviewed in the preceding subsections, suggesting that the boundary conditions of trough are negligible when each side of the trough is longer than $4a$. Let us examine now the experimental results.

Figure 8 is a universal plot showing the dichroic data derived from various samples with different values of the deposition velocity v_d, the substrate width a and the relative position $|x|/a$ on the substrate. Here, the dichroic ratio R at J-band peak (590 nm) is plotted against the positive eigenvalue $\mathbf{G}$ of velocity gradient tensor G near the substrate, where G is given by Eqs. (29) and (30).

It is evident from Fig. 8 that G is the predominant factor in governing the dichroic behaviour. R is approximately 1.3 for $G \rightarrow 0$. This fact favors the Bingham fluid approximation, since the c-parameter tends to $\tau_0/(2k_BT)>0$ for $G \rightarrow 0$ in the Bingham case according to Eq. (42), while in the Newtonian case c $\rightarrow 0$ for $G \rightarrow 0$, resulting in the random angular distribution with R=1 (cf. Eq. (30)). R increases with G and then tends to saturate at a constant value ($\approx$2.3) as predicted by the present model: $R \rightarrow \cot^2\alpha$ for c $\rightarrow \infty$. The dashed line in Fig. 8 is the best least-squares fit calculated using Eqs. (38) and (41). The corresponding values of α, ζ' and τ_0 are given in Table 1.

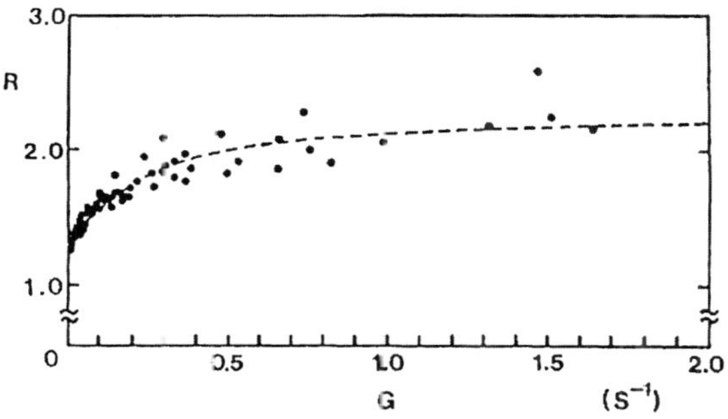

Fig. 8 Universal plot: Dependence of the dichroic ratio R at 590 nm on the large eigenvalue G of the velocity gradient tensor **G**. The dashed line is the best, least-squares fit, where the values of α, ζ' and τ_0 are listed in Table 1. (Reproduced from Ref. [22] with modification by permission of the Physical Society of Japan.)

The coincidence of the model and the experiment is satisfactory, suggesting that the Bingham fluid approximation is effective in describing the behaviour of the monolayer at the air-water interface. The value of angle α=33.5° is comparable to that of about 30° estimated from the ESR line shape analysis mentioned in §2.2 and is also compatible with the structure of J-aggregate according to Nakahara et al. deduced from the analysis of J-band peak position using Kuhn's extended dipole model.

Table 1: The values of parameters, α, ζ and τ_0, estimated using the best least-squares method.

Name of parameter	Estimated value
Tilt angle of the transition dipole moment	α=33.5°
Newtonian rotatory coefficient	$\zeta'=5.00\times10^{-20}$ kg m^2 s^{-1}
Bingham yield value	$\tau_0=6.22\times10^{-21}$ kg m^2 s^{-2}

The approximate size of the J-aggregate can be estimated using the values of ζ' and τ_0 listed in Table 1, which represent the Bingham rotatory friction coefficient $\zeta(G)=\zeta'+\tau_0/G$ (Eq. (39)). If we assume $\eta'=3.64\times10^{-7}$ kg s^{-1} referring to the viscoelastic data of icosanyl alcohol CH_3-$(CH_2)_{19}$-OH reported by Patil et al., we find $b\approx400$ nm using $b\approx(\zeta'/\eta')^{1/2}$ (cf. Eq. (43)), which corresponds to a length of ~10^2 DS molecules aligned with a center-to-center distance of ~ 1 nm.

Dependence of R on the Position on Substrate

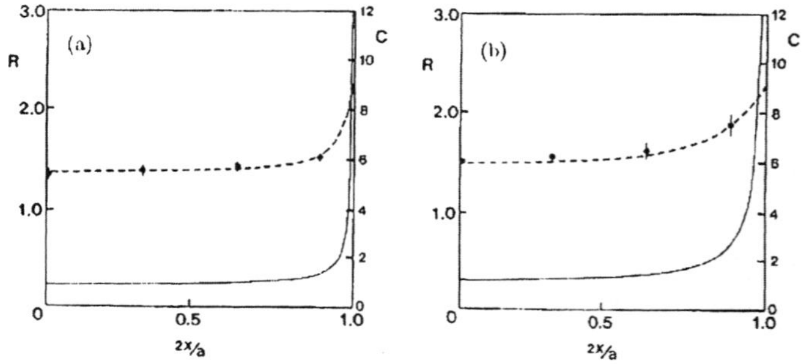

Fig. 9 Dependence of the dichroic ratio R at 590 nm on the relative position $2|x|/a$ on the substrate. (a) v_d=20 mm/min, and (b) 80 mm/min. The substrate width is a=26 mm. The dashed lines are the calculated curves using the set of parameter values listed in Table 1. The corresponding c-parameter is also given by the solid lines. (Reproduced from Ref. [22] by permission of the Physical Society of Japan.)

The flow orientation parameter c for the Bingham case is given by Eq. (42).

As mentioned in §3.5.1, the effect of the relative position is represented by the dimensionless term $(1-4x^2/a^2)^{-1}$, which increases slowly when $2|x|/a \ll 1$, but tends to infinity when $2|x|/a \rightarrow 1$.

Figure 9 shows the experimental data of R plotted against $2|x|/a$. The values of α, ζ' and τ_0 in Table 1 are found to lead to a satisfactory coincidence of the model and the experiment. The dashed and the solid lines show the calculated curves of R and c using Eqs. (38) and (42), respectively.

Dependence of R on the Substrate Width

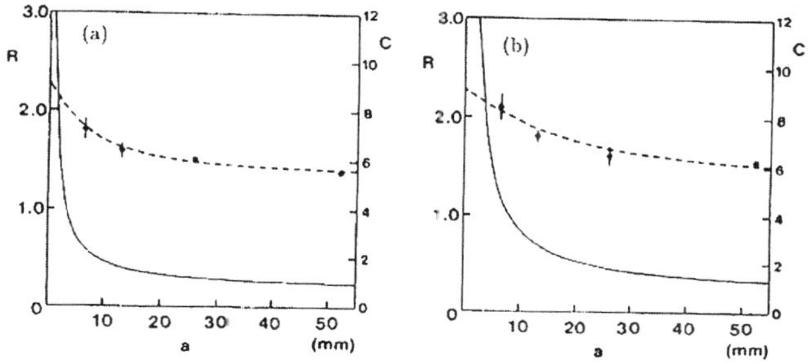

Fig. 10 Dependence of the dichroic ratio R at 590 nm on the substrate width a (for $2|x|/a<<1$). (a) v_d=80 mm/min, and (b) 200 mm/min. The dashed lines are the calculated curves using the set of parameter values listed in Table 1. The corresponding c-parameter is also given by the solid lines. (Reproduced from Ref. [22] by permission of the Physical Society of Japan.)

The flow orientation parameter c around the center of the substrate ($2|x|/a<<1$) is derived from Eq. (42) as,

$$c = \left(\frac{2}{\pi}\right)\left(\frac{\zeta' v_d}{ak_B T}\right) + \frac{\tau_0}{2k_B T}, \tag{45}$$

where the term $1/a$ represents the effect of the width of the substrate a. Therefore, the degree of orientation decreases with a.

The experimental data of R are plotted against a in Fig. 10. A satisfactory coincidence is attained between the model and the experiment using the values of α, ζ' and τ_0 given in Table 1. The dashed and the solid lines show the calculated curves of R and c using Eqs. (38) and (45), respectively.

Figure 11 shows the experimental data of R plotted against the deposition velocity v_d, where a and $2|x|/a$ are kept constant. The cases of $2|x|/a<<1$ are exemplified. The dichroic ratio R increases with v_d and saturates at a constant value as expected from Eq. (45). For $v_d \to 0$, R

remains larger than unity, reflecting the behaviour of the Bingham fluid. The values of α, ζ' and τ_0 in Table 1 are found to result in a good coincidence. The solid line refers to the flow orientation parameter c calculated using Eq. (45), which indicates that c varies linearly with v_d.

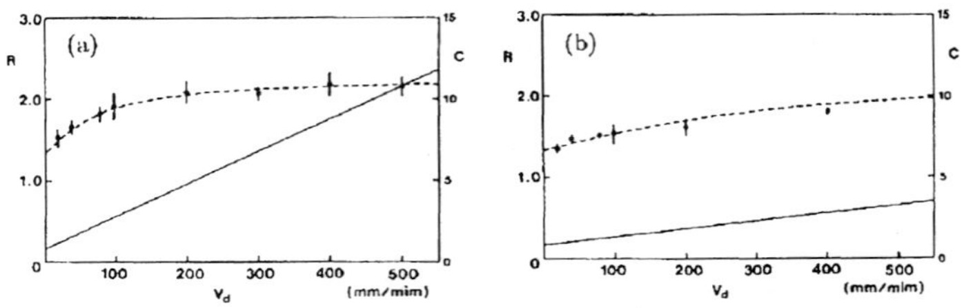

Fig. 11 Dependence of the dichroic ratio R at 590 nm on the deposition velocity v_d (for $2|x|/a<<1$). (a) a=6.5 mm, and (b) 26 mm/min. The dashed lines are the calculated curves using the set of parameter values listed in Table 1. The corresponding c-parameter is also given by the solid lines. (Reproduced from Ref. [22] by permission of the Physical Society of Japan.)

The dashed line shows the curve of R drawn by substituting c for Eq. (38).

3.6.2. Law of Similarity

The c-parameter involves three dimensionless terms as mentioned in §3.5.1: (1) $\zeta' v_d/(ak_BT)$, (2) $(1-4x^2/a^2)^{-1}$ and (3) $\tau_0/(2k_BT)$ (cf. Eq. (42)), leading to a law of similarity governed by the ratio v_d/a, since ζ' and τ_0 are characteristic of the monolayer and T was kept constant throughout the experiment. Table 2 shows five sets of the dichroic data I-V, each sharing the same value of v_d/a. R_{exp} and $\overline{R}_{exp}$ denote the experimental values of the dichroic ratio at the center of the substrate and the average of them, respectively, and R_{cal} refers to the calculation using the values of α, ζ' and τ_0 given in Table 1. The competitive relation between v_d and a has been actually observed, indicating that the law of similarity holds as predicted.

Table 2: The law of similarity

	v_d (s^{-1})	v_d (mm/min)	a (mm)	R_{exp}	$\overline{R}_{exp}$	R_{cal}
I	1.282×10^{-2}	10	13	1.27±0.01	1.33±0.06	1.369
		20	26	1.37±0.07		
		40	52	1.31±0.04		
II	2.564×10^{-2}	10	6.5	1.38±0.03	1.43±0.06	1.411
		20	13	1.44±0.07		
		40	26	1.47±0.05		
		80	52	1.39±0.02		
III	5.128×10^{-2}	20	6.5	1.52±0.10	1.52±0.07	1.491
		40	26	1.53±0.07		
		80	52	1.51±0.03		
IV	1.282×10^{-1}	50	6.5	1.63±0.01	1.62±0.06	1.692
		100	13	1.64±0.04		
		200	26	1.61±0.10		
		400	52	1.62±0.03		
V	2.564	100	6.5	1.91±0.16	1.86±0.12	1.898
		200	13	1.80±0.03		
		400	26	1.81±0.04		

4. APPLICATION TO COMPLICATED CASES

We have reviewed in §3 the simple case of LB deposition onto a single substrate, flat in shape, using a sufficiently large trough. Within the same framework of the analytic approach, the model can be extended to the more complicated cases, of which we will refer to batch production using arrays of substrates, deposition processes in small troughs, and those associated with anomalies such as peeling-off, slipping, and expansion-compression of the monolayer.

4.1. Batch Production of LB Films

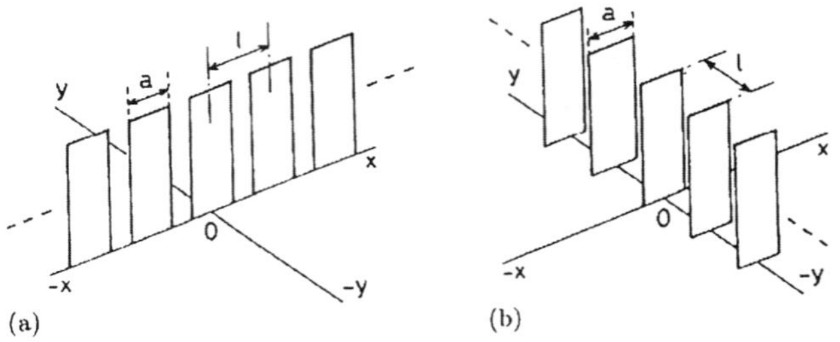

Fig. 12 Infinite arrays of flat substrates aligned (a) side by side, and (b) face to face. *a* and *l* denote the substrate width and the center-to-center. (Reproduced from Ref. [40] by permission of Elsevier.)

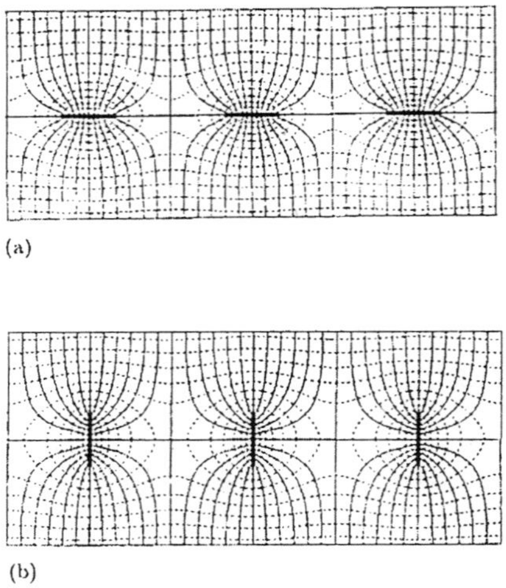

Fig. 13 Schematic representations of (a) the side-by-side, and (b) the face-to-face cases given by the potential functions W_{ss} and W_{ff}, respectively. The solid and the dashed lines refer to the streamlines and the equipotential lines for l/a=3, respectively. The thick rods denote the substrates. (Reproduced from Ref. [40] by permission of Elsevier.)

When the LB films are prepared employing more than one substrate, the velocity potential governing the flow orientation effect is obtainable as the superposition of the potential functions, each analogous to Eq. (7) as representing a substrate [39], [40].

4.1.1. Side-by-side Alignment of Substrates

For an infinite array of flat substrates aligned side by side on the x-axis as shown in Fig. 12(a), the complex velocity potential $W_{ss}(z)$ is formally written using $W(z)$ given by Eq. (7) as a sum [39], [40],

$$W_{ss}(z) = \sum_{k=-\infty}^{\infty} W(z+kl) = -\frac{v_d}{\pi}\int_{-a/2}^{a/2}\left\{\sum_{k=-\infty}^{\infty}\log(z+kl-x')\right\} dx', \tag{46}$$

where l ($\leq a$) is the center-to-center distance between the adjacent substrates.[3]

Equation (46) leads to a potential function,

$$W_{ss}(z) = -\frac{v_d}{\pi}\int_{-a/2}^{a/2} log\left[sin\left\{\frac{\pi(z-x')}{l}\right\}\right] dx', \tag{47}$$

and the positive eigenvalue of velocity gradient tensor **G** near the substrate as,

$$G_{ss} = \frac{v_d}{\pi}\left[\cot\left\{\frac{\pi(x+a/2)}{l}\right\} - \cot\left\{\frac{\pi(x-a/2)}{l}\right\}\right], \tag{48}$$

showing that $G_{SS} \rightarrow \infty$ for $|x| \rightarrow a/2 \pm kl$, while Eq. (30) for the single-substrate case is formally rewritten as,

$$G_{ss} = \frac{v_d}{\pi}\left[\left\{\frac{\pi(x+a/2)}{l}\right\}^{-1} - \left\{\frac{\pi(x-a/2)}{l}\right\}^{-1}\right] \tag{49}$$

Equations (48) and (49) indicate that $G_{SS}<G$ for $|x|<a/2$ since $\cot x/x^{-1}<1$ for any real, non-zero x. The flow orientation effect for the side-by-side case is therefore weaker than that for the single-substrate case.

---

[3] Equation (46) cannot be evaluated since it involves a diverging constant. This difficulty is avoided if it is once differentiated. The summation involved in dW_{ss}/dz is easily carried out, leading to Eq. (47).

The streamlines and the equipotential lines are given by Im W_{ss}=const and Re W_{ss}=const, respectively, for the present case of the side-by-side array. They are shown in Fig. 13(a), in which we recognize stagnation points located at $x=\pm l/2, \pm 3l/2, \pm 5l/2, \ldots$, forming a progression along the x-axis, in contrast to the flow around one single substrate as shown in Fig.6(a).

4.1.2. Face-to-face Alignment of Substrates

When the substrates are aligned face to face on the y-axis as shown in Fig. 12(b), the velocity potential is formally written using Eq. (7) as [39], [40],

$$W_{ff}(z) = \sum_{k=-\infty}^{\infty} W(z+ikl) = -\frac{v_d}{\pi}\int_{-a/2}^{a/2}\left\{\sum_{k=-\infty}^{\infty}\log(z+ikl-x')\right\} dx', \tag{50}$$

which is reduced to an expression as,

$$W_{ff}(z) = -\frac{v_d}{\pi}\int_{-a/2}^{a/2}\log\left[\sinh\left\{\frac{\pi(z-x')}{l}\right\}\right] dx', \tag{51}$$

by employing some manipulations. The positive eigenvalue of the velocity gradient tensor **G** near the substrate is given as,

$$G_{ff} = \frac{v_d}{\pi}\left[\coth\left\{\frac{\pi(x+a/2)}{l}\right\} - \coth\left\{\frac{\pi(x-a/2}{l}\right\}\right], \tag{52}$$

which indicates that $G_{ff}>G$ for $|x|<a/2$, since $\coth x/x^{-1}>1$ for any real, non-zero x (cf. Eq. (49)). The flow orientation effect for the face-to-face case is therefore stronger than that for the single-substrate case.

The streamlines and the equipotential lines, given by Re W_{ff}=const and Im W_{ff}=const, respectively, of the face-to-face case are shown in Fig. 13(b). The stagnation points are now located at $y=\pm l/2, \pm 3l/2, \pm 5l/2, \ldots$.

4.1.3. Comparison with Experimental Results

The predictions for the batch production of LB films have been examined employing finite arrays, involving two up to nine substrates aligned either side-by-side or face-to-face, as representing the corresponding cases of infinite arrays [47], [48].

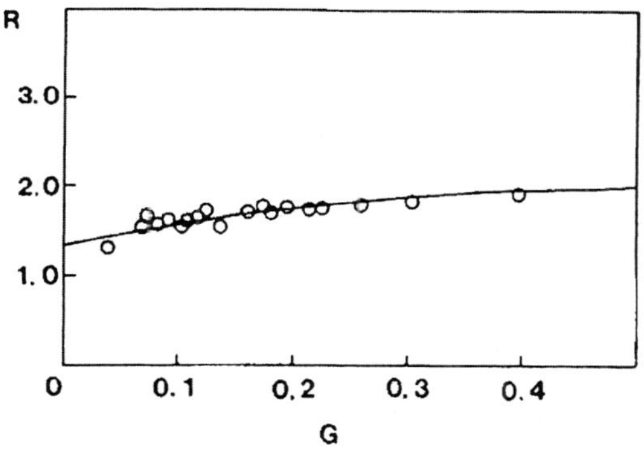

Fig. 14 Dependence of the dichroic ratio R at 590 nm on the positive eigenvalue G_{ss} of the velocity gradient tensor in the side-by-side case. The solid line is the theoretical curve for the infinite array calculated using the set of parameter values listed in Table 1. (Reproduced from Ref. [48] by permission of the American Institute of Physics.)

A mixture of merocyanine DS and arachidic acid C_{20} with [DS] : [C_{20}]=1 : 2 was used as the film-forming material as in the experiment for the single-substrate deposition reviewed in §3.6. The chemical structures of DS and C_{20} are given in Fig. 1. When an array involved three or more substrates, the data were obtained on those in the middle of it. For an array involving only two substrates, the data were from the inner surfaces of them. For further details of the experiment, the readers are referred to the original reports [47], [48].

Figure 14 shows the dichroic ratio R plotted against G_{ss} given by Eq. (48), which is assumed to represent the actual finite arrays of the side-by-side alignment. The open circles refer to the data taken at the center of the substrates x=0 with different values of v_d, a and l. The solid line is the theoretical curve calculated using Eqs. (38), (41) and (48) with $\zeta'=5.0\times10^{-20}$ kg m^2 s^{-1}, $\tau_0=6.2\times10^{-21}$ kg m^2 s^{-2} and $\alpha=33.5^\circ$ given in Table 1, i.e., the same set as estimated for the single-substrate case. As seen in the figure, the dependence of R on G coincides well with that predicted from the flow-orientation model.

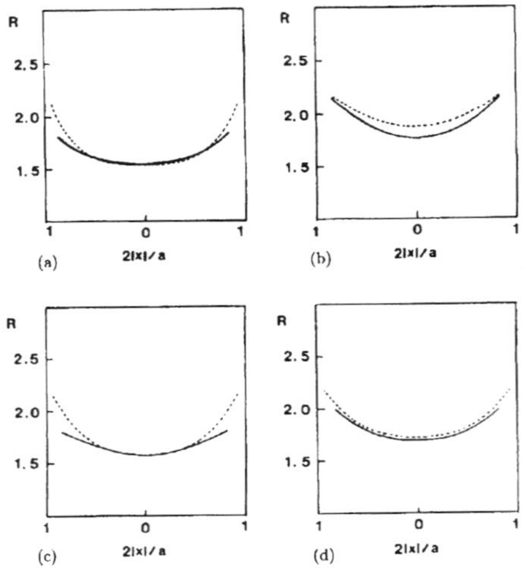

Fig. 15 Dependence of the dichroic ratio R at 590 nm on the relative position $2|x|/a$ in the side-by-side case. The solid line is the theoretical curve for the infinite array calculated using the set of parameter values listed in Table 1. (Reproduced from Ref. [48] by permission of the American Institute of Physics.)

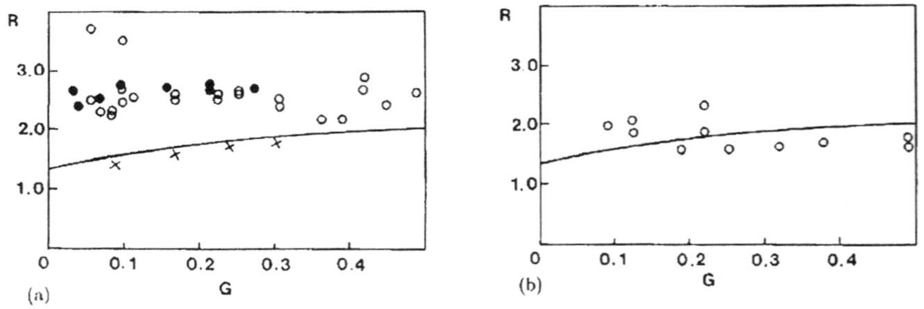

Fig. 16 Dependence of the dichroic ratio R at 590 nm on the positive eigenvalue G_{ff} of the velocity gradient tensor in the face-to-face case. The solid lines are the theoretical curves for the infinite arrays calculated using the set of parameter values listed in Table 1. (a) Open circles, solid circles and crosses are values for $l/a=1$, 2 and 6, respectively. The data for $2|x|/a=0$ are shown. (b) Open circles are values for $l/a=1$. The data for $2|x|/a=5/6$ are referred to. (Reproduced from Ref. [48] by permission of the American Institute of Physics.)

The dichroic ratio R is predicted to be an increasing function of $|x|/a$ according to Eq. (48). Figure 15 shows R plotted against $2|x|/a$. The solid lines are the experimental curves. The observed tendency of increasing R towards the edge of the substrate is well reproduced by the dashed curves calculated using the same set of parameter values as given in Table 1.

Dichroic Behaviour in Face-to-face Alignment

Figure 16 shows the dichroic ratio R plotted against G_{ff} given by Eq. (52) assumed to represent the finite face-to-face case. The open circles, solid circles and crosses are the R-values at x=0 for l/a=1, 2 and 6, respectively. The solid line is the theoretical curve calculated using Eqs. (38), (41) and (52) with the values given in Table 1. R for l/a=1 and 2 do not fit with the calculated curves, but are independent of G_{ff}, while the behaviour of R for l/a=6 is well reproduced by the curve.

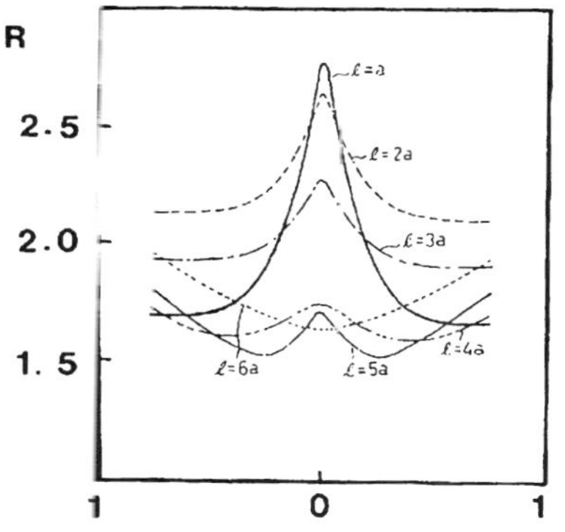

Fig. 17 Dependence of the dichroic ratio R at 590 nm on the relative position on the substrate $2|x|/a$ in the face-to-face case with v_d=100 mm/min. When $l/a \leq 5$, R has a peak at the center of the substrate. When $l/a \leq 6$, the dependence of R on $|x|/a$ is consistent with the theory. (Reproduced from Ref. [48] by permission of the American Institute of Physics.)

The dependence of R on $|x|/a$ is shown in Fig. 17. The actual behaviour of R differs from Eq. (52) which predicts an increasing function of $|x|/a$. For $l/a \leq 3$, R is independent of G_{ff}, having a sharp peak at x=0 with its skirt going downward as $x \rightarrow \pm a/2$. This tendency is observed for the whole range of the deposition velocity v_d = 30 – 400 mm/min.

The peak height decreases with increasing relative distance l/a between the adjacent substrates. At l/a=6, R comes to be an increasing function, showing a G_{ff}-dependence as seen in Fig. 16. The curves for l/a=4 and 5 represent the intermediate between both extremes, characterized with two minima located symmetrically with respect to the center (x=0).

Unsteady Flow in the Face-to-face Case

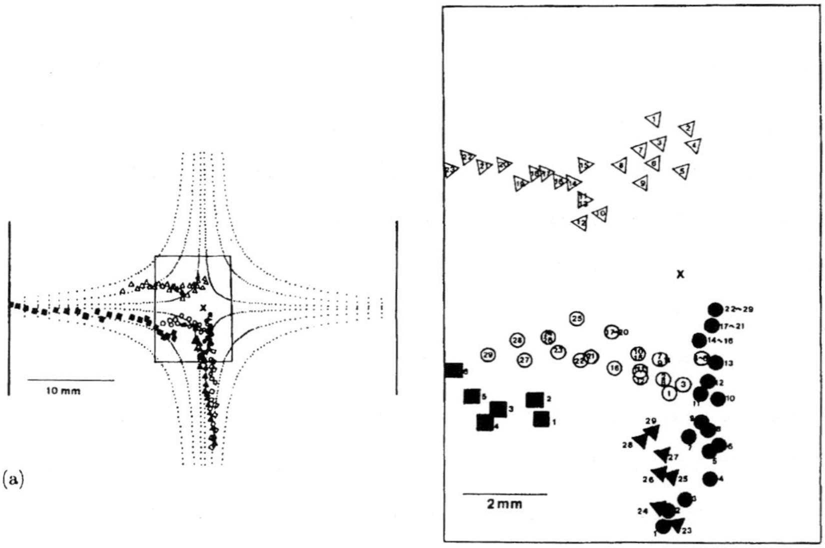

(a)

(b)

Fig.18 (a) Traces of six different particles recorded during the deposition with the array of substrates aligned face to face. The positions of the particles are written every 0.5 s at the same time. The dots refer to the theoretical traces for the interval of 0.5 s calculated using Eq. (50). The solid rods show the adjacent substrates. The cross denotes the stagnation point. (b) The traces of the particles in the region near the stagnation point. The numerical figure added to each point correspond to the elapsed time from the onset of the measurement, e.g., 10 denotes 0.5×10=5 s. (Reproduced from Ref. [48] by permission of the American Institute of Physics.)

We have seen that a significant discrepancy between the model and the experiment appears in the face-to-face case when $l/a<6$, while the model for the side-by-side case has led to a satisfactory coincidence. This discrepancy may be related with the difference between the both cases in the occurrence of the stagnation points. They are located between the edges of the adjacent substrates in the side-by-side case (Fig.13(a)), but directly faced by the substrate surfaces in the face-to-face case (Fig.13(b)), resulting in certain influence.

In order to examine this, the flow of monolayer on the water surface has been observed by taking photographs using the stroboscope lamp. After the DS-C_{20} mixed monolayer was compressed, tiny lumps of sulphur powder were placed on it and traced by successively taking photographs at a constant time-interval during the deposition. The photographs were then transformed into the top-view pictures and the displacement of each particle was traced.

Figure 18(a) shows six different particles successively traced in one and the same set of photographs as representing the monolayer flow. The dotted lines are the theoretical traces of particles calculated with an interval of 0.5 s using Eq. (51). The flow near the stagnation point is shown in Fig. 18(b) in an enhanced scale. In this region, each particle moves unsteadily here and there like random walk. Consequently, the path line or the trajectory of the particle does not coincide any more with the theoretical streamline.

It is therefore shown that there is an actual discrepancy between the flow predicted by the ideal fluid approximation and the observed one. The influence of the anomalous motion as reviewed here should decrease with increasing l. The results shown in Fig. 16 suggest that the influence becomes small enough for $l/a \leq 6$ in comparison to the typical flow orientation effect as described in §3. It should be noted that the maximum of R observed at the center of substrate for smaller values of l/a remains to be quantitatively explained.

4.2. Deposition in Small Troughs

As reviewed in §3.6, the model and the experiment have well coincided with each other when the trough size is large enough to allow a space for four times the substrate $4a$. Further, we have seen in the preceding subsection that the substrate-to-substrate distance of $6a$ plays an important role in the face-to-face case of LB batch production. Let us examine now the LB deposition in small troughs, i.e., the cases when these $4a$- and $6a$-rules may be violated. We will examine two different types of rectangular troughs. They are the two-side compression and the one-side compression types, each having a pair of moving barriers with a pair of fixed side-walls or one single barrier with a pair of side walls and an end wall. The substrate is assumed to be centered on the bisector between the side walls, and two cases for each type will be dealt with:

(1) the transversal position of a substrate with its surface facing the barrier, and (2) the longitudinal position where the substrate surface faces the side wall.

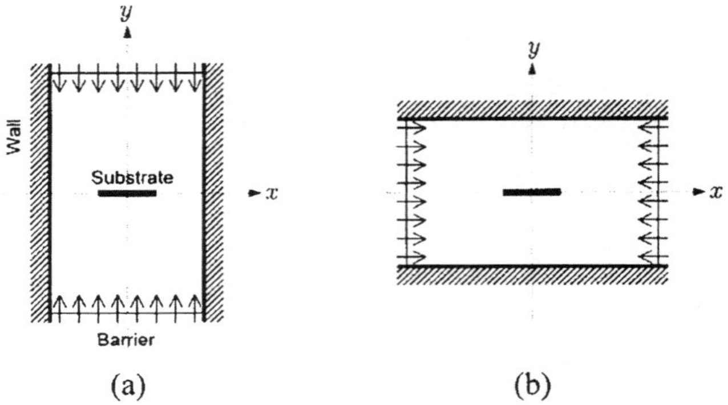

Fig. 19 Top-view figures of various small-trough configurations for the two-side compression. (a) and (b) refer to the cases with the substrate in transversal and longitudinal positions, respectively.

4.2.1. Two-side Compression Cases

Let l and l_l be the width and the length of the monolayer in a trough, where l and l_l correspond to the wall-to-wall and the barrier-to-barrier distances, respectively. The water-wall boundary condition is that the wall front itself represents a streamline. It is evident from Figs. 13(a) and

(b) that the water-wall boundary condition is satisfied by both W_{ss} and W_{ff} derived for the side-by-side and the face-to-face cases, respectively, if each bisector between the adjacent substrates is identified with a wall front. The water-barrier boundary condition, on the other hand, is that the barrier front itself represents an equipotential line. Although the barriers sweep the monolayer towards the substrate as the deposition proceeds, the associated decrease in l_l will be neglected for simplicity. The boundary condition is instead satisfied by assuming appropriate fictive sources and sinks outside of the $l \times l_l$ monolayer area with a substrate at its center.

The velocity potential $W_{ttr}(z)$ for the transversal position as shown in Fig. 19(a) can be formally written using $W_{ss}(z)$ for the side-by-side case (Eq. (47)) as a superposition,

$$W_{ttr}(z) = \sum_{k=-\infty}^{\infty}(-1)^k W_{ss}(z + ikl_l), \tag{53}$$

which satisfies the water-barrier boundary condition due to the $2l_l$-periodicity along the y-axis. Here, the above W_{ttr} corresponds to a two-dimensional array composed of the alternate stacking of the side-by-side-type sinks and sources with a constant interval l_l along the y-axis which coincides with the longitudinal direction of the trough. The trough is therefore represented by an $l \times l_l$ area surrounding a sink. Equation (53) is explicitly written as,

$$W_{ttr}(z) = -\frac{v_d}{\pi}\sum_{k=-\infty}^{\infty}(-1)^k \int_{-a/2}^{a/2} \log\left[\sin\left\{\frac{\pi(z + ikl_l - x')}{l}\right\}\right] dx', \tag{54}$$

and the positive eigenvalue of velocity gradient tensor near the substrate is given by,

$$G_{ttr} = \frac{v_d}{l}\sum_{k=-\infty}^{\infty}(-1)^k\left[\frac{\sin\{2\pi(x + a/2)/l\}}{\cosh(2\pi kl_l/l) - \cos\{2\pi(x + a/2)/l\}} - \frac{\sin\{2\pi(x - a/2)/l\}}{\cosh(2\pi kl_l/l) - \cos\{2\pi(x - a/2)/l\}}\right], \tag{55}$$

where the term for k=0 at the right-hand side coincides with G_{ss} given by Eq. (48) as a matter of consequence.

Each term at the right-hand side of Eq. (55) involves a hyperbolic function $\cosh(2\pi k l_t/l)$ evaluated as $\exp(2\pi|k|l_t/l)/2$ when $|k|l_t/l \rightarrow \infty$, while others are the trigonometric functions oscillating in the range of ±1. Equation (55) is therefore approximated for $2\pi|k|l_t/l >> 1$ as,

$$G_{ttr} \approx G_{ss} + \frac{2v_d}{l}\sum_{k=-\infty}^{\infty}(-2e^{-2\pi l_t/l})^k\left[\sin\left\{\frac{2\pi(x+a/2)}{l}\right\} - \sin\left\{\frac{2\pi(x-a/2)}{l}\right\}\right]$$

$$= G_{ss} - \frac{4(v_d/l)e^{-2\pi l_t/l}}{1+2e^{-2\pi l_t/l}}\left[\sin\left\{\frac{2\pi(x+a/2)}{l}\right\} - \sin\left\{\frac{2\pi(x-a/2)}{l}\right\}\right]. \quad (56)$$

Let us estimate the relative contribution of each term in Eq. (56) to G_{ttr}. The main contribution comes from $G_{ss} \geq (2v_d/l)\cot\{\pi a/(2l)\}$ unless l/a=1. The second term has a negative value to weaken the orientation effect, however, even for an extreme case of l_t/l=1 (i.e., a square trough), $e^{-2\pi}=1.87\times10^{-3}$ and the resultant decrease in G_{ttr} is less than 1% for $l/a \geq 2$.

It is therefore shown that the deposition with a substrate in transversal position can be identified in its orientation effect with that using a side-by-side array of substrates in any realistic case of $l_t/l \geq 1$ and $l/a \geq 2$. As a consequence, the streamlines and the equipotential lines should be similar to those shown in Fig. 13(a), with the pair of adjacent bisectors being regarded as the trough walls. Here, $l_t/l \rightarrow \infty$ and l/a=1 refer to an extreme case when the substrate and the trough shares the same width and the trough has an infinite length. Then, the flow orientation effect ceases to exist, and the velocity potential is given by,

$$W(z) = \pm i v_d z, \quad (57)$$

for the upper and the lower half planes (y>0 and <0), respectively [38].

Equation (57) suggests that the monolayer can be deposited without any deformations if we use a substrate sharing the same width with the trough in transversal position, since all the elements of velocity gradient tensor vanish due to $d^2W/dz^2 = 0$. It is noted that this situation has been realized in Kumehara-Miyata method [49] employing a pair of sliding walls running parallel to each other at a distance equal to the substrate width.

Longitudinal Case

For the longitudinal case as shown in Fig. 19(b), the velocity potential $W_{tlo}(z)$ can be formally written using $W_{ff}(z)$ for the face-to-face alignment given by Eq. (51) as,

$$W_{tlo}(z) = \sum_{k=-\infty}^{\infty}(-1)^k W_{ff}(z+kl_l), \tag{58}$$

by which the water-barrier boundary condition is satisfied due to the $2l_l$-periodicity along the x-axis. Equation (58) refers to a two-dimensional array composed of the alternate stacking of the face-to-face-type sinks and sources along the x-axis with an interval l_l. The longitudinal direction of the trough is now represented by the x-axis. $W_{tlo}(z)$ is explicitly written as,

$$W_{tlo}(z) = -\frac{v_d}{\pi}\sum_{k=-\infty}^{\infty}(-1)^k \int_{-a/2}^{a/2} \log\left[\sinh\left\{\frac{\pi(z+kl_l-x')}{l}\right\}\right] dx', \tag{59}$$

which leads to the positive eigenvalue of velocity gradient tensor near the substrate expressed by,

$$G_{tlo} = \frac{v_d}{l}\sum_{k=-\infty}^{\infty}(-1)^k\left[\frac{\coth\{2\pi(x+a/2)/l\}\coth(2\pi kl_l/l)+1}{\coth\{2\pi(x+a/2)/l\}+\coth(2\pi kl_l/l)}\right.$$
$$\left.-\frac{\coth\{2\pi(x-a/2)/l\}\coth(2\pi kl_l/l)+1}{\coth\{2\pi(x-a/2)/l\}+\coth(2\pi kl_l/l)}\right]. \tag{60}$$

where the k=0 term at the right-hand side coincides with G_{ff} in Eq. (52) as expected.

Let us examine the effect of $k \neq 0$ terms at the center of the substrate. For x=0, Eq. (60) is reduced as,

$$G_{tlo} = G_{ff} + \frac{4v_d}{l} \sum_{k=-\infty}^{\infty} \frac{\coth(\pi a/l)\{\coth^2(2\pi k l_t/l) - 1\}}{\coth^2(2\pi k l_t/l) - \coth^2(\pi a/l)}, \tag{61}$$

which indicates that each term in the summation makes a negative contribution to G_{tlo} unless $k<a/(2l_l)$ is satisfied. This is however an impossible condition: even for k=1, for instance, $l_l<a/2$, i.e., the substrate width is twice or more larger than the trough length. Each $k \neq 0$ term involves a factor $\{\coth^2(2\pi kl_l/l)-1\}$ which decreases, tending to zero for $kl_l \rightarrow \infty$. In the extreme case of a square trough $l=l_l$, $\{\coth^2(2\pi kl_l/l)-1\} = 1.40\times10^{-5}$ for k=1, showing that the contributions from the $k \neq 0$ terms are negligible, and therefore $G_{tlo} \approx G_{ff}$ for any realistic case.

The deposition with a substrate in longitudinal position is therefore found to be identifiable in its orientation effect with that using a face-to-face array of substrates. Consequently, the streamlines and the equipotential lines should be similar to those shown in Fig. 13(b), where the pair of adjacent bisectors correspond to the trough walls. The longitudinal position or the substrate surface facing the wall is therefore as effective as the face-to-face alignment to enhance the degree of orientation, in contrast to the transversal position. This is consistent with the experimental results reported so far: the higher ordering is found in LB films deposited with the substrate surface held parallel to the wall than otherwise [50], [51].

According to the present model, G_{tlo} for given widths a and l is predicted to be insensitive to the trough length l_l, quickly saturating towards the upper limit identical with G_{ff}. It is therefore suggested that the use of very long trough is not necessarily an effective measure to enhance the degree of orientation.

4.2.2. One-side Compression Cases

So far we have dealt with the deposition in the central area of water surface using a rectangular trough having a pair of barriers facing each other to compress the monolayer, i.e., the two-side compression cases with the point symmetry. While LB troughs for the two-side compression are available, most of the commercial apparatuses are of single-barrier type for the one-side compression.

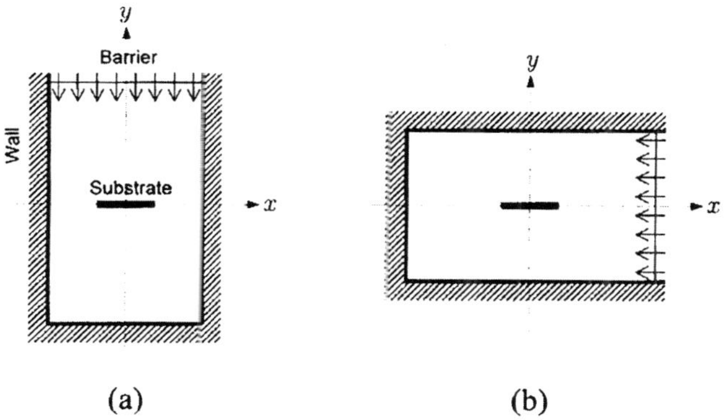

Fig. 20 Top-view figures of various small-trough configurations for the one-side compression. (a) and (b) refer to the cases with the substrate in transversal and longitudinal positions, respectively.

Let us examine the deposition using this type of troughs. The trough has now three fixed walls, i.e., an end wall besides a pair of side walls, and the point symmetry does not hold any more with respect to the center of the trough. Here, l denotes now the separation between the side walls, while l_e and l_b are defined as the distances from the substrate center to the end wall and the barrier front, respectively, with $l_e + l_b$ being the total length of the trough.

Transversal Case

The transversal case as shown in Fig. 20(a) is inevitably associated with an asymmetry of the barrier and the end-wall sides of substrate. The positive eigenvalue of velocity gradient tensor should therefore be separately obtained for each surface. Further, it would be difficult to simulate this asymmetry by introducing fictive sources and sinks outside of the $l \times (l_e + l_b)$ area surrounding the substrate. The eigenvalues are however plausibly evaluated using the expressions for the two-side cases Eqs. (55) and (60) as,

$$G^{b}_{otr} \approx (G_{ttr})_{l_t=2l_b}, \quad \text{and,} \quad G^{e}_{otr} \approx (G_{tlo})_{l_t=l,l=2l_e} \tag{62}$$

where G^{b}_{otr} and G^{e}_{otr} refer to the barrier side and the end-wall side, respectively, and the suffixes denote the replacements of variables.

When l_b>>l>>l_e and the substrate width a is appreciably smaller than l, Eq. (62) can further be approximated as,

$$G^{b}_{otr} \approx G_{ss}, \quad \text{and} \quad G^{e}_{otr} \approx (G_{ff})_{l=2l_e}, \tag{63}$$

using Eqs. (48) and (52) for the infinite side-by-side and face-to-face arrays given in the preceding subsections, respectively.

Longitudinal Case

The velocity potential for the longitudinal case as shown in Fig. 20(b) can be written using a fictive potential as,

$$W_{olo}(z) = \{W_{tlo}(z) + W_{tlo}(z + 2l_e)\}_{l_t=2l_b}, \tag{64}$$

where $W_{tlo}(z)$ is the longitudinal-case potential for the two-side compression given by Eq. (59). The x-axis corresponds to the longitudinal direction in this case, and the second term is a fictive sink located at x=$-2l_e$. Thus, the end wall at x=$-l_e$ itself represents a streamline. The positive eigenvalue of velocity gradient tensor is written as,

$$G_{olo}(x) = G_{tlo}(x) + G_{tlo}(x + 2l_e), \tag{65}$$

using Eq. (60) for the two-side compression case.

The degree of orientation is more or less asymmetric with respect to the center of substrate x=0. The most prominent asymmetry is predicted for the extreme case of l_e=a/2 when one of the edges contacts the end wall. $W_{olo}(z)$ is now equivalent to the longitudinal-case potential for the two-side compression with a fictive substrate of doubled width $2a$ centered at $x=-l_e$, of which the right half represents the real substrate with its left-side edge corresponding to the center of the fictive substrate. As a consequence, the degree of orientation increases with x. This increasing tendency weakens with increasing l_e, since the second term $\{W_{tlo}(z+2l_e)\}_{l_f=2l_h}$ in Eq. (64) approaches a potential for a uniform flow from right to left $-(a/l)v_d z$ which makes no contribution to the velocity gradient tensor, i.e., the second term at the right-hand side of Eq. (65) tends to vanish, leading to $G_{olo}=G_{tlo}$ for $l_e \to \infty$.

4.3 Anomalous Deposition

We have so far dealt with the potential functions given by Eq. (7) and its superpositions, where the monolayer on the water surface, assumed as an ideal fluid, is transferred without any expansion or contraction. This situation can be characterized with the deposition ratio of r_d=1. Here, r_d denotes the ratio of the area occupied by the monolayer on the water surface to that after the transfer onto the substrate.

r_d>1 and r_d<1 may correspond to the contraction and the expansion, respectively, and r_d <0 refers to the case when the monolayer once deposited is peeled off from the substrate. In the following, a brief review will be given of the expressions derived to describe the anomalous deposition processes with $r_d \neq 1$ within the framework of the analytic model shown in §3 [39], [40].

The monolayer must be compressed up to the solid-like state prior to the transfer, where the incompressibility condition is inevitably violated. This occasionally induces an effect referred to as the compression orientation, which will also be touched upon here, although the process in itself is regular rather than anomalous.

4.3.1. Asymmetric Deposition

Let us consider an anomalous deposition in which the monolayer is normally deposited at the front of substrate but peeled off at the rear. The potential function $W_a(z)$ for this case can be written as,

$$W_a(z) = \lim_{\varepsilon \to +0} \{W(z+i\varepsilon) - W(z-i\varepsilon)\}, \tag{66}$$

using $W(z)$ of the single-substrate case given by Eq. (7), where ε is a real number. Equation (66) refers to a pair of substrates held together with an infinitesimal distance. The upper and the lower substrates are characterized with the normal deposition with r_d=1 and the complete peeling-off with r_d=−1, respectively.

After some manipulations, we derive,

$$W_a(z) = i\frac{v_d}{\pi}\int_{-a/2}^{a/2} \frac{\sqrt{a^2/4 - x'^2}}{z - x'}\, dx' = iv_d\left(z + \sqrt{z^2 - a^2/4}\right), \tag{67}$$

which satisfies $v_y \to -v_d$ for $y \to 0$ at both front and rear of the substrate. The streamlines and the equipotential lines of $W_a(z)$ are shown in Fig. 21(a).

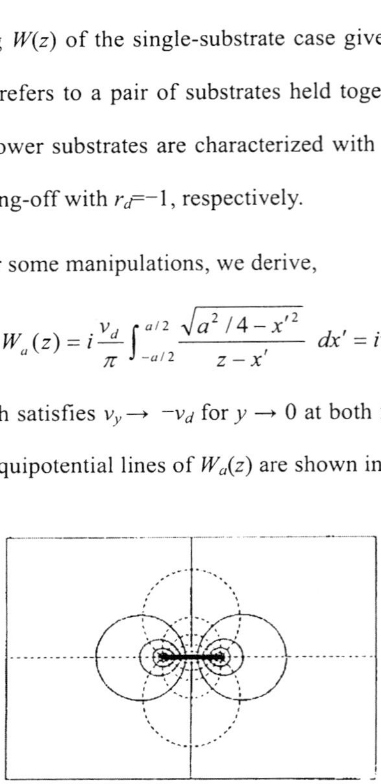

(a)

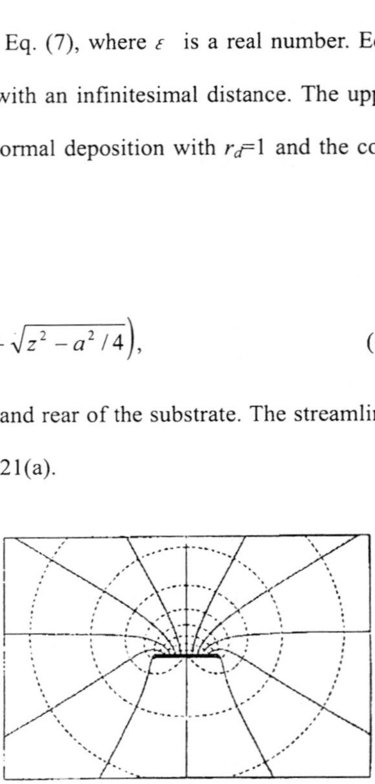

(b)

Fig. 21 Complex velocity potential functions (a) W_a and (b) $(W+W_a)/2$ shown schematically for the asymmetric cases. The solid and the dashed lines refer to the streamlines and the equipotential lines, respectively. The thick rods denote the substrates. (Reproduced from Ref. [40] by permission of Elsevier)

The positive eigenvalue of the velocity gradient tensor near the substrate is given by,

$$G_a = \left(\frac{2v_d}{a}\right)\left(1-\frac{4x^2}{a^2}\right)^{-3/2}, \tag{68}$$

which shows stronger divergence at $|x|=a/2$ than that for the single-substrate case given by Eq. (30).

Linear combinations of $W(z)$ and $W_a(z)$, given by Eqs. (7) and (67), respectively, may express more realistic cases. The arithmetic average $(W+W_a)/2$, for instance, can be related to the normal deposition $r_d=1$ at the front with complete slipping of the monolayer $r_d=0$ at the rear. The streamlines and the equipotential lines of $(W+W_a)/2$ are given in Fig. 21(b).

4.3.2. Fractional Deposition Ratios

When the monolayer is transferred from the air-water interface onto the substrate, the y-axis is turned by 90° to be identical with the dipping direction as mentioned in §3.2.2. The angular distribution $P(\theta)$ has been so far assumed to be frozen at the transfer. For the deposition ratio of $r_d=\pm 1$, this assumption is compatible with the ideal fluid approximation, i.e., the area of monolayer is incompressible.

In the case of fractional values of r_d, however, either the uniform coverage of the substrate or the invariance of the area of monolayer is violated as a consequence, or even the both may not hold any more. Of many possibilities, some examples for $0<r_d<1$ will be reviewed here [39], [40].

Discontinuous Deposition

If the invariance of the area is conserved, the monolayer should be disrupted into fragments, which can be patches of fractal nature or may retain certain regularity in size and shape. The latter can be exemplified by two cases, each referring to horizontal or vertical stripes.

In the horizontal case, $P(\theta)$ given by Eq. (33) could remain valid, with the coverage identical with r_d. The vertical case is, however, inevitably associated with the change in angular orientation. The monolayer expands by a factor $1/r_d$ in the dipping direction and contracts by r_d in the horizontal direction. As a consequence, the distribution function $P_v(\theta)$ would be expressed as,

$$P_v(\theta) \propto \frac{1}{1-(1-r_d^4)\cos^2\theta} \exp\left\{-\frac{1}{1-(1-r_d^4)\cos^2\theta}\left(\frac{\zeta G}{k_B T}\right)\sin^2\theta\right\}, \tag{69}$$

by assuming a projection with the expansion and contraction along the vertical and horizontal axes, respectively [39], [40]. The factor $1/\{1-(1-r_d^4)\cos^2\theta\}$ involved in Eq. (69) has a maximum at $\theta=0$ for $0<r_d<1$. The distribution $P_v(\theta)$ is therefore predicted to be sharper than $P(\theta)$ given by Eq. (33) for the normal deposition.

Stretched Deposition

The uniform coverage for $r_d<1$ apparently contradicts the assumption of ideal fluid. The monolayer is stretched along the dipping direction, and the particles may be rearranged at the transfer, resulting in an angular distribution function totally different from Eq. (33).

If we compare the stretched transfer to a projection with the vertical axis expanded by a factor $1/r_d$, we derive an angular distribution function,

$$P_s(\theta) \propto \frac{1}{1-(1-r_d^2)\cos^2\theta} \exp\left\{-\frac{1}{1-(1-r_d^2)\cos^2\theta}\left(\frac{\zeta G}{k_B T}\right)\sin^2\theta\right\}, \tag{70}$$

as a formal expression for the stretched deposition [39], [40]. The above $P_s(\theta)$ is predicted to be sharper than $P(\theta)$ without stretching, since the factor $1/\{1-(1-r_d^2)\cos^2\theta\}$ involved has a maximum at $\theta=0$.

The stretched transfer may actually take place due to the decrease in surface pressure towards the substrate. Indeed, this effect has been discussed for monolayers with larger compressibilities [50]. It is therefore indispensable to refer to the deposition ratio r_d as an index of prime importance in any attempt to clarify quantitative aspects of the transfer process.

4.3.3. Effect of Compression Process

The monolayer at the air-water interface is compressed prior to the deposition onto the substrate. In this stage, the monolayer is compressible as a matter of consequence and subjected to deformation which may orient the elongated particles involved in it [51]. Two extremes, the solid and the fluid models, will be exemplified assuming the uniaxial compression of monolayer along the y-axis with the width held constant.

Solid Model

If the monolayer is characterized as a plastic solid, the Brownian motion of the particle will be negligible during the compression process, and the angular distribution can be compared to a parallel projection with a contraction by r_c along the y-axis, where $r_c \equiv A_i/A_f$ denotes the ratio between the initial and the final areas, A_i and A_f, respectively.

Let us assume that the monolayer is isotropic in the initial state. The angular distribution is then governed by r_c alone and written as,

$$P_{sol}(\theta) = \left(\frac{r_c}{\pi}\right)\frac{1}{1+(r_c^2-1)\cos^2\theta}, \tag{71}$$

referring to Eq. (70). Here, the initial value A_i may be identified as the limiting area A_0 representing the onset of the solid-like phase. The preferred direction corresponds to the x-axis, since Eq. (71) has a maximum at $\theta=\pi/2$ when $r_c>1$.

Fluid Model

When the monolayer is characterized as a fluid, the flow of monolayer can be represented by that shown in Fig. 5(d), i.e., $v_x=0$ and $v_y=v_0-gy$ with $g>0$, where $g\equiv-(1/A)(dA/dt)$ is an experimentally adjustable parameter known as the compression rate. The velocity gradient tensor is then written as,

$$\mathbf{G}_{fl}=\begin{pmatrix} 0 & 0 \\ 0 & -g \end{pmatrix}, \tag{72}$$

which leads to an expression by using Eq. (18) as,

$$\dot{\theta}=\frac{1}{2}g\sin 2\theta, \tag{73}$$

having a maximum $\dot{\theta}_{\max}=g/2$ at $\theta=\pi/4$. We obtain a diffusion equation for this case by substituting Eq. (73) for Eq. (25). If we employ the local thermal equilibrium approximation, the angular distribution function is derived as,

$$P_{fl}(\theta)=\left(\frac{e^{c_{fl}}}{\pi I_0(c_{fl})}\right)\exp(-2c_{fl}\cos^2\theta), \tag{74}$$

using a similar procedure as shown in §3.2.2. Here, I_0 is the modified Bessel function of the zeroth order, and c_{fl} is defined as,

$$c_{fl}=\frac{\zeta\dot{\theta}_{\max}}{2k_BT}=\frac{\zeta g}{4k_BT}, \tag{75}$$

where ζ denotes the friction coefficient. Equation (74) shows that the preferred direction coincides with the x-axis.

If the monolayer is a Newtonian fluid, the compression orientation is of transient nature, and should vanish sooner or later after the compression. In contrast, if the monolayer is a Bingham fluid characterized with $\zeta=\zeta'+\tau_0/\dot{\theta}_{max}$, the effect is frozen as long as the flow torque remains below τ_0. The frozen orientation can be transferred onto the substrate by using the stamping or horizontal-lifting method of Nakahara and Fukuda [52]. In the case of vertical dipping technique, however, the flow orientation effect is set to work at the onset of the deposition, and the frozen distribution begins to melt away. There arises a question as to how the initial orientation in the monolayer at the air-water interface is lost during the deposition process. This will be quantitatively examined in §5 by the numerical calculation based on a more precise model.

5. NUMERICAL CALCULATION BASED ON A MORE PRECISE MODEL

Although the analytical model of flow orientation have been found to agree well with the dichroic behaviour of the mixed LB films of a merocyanine dye DS and a fatty acid C_{20} (cf. Fig. 1) for the case of one single substrate in a large trough, it is not always adequate for explaining the experimental results as seen in §4.1 for the face-to-face case of batch production of LB films, where the observed discrepancy remains to be quantitatively explained.

To know the limits inherent in this model, it is necessary to examine to what extent the local thermal equilibrium and the conservation of in-plane anisotropy on the transfer are relevant to the actual LB deposition process. Indeed, the dynamic aspects of LB deposition process have been referred to as qualitatively explaining the experimental observations [50]. In this respect, it will be useful to examine the dynamic aspects of the process based on a more precise model.

The angular distribution function $P(\theta)$ have been numerically calculated without assuming the thermal equilibrium, introducing a Markov-type integral expression [43] of the flow orientation process with an intermediate meniscus region allowed for between the water surface and the substrate [53], [54]. A good coincidence of the calculation with the experiment has been obtained using the DS-C_{20} 1 : 2 mixed LB films. The results will be reviewed in this section.

5.1. Flow Orientation Effect as a Markov Process

5.1.1. Integral Expression of Angular Distribution Function

As the angular distribution function for the long axis of the rod-like particle, let $P(\theta,t)\,d\theta$ be the probability for the angle of the long axis of the particle θ to lie at time t in a range between θ and $\theta+d\theta$. Then, by employing the Markovian transition probability density function $Tr(\Delta\theta, \Delta t; t)$, the angular distribution function is given by [43], [53], [54],

$$P(\theta,t+\Delta t)=\int_{-\pi/2}^{\pi/2}P(\theta-\Delta\theta,t)Tr(\Delta\theta,\Delta t;t)d(\Delta\theta),$$

where,

$$Tr(\Delta\theta,\Delta t;t)=\sum_{n=-\infty}^{\infty}Tr'(\Delta\theta+n\theta,\Delta t;t),$$

$$Tr'(\Delta\theta,\Delta t;t)=\frac{1}{\sqrt{4\pi D(t)\Delta t}}\exp\left[-\frac{\{\Delta\theta-(d\theta/dt)\Delta t\}^2}{4D(t)\Delta t}\right]. \tag{77}$$

In Eq. (77), $D(t)$ is the rotatory diffusion coefficient, and $d\theta/dt$ is the angular velocity around the center of the particle (x,y) defined as a function of time t. Equation (76) is an integral expression equivalent to the differential equation,

$$\frac{\partial P}{\partial t}=-\frac{\partial}{\partial\theta}\left(\frac{d\theta}{dt}P\right)+D(t)\frac{\partial^2 P}{\partial\theta^2}, \tag{78}$$

which is identical with Eq. (25). [4]

According to the Einstein relation, the diffusion coefficient for an isotropic Bingham fluid is

---

[4] The equivalence of the integral and the differential forms, Eqs. (76) and (78), respectively, is demonstrated in Ref. [54].

written as,

$$D(t) = \frac{k_B T}{\zeta' + \{\tau_0 /(d\theta / dt\}_{max}}, \tag{79}$$

where k_B is the Boltzmann constant, T is the absolute temperature, and ζ' and τ_0 a re the Newtonian rotatory friction coefficient and the Bingham yield value, respectively. In Eq. (79), $(d\theta/dt)_{max}$ is the maximum shear rate given as the maximum value of the angular velocity $d\theta/dt$ that is explicitly written as Eq. (18) in §3.2.1:

$$\frac{d\theta}{dt} = \frac{G_{xx} - G_{yy}}{2} \sin 2\theta + G_{yx} \cos 2\theta + (G_{xy} - G_{yx}) \cos^2 \theta, \tag{80}$$

where G_{xx}, G_{xy}, G_{yx} and G_{yy} are the afore-mentioned elements of the velocity gradient tensor **G**.

5.1.2 Velocity Vectors on the Water Surface and the Substrate

On the water surface, the flow of a monolayer is governed by Eq. (7) as $W(z) = -(v_d / \pi) \int_{-a/2}^{a/2} \log(z - x')dx'$. On the substrate, the monolayer should move either upward or downward together with the substrate. The velocity potential is therefore given by,

$$W_s(z) = iv_d z, \tag{81}$$

to satisfy the condition of the velocity vector parallel to the dipping direction, where all the second derivatives vanish, indicating the absence of the deformation-orientation effect. It is noted that Eq. (81) is identical with Eq. (57) given in §4 as representing the potential for the monolayer at the air-water interface [38] of the Kumehara-Miyata method [49].

To connect the streamlines on the water surface with the substrate, two different models have been examined as the approximations [53], [54]. Model I refers to the abrupt connection of the streamlines on the water surface and those on the substrate satisfying a condition,

$$\tan\theta_s = \tan\theta_w - \tan\theta_t, \tag{82}$$

where the suffices s and w refer to the substrate and the water surface regions, respectively, and θ_t is the angle between the streamline and the substrate normal immediately before the transfer as shown in Fig. 22(a).

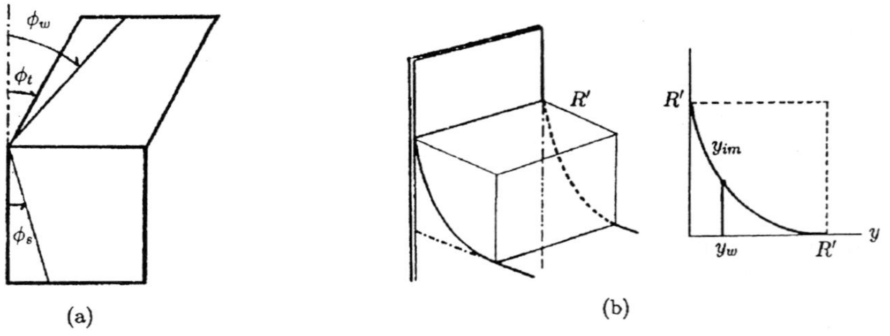

Fig. 22 (a) Model I, the abrupt connection model: The intermediary region is represented by a line. ϕ_s and ϕ_w are the angles between the long axis of the particle and the y-coordinate on the water surface and on the substrate, respectively. ϕ_t is the angle between the streamline and the substrate normal immediately before the transfer. (b) Model II, the smooth connection model: The intermediary region is represented by one fourth of a cylindrical surface. l_{im} and y_w are the y-coordinate for the intermediary region and the corresponding coordinate for the water surface region. R' denotes the radius of the cylinder. (Reproduced from Ref. [54] by permission of the American Institute of Physics.)

Model II exemplifies the smooth connection, assuming 1/4 of a cylinder surface as the intermediate region as shown in Fig. 22(b). Here, each streamline is determined in such a way that its projection onto the x,y-plane should coincide with a streamline $V(x,y)$=const given using Eq. (10) for the ideal-fluid case. Accordingly, the y-equivalent coordinate y_{im} is defined as,

$$y_{im} = R'\arccos\left(1 - \frac{y_w}{R'}\right) - \left(\frac{\pi}{2} - 1\right)R', \tag{83}$$

for $|y| \leq R'$, where R' denotes the radius of the cylinder representing the meniscus in its steady state [53], [54]. The stream function $V(x,y)$ given by Eq. (10), is then smoothly connected up to the second-order derivative, while conserving the incompressibility condition, $V_w(x,y_w)=V_{im}(x,y_{im})$. The flow derived above is, however, no longer irrotational, and the equipotential lines cannot be defined in the intermediate region. It is noted that, if the above procedure is extended to Re $W(z)\equiv U(x,y)$ instead, the irrotationality is conserved at the expense of the incompressibility.

The numerical calculation has clarified that Model I is too simple to reproduce the main features of the dichroic behaviour as reviewed in §3.6 [54]. In the following, the results on Model II will be shown.

5.1.3. Procedure of Computer Simulation

The angle θ is here regarded as a discrete variable, whose allowable values are $\theta=0$, π/N, $2\pi/N$, ... , $(J-1)\pi/N$, ... , $(N-1)\pi/N$ ($J=1-N$) [53], [54]. The Jth division is defined as the region from $\{(J-1)\pi/N\}-(\pi/2N)$ to $\{(J-1)\pi/N\}+(\pi/N)$ around $\theta=(J-1)\pi/N$. We recognize the transition probability matrix Tr as,

$$Tr \to \begin{pmatrix} 1 & & 0 \\ & 1 & \\ & & \ddots & \\ 0 & & 1 \end{pmatrix}, \quad \text{for } \Delta t \to 0.$$

It is assumed that the off-diagonal elements Tr_{ij} between the distant divisions can be neglected if we take a sufficiently small Δt. $m=3$ was taken as appropriate for the condition. $Tr_{ij}=0$ for $|i-j|\geq m$. According to Eq. (77), $Tr(\Delta\theta, \Delta t)$ is a Gaussian-like distribution whose deviation is approximated as $\sigma=\sqrt{2D(t)\Delta t}$. So, the appropriate Δt is given by $\Delta t=\pi^2/(2N^2D(t))$, which corresponds to a condition $\sigma=\pi/N$ [53], [54]. When the particle approaches the substrate, $D(t)$ decreases according to Eq. (79) and a difficulty arises in the computer calculation.

Consequently, the lower limit of $D(t)$ is set at 0.002 s^{-1}, below which the effect of rotatory diffusion is neglected.

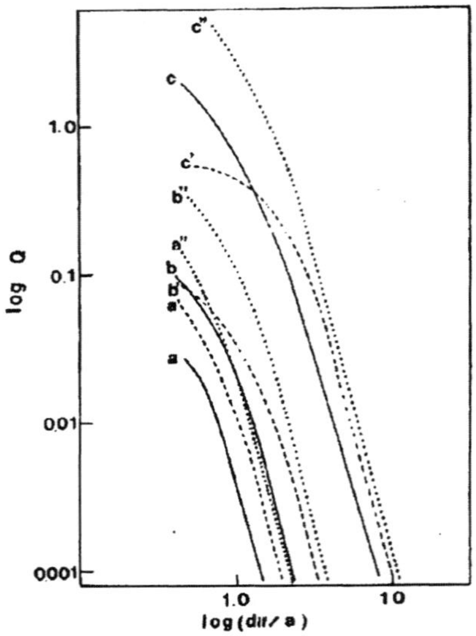

Fig. 23 The dependence of the convergence of $\langle \sin^2\phi \rangle$ on the flow distance d_{if}. log $\{Q = |(\langle\sin^2\phi\rangle - \langle\sin^2\phi\rangle_0)/\langle\sin^2\phi\rangle_0|\}$ is plotted against $\log(d_{if}/a)$. Lines a-a", b-b", and c-c" are the results for v_d= 30, 120, and 400 mm/min, respectively. Various kinds of lines represent the difference in the initial state: $P(\phi)= 1/N$ (solid lines); $P(\phi)=\delta(0)$ (dashed lines), and $P(\phi)=\delta(\phi/2)$ (dotted lines). (Reproduced from Ref. [54] by permission of the American Institute of Physics.)

5.2. Comparison with Experimental

5.2.1. Dependence of P on the Flow Distance dif

The dependence of P on the flow distance d_{if} was examined by drawing the particle on a streamline running toward the center of the substrate [54]. Here, d_{if} denotes the distance between the initial point and the final point. As a measure of the degree of convergence, the relative deviation in anisotropy is defined as,

$$Q = \left| \frac{< \sin^2\theta > - < \sin^2\theta >_0}{< \sin^2\theta >_0} \right|, \tag{84}$$

where,

$$< \sin^2\theta >= \int_0^\pi P(\theta)\sin^2\theta\, d\vartheta = \sum_{k=0}^{N} P\left(\frac{k\pi}{N}\right)\sin^2\left(\frac{k\pi}{N}\right) / \sum_{k=0}^{N} P\left(\frac{k\pi}{N}\right). \tag{85}$$

Here, $<\sin^2\theta>_0$ corresponds to the value for $d_{if}=\infty$. Figure 23 shows log (Q) plotted against log(dif/a) for various vd ranging from 30 to 400 mm/min and for various initial states [54]. Lines a-a", b-b", and c-c" are the results for v_d= 30, 120, and 400 mm/min, respectively. Various kinds of lines represent the difference in the initial state: $P(\theta)=1/\pi$ (solid lines), $P(\theta)=\delta(\theta)$ (dashed lines), and $P(\theta)=\delta(\theta-\pi/2)$ (dotted lines), each corresponding to the isotropic, the complete orientation to the vertical or the horizontal axis. From Fig. 23, when $d_{if}\geq 5a$, $Q\leq 0.02$ even for v_d=400 mm/min.

As each particle keeps memory of the flow history on the water surface in this numerical calculation, the final distribution function depends on the initial conditions of the particle [54]. As seen in Fig. 23, the influence of the initial state on the final distribution increases with increasing v_d for the same d_{if}. When v_d=30 mm/min, the difference in $<\sin^2\theta>$ for various initial states remains within 2% for $d_{if}=a$. When v_d=400 mm/min, however, dif must be $5a$ to attain the same level of convergence; running with larger velocity, the particle needs to move a longer distance before it loses the information of the initial state. Therefore, the values for $d_{if}\geq 5a$ can be regarded as nearly saturated for $v_d\leq$400 mm/min, and the values for $d_{if}=5a$ are mainly referred to in the following as representing the tendencies seen in the results.

5.2.2. Optimization of Parameter Values

The values of ζ', τ_0 and the deviation angle α were optimized for a=13 mm, R'=3 mm, $d_{if}\geq 5a$, and v_d=20 – 400 mm/min [54]. In the first stage, the position dependence of the dichroic ratio

R was examined in order to identify the region in the parameter space where the increasing R toward the edge of the substrate can be reproduced.

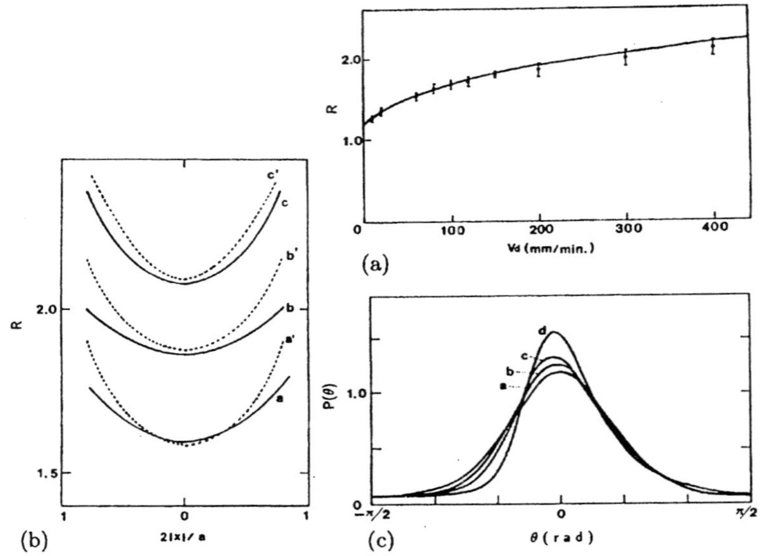

Fig. 24 (a) The dependence of the dichroic ratio R on the dipping velocity. The solid line is the calculated R and the experimental data with the estimated errors are also shown. (b) The dependence of R on the relative substrate position $2|x|/a$. Solid lines are experimental results and dashed lines are calculated R. a and a' are for v_d=60 mm/min, b and b' are for v_d=150 mm/min, and c and c' are for v_d=300 mm/min. (c) Calculated angular distribution function for v_d=120 mm/min. a is for $2|x|/a$=0, b is for $2|x|/a$=1/4, c is for $2|x|/a$=1/2, and d is for $2|x|/a$=3/4. (Reproduced from Ref. [54] by permission of the American Institute of Physics.)

In the second stage, the position on the substrate was fixed at its center x=0, and the remaining 3-dimensional region of ζ', τ_0 and α was examined using the least-squares method [54]. Figure 24(a) shows thus obtained curve of R as a function of the dipping velocity. The experimental data with the estimated errors are also given in the figure.

The optimized values of ζ', τ_0 and α are listed in Table 3 together with those obtained using the local thermal-equilibrium approximation. The values of each parameter fall in the same order of magnitude. The largest difference, nearly a factor of 2, is seen in τ_0, while the values of α differ only by 10% or less.

Table 3 List of optimized parameters ζ', τ_0 and α.

Parameter	Numerical calculation	Local thermal-equilibrium Approximation in §3.6
ζ' (kg m^2 s^{-1})	8.0×10^{-20}	5.00×10^{-20}
τ_0 (kg m^2 s^{-2})	3.5×10^{-21}	6.22×10^{-21}
α (deg)	31	33.5

Figure 24(b) shows the dichroic ratio R plotted as a function of the relative substrate position $2|x|/a$. The dashed lines a', b', and c' are calculated for v_d=60, 150, and 300 mm/min, respectively, using the parameter values in Table 3. The corresponding experimental data are shown by the solid lines a, b, and c. Although the increasing tendency of R toward the edge is reproduced, the difference between the calculated R and the observed value increases as $2|x|/a \rightarrow 1$. The examples of the final distribution functions are shown in Fig. 24(c).

The agreement between the calculated and the observed results may however be regarded as satisfactory in view of the crude approximation with one single R'-value adopted for the entire meniscus region. It is noted that the radius of curvature of the meniscus may be more precisely represented by a decreasing function of $2|x|/a$ as is experimentally observed.

5.2.3. Law of Similarity

Within the local thermal equilibrium approximation, the angular distribution function is given by a universal function Eq. (33) of the c-parameter. It is then of interest to examine whether this relation survives even when the approximation is removed. At the substrate center x=0, the c-parameter is written as,

$$c_0 \equiv (c)_{x=0} = \frac{\zeta'(4/\pi)(v_d/a)+\tau_0}{2k_BT} = \frac{2}{\pi}\frac{v_d}{Da}, \tag{86}$$

by using Eq. (41). Figure 25(a) shows $<\sin^2\theta>$ calculated for x=0 as a function of c_0 for the following range of variables: $\zeta'=5.0\times10^{-20} - 2.0\times10^{-19}$ kg m^2 s^{-1}, $\tau_0=1.0\times10^{-21} - 5.0\times10^{-21}$ kg m^2 s^{-2}, a=$6.5\times10^{-3} - 1.3\times10^{-2}$ m, and $v_d=5.0\times10^{-4} - 7.0\times10^{-3}$ m s^{-1}. The result in Fig. 25(a) shows that, when plotted as a function of c_0, the resultant dichroic ratio falls in a rather narrow range, indicating that the similarity law is still approximately valid: It must, however, be noted that the line for the local thermal-equilibrium case does not overlap the present region except for the limit of $c_0 \to 0$.

We have seen that $<\sin^2\theta>$ given by this numerical calculation for x=0 is determined almost uniquely by c_0 defined by Eq. (86). If we assume the same functional form as Eq. (35) for the dependence of $<\sin^2\theta>$ on the c-parameter, the behaviour shown in Fig. 25(a) can be reproduced by taking,

$$<\sin^2\theta> \approx \frac{1}{2}\left(1-\frac{I_1(c_0')}{I_0(c_0')}\right), \tag{87}$$

while,

$$c_0' = 0.7c_0^{0.7}. \tag{88}$$

In Eq. (88), the coefficient and the exponent of c_0 are both smaller than unity. This indicates that the calculated distribution function is less sensitive to the variation of c_0 than in the local thermal-equilibrium approximation. Since c_0 corresponds to the maximum shear stress immediately before the transfer, it is natural that the present distribution function, which reflects the whole flow process, is less sensitive to the c_0-value compared to that based on the local thermal equilibrium.

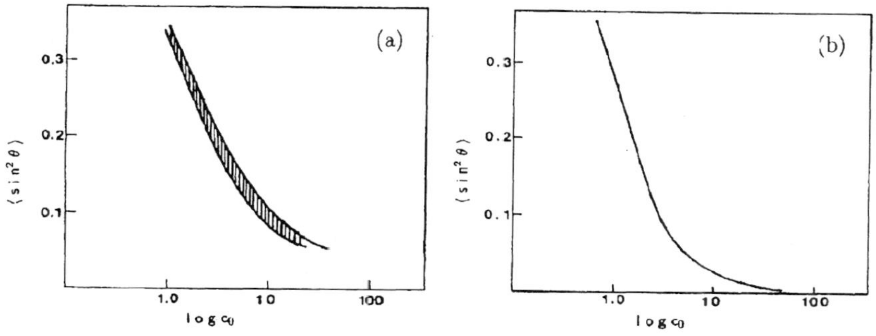

Fig. 25 (a) The relation between calculated $<\sin^2\theta>$ and the dimensionless number c_0. (b) The dependence of $<\sin^2\theta>$ on c_0 for the local thermal-equilibrium approximation. (Reproduced from Ref. [54] by permission of the American Institute of Physics.)

5.3. General Remarks

We have reviewed the results of numerical calculation of the angular distribution function based on the integral form of the rotatory diffusion equation in the LB deposition process for the case of a single substrate in a large trough. Allowing for the role of the meniscus region, the streamlines are smoothly connected by assuming a cylindrical shape for it.

The final angular distribution function depends on the flow distance d_{if} over which the particles traverse. When the dipping velocity is increased, the larger d_{if} is required to obtain satisfactorily saturated distribution functions. For $v_d \leq 400$ mm/min, it is found that dif should be equal to or larger than $5a$ for the memory of the initial state to be almost lost by the flow orientation. This $5a$-rule is quite compatible with the $4a$-rule empirically obtained for the single-substrate case in §3.6, and also consistent with the $6a$-rule for the face-to-face case of LB batch production in §4.1.2. Further, the $5a$-rule will serve as a quantitative criterion when we discuss the compression orientation effect in LB films prepared using the vertical dipping method.

The viscoelastic parameters have been estimated by comparing the experimental and the calculated dipping-velocity dependence of the dichroic behaviour, which results in the Newtonian friction coefficient $\zeta'=8.0\times 10^{-20}$ kg m^2 s^{-1}, the Bingham yield value $\tau_0=3.5\times 10^{-21}$ kg m^2 s^{-2}, and the angle between the transition dipole moment and the long axis of the rod-like particle $\alpha=31^\circ$. Using these parameters, the final distribution function has satisfactorily reproduced the experimentally observed dichroic behaviour as to the dependence on the dipping velocity and as to the dependence on the substrate position.

Within the local thermal approximation, the angular distribution function is analytically determined by the c_0 parameter. Also in the numerical calculation, it is found that the dimensionless number $c_0=\{\zeta'(4/\pi)(v_d/a)+\tau_0\}/2k_BT$ leads to a law of similarity, i.e., the final distribution function is approximately determined by c_0 alone regardless of different combinations of ζ', τ_0, a and v_d.

Although the present analysis gives a fair degree of quantitative agreement with experiment, there remain several points to be improved. The meniscus has been represented by 1/4 of the cylinder surface, but the real meniscus may be more complicated as already mentioned. Especially, to take care of the edge of the substrate, a better model for the meniscus should be considered, and the velocity vector on the water surface should probably be modeled on a more general framework than the ideal-fluid approximation. Nevertheless, the present numerical calculation is expected to serve as a way to describe the flow orientation effect in monolayers at the air-water interface as well as LB multilayers prepared under various conditions such as the compression-orientation scheme.

6. FLOW ORIENTATION WITH ROTATING DISKS

This section is devoted to review the rotating-disk method, which has been developed aiming at the extended feasibility of controlling the in-plane orientation of mesoscopic domains in the monolayers beyond the scope of the flow orientation model summarized in the preceding sections. A phenomenological, but analytical model to be shown here will figure out the important factors of the rotating-disk method and provide the guidelines for its applications.

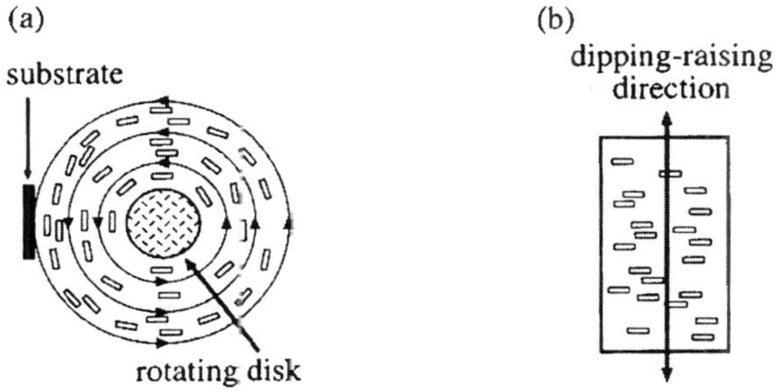

Fig. 26 Schematic illustration of the "rotating disk method". (a) A rotating disk generates a concentric monolayer flow at an air-water interface. This monolayer flow orients rod-like particles preferentially parallel to the streamlines. In the LB deposition process, a solid substrate is dipped and raised across the monolayer with keeping its surface parallel to the flow lines. (b) Molecular alignment within the plane of an LB film deposited with this method.

6.1. Rotating-disk Method

After the flow orientation model was proposed by Minari et al. [21], [22], it has been widely recognized that the in-plane anisotropy can be controlled in LB films of such compounds as self-aggregating molecules and rod-like polymers [50], [51], [55]-[60]. The law of similarity described in §3.5.1, for instance, serves as guidelines for this purpose. The possible range of controlling is, however, sometimes restricted, e.g., the enhanced deposition velocity, predicted to result in the higher anisotropy, is limited by the normal drainage during the transfer process.

For the enhanced controllability, Mingotaud et al. [61] have developed the "rotating-disk method", which employs an LB trough with a Teflon disk put in contact with the monolayer and rotated to generate a monolayer flow, which is expected to induce the in-plane orientation of the monolayer. They have successfully proved this idea by using a substituted polyacrylate bearing a mesogenic side-group and a triazolemioporphyrazine as film-forming materials.

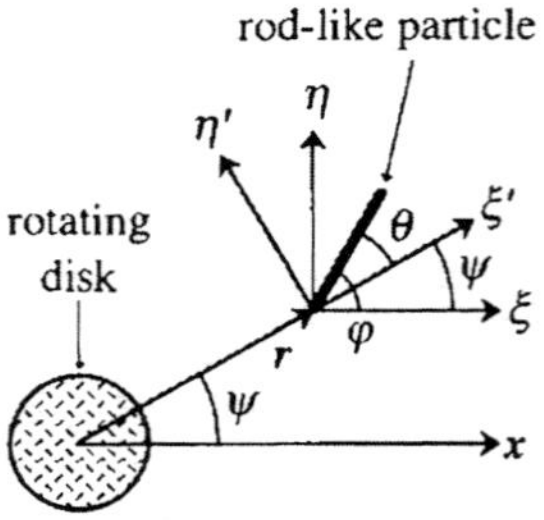

Fig. 27 Definition of the angles ψ, φ and θ.

The orientation mechanism of the rotating-disk method is qualitatively explained as follows [61]: The disk generates around it a concentric monolayer flow characterized by an angular velocity of the monolayer around its rotary axis $\Omega(r)$ which decreases towards its periphery, i.e., $d|\Omega(r)|/dr<0$ with r being the distance from the axis. The velocity gradient causes shearing and mesoscopic rod-like domains are preferentially oriented along the streamlines (Fig. 26(a)). Thus generated in-plane orientation is transferred at an appropriate deposition velocity (Fig. 26(b)). Since the disk-generated shear rate and the deposition velocity are independent of each other, we can choose the more effective of the conventional and the rotating-disk methods for a given compound.

Two stages are involved in the orientation mechanism: the flow generation in the monolayer by the disk and the orientation of the mesoscopic rod-like domains by the flow-induced shear.

6.2. Monolayer Flow Generated by a Rotatory Disk

6.2.1. Observation of Actual Flow

Let us start with the first stage to make a simple model of the disk-generated flow. First of all, we should briefly review the remarks from the experiments. The observation of the flow was performed for monolayers having dimensions of 38 cm×38 cm [62]. A disk with a radius r_{disk}=2.5 cm placed at the center was rotated with an angular velocity of 0.63 – 6.3 rad/s. Mixtures of an alkylated merocyanine and beheric acid (DS and C_{22} in Fig. 1, respectively) were used for model compounds with varying mixing ratio to modify the viscoelastic properties of monolayers. To measure the flow velocity, dr/dt and $\Omega \equiv d\psi/dt$ in the cylindrical coordinate with its origin fixed at the center of the disk (Fig. 27), grids for transmission electron microscopy were put on the monolayers as floats whose trajectories were digitally recorded for analysis.

The actual flow was found to be affected by the boundary of the monolayers. With the 1 : 1 mixture of DS and C_{22}, the flow was approximately concentric at a surface pressure of 15 mN/m, but there appeared a macroscopically inhomogeneous structure in it at 25 mN/m. In the latter case, a clear boundary was observed to divide the monolayer into two, the inner and the outer parts. This boundary seemed frictionless, with the inner part freely rotating with the disk, while the outer part was immobilized by the trough wall. With the DS-C_{22} 1 : 2 and 1 : 4 mixtures, and with pure C_{22} as well, the flow was approximately concentric even when the surface pressure was increased up to 25 mN/m, although vortices were seen at the corners of the monolayers. dr/dt and $d\Omega/dt$ were therefore practically zero for the 1 : 2 and 1 : 4 mixtures.

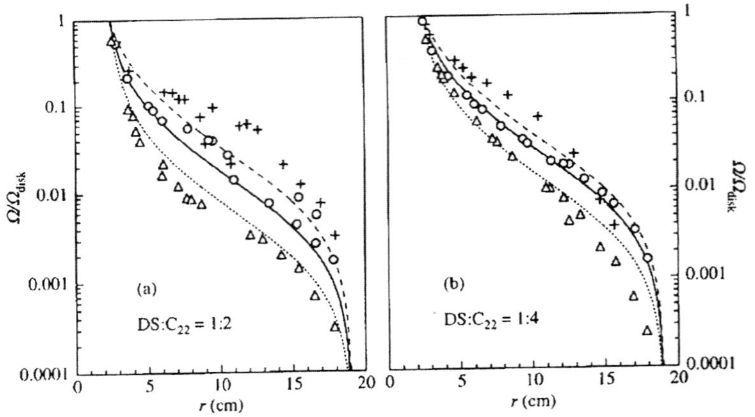

Fig. 28 Typical r-dependence of the normalized angular velocity, Ω/Ω_{disk}, of the 1 : 2 (a) and 1 : 4 (b) mixed monolayers of DS and C_{22}. The rotation speed of the disk, Ω_{disk}, is 6.3 (△), 2.1 (○) or 0.63 rad/s (+). Simulated curves calculated by Eq. (100) with K=0.4 and 1.0 s^{-1}, respectively, for the 1 : 2 and 1 : 4 mixtures are also indicated. These curves correspond to Ω_{disk} of 6.3 (dotted lines), 2.1 (solid lines) and 0.63 rad/s (broken lines). It should be noted that the simulated curves for Ω_{disk}=0.63 rad/s are not so much deviated from the one given by Eq. (98).

Figures 28(a) and (b) shows the data of $\Omega(r)$ reported for the DS-C_{22} 1 : 2 and 1 : 4 mixtures, respectively (both at 25 mN/m) [62]. The striking features of the observed velocity field are the nonlinear dependence of $\Omega(r)$ upon the angular velocity of the Ω_{disk} and the steep decrease in $\Omega(r)$ in the small-r region. The observed finite velocity in the larger-r region is quite natural if the plasticity of the monolayer is small. It should be noted that similar results are obtained for pure C_{22} monolayers.

6.2.2. Flow in Infinite Monolayer

Let us first consider an infinitely large monolayer. In this case, the monolayer flow should be concentric, as long as the phenomenon is regular, since the whole system is centrosymmetric.

Here, the viscosity of the monolayer is essential for the generation of the flow, and the ideal-fluid approximation reviewed in §3 is no more applicable. With a concentric flow, the velocity gradient tensor defined by Eq. (16) is given by,

$$\mathbf{G}_{\mathrm{rd}}=\begin{pmatrix} -\dfrac{\partial}{\partial x}y\Omega & -\dfrac{\partial}{\partial y}y\Omega \\ \dfrac{\partial}{\partial x}x\Omega & \dfrac{\partial}{\partial y}x\Omega \end{pmatrix}=\begin{pmatrix} -\dfrac{g_{rd}}{2}\sin 2\psi & -\Omega-g_{rd}\sin^2\psi \\ \Omega+g_{rd}\cos^2\psi & \dfrac{g_{rd}}{2}\sin 2\psi \end{pmatrix}, \tag{89}$$

where g_{rd} is a value defined as,

$$g_{rd}=r\frac{d\Omega(r)}{dr}, \tag{90}$$

If we rotate the local Cartesian coordinate (ξ,η) by the angle ψ to make ξ'-axis parallel to the radius vector r (Fig. 27), $\mathbf{G}_{\mathrm{rd}}$ can be written as,

$$\mathbf{G}_{\mathrm{rd}}=\begin{pmatrix} 0 & -\Omega-\dfrac{g_{rd}}{2} \\ \Omega+\dfrac{g_{rd}}{2} & 0 \end{pmatrix}+\begin{pmatrix} 0 & \dfrac{g_{rd}}{2} \\ \dfrac{g_{rd}}{2} & 0 \end{pmatrix}, \tag{91}$$

which shows that a concentric flow field is a "simple shear" field, which is the sum of a "pure rotation" field with a velocity of $(\Omega+g_{rd}/2)$ and a "pure shear" field with a shear rate of g_{rd} [63] .

The r-dependence of Omega in a concentric flow is described by the simple torque equation: an equilibrium must be held between the torques applied from the inner and the outer parts, $M(r)$ and $-M(r+dr)$, respectively, to a thin ring $[r, r+dr]$ of the monolayer [64]. The torque is the product of the radial distance r, the circumference of the ring $2\pi r$ and the shear stress σ, of which σ is a function of the shear rate g_{rd} and the surface density of the monolayer ρ. The moment is therefore expressed as,

$$M = -2\pi r^2 \sigma(g_{rd}, \rho) = \text{const.} \tag{92}$$

Another condition is the equilibrium in a small rectangular part in the thin ring between the force due to the r-dependence of the surface pressure f and the centrifugal one [63], i.e.,

$$\frac{df}{dr} = \rho(f) r \Omega^2. \tag{93}$$

Let us simplify the physical situation described by the above simultaneous equations, Eqs. (90), (92) and (93), before stepping into a finite system. We begin with assuming a constant surface viscosity η_{L} of the monolayer. Then, σ is proportional to g_{rd} as,

$$\sigma = \eta_{\mathrm{L}} g_{rd} = \eta_{\mathrm{L}} r \frac{d\Omega}{dr}, \tag{94}$$

and then Eq. (92) can readily be solved as,

$$\Omega(r) = \Omega_{disk} \left(\frac{r_{disk}}{r} \right)^2, \tag{95}$$

because $\Omega(r_{disk}) = \Omega_{disk}$ and $\Omega(r \to \infty) \to 0$. Consequently, Eq. (93) can be integrated as,

$$f = \left(\frac{\rho_0}{\varepsilon} + f_\infty \right) \exp\left(-\frac{\varepsilon \, \Omega_{disk}^2 \, r_{disk}^4}{2r^2} \right) - \left(\frac{\rho_0}{\varepsilon} \right), \tag{96}$$

if we expand ρ up to the first order as $\rho \to \rho_0 + \varepsilon f_\infty$ where f_∞ is the surface pressure at $r \to \infty$. The surface pressure decreases with decreasing r and takes its minimum at $r = r_{disk}$. The domain of variability $\Delta f\,(>0)$ of f is estimated as,

$$\begin{aligned} \Delta f &= f_\infty - f(r_{disk}) = (\rho_0 \varepsilon^{-1} + f_\infty) \left\{ 1 - \exp\left(-\frac{1}{2} \varepsilon \, \Omega_{disk}^2 r_{disk}^2 \right) \right\} \\ &< \frac{1}{2} (\rho_0 + f_\infty \varepsilon_0) \Omega_{disk}^2 r_{disk}^2 = \frac{1}{2} \rho(f_\infty) \Omega_{disk}^2 r_{disk}^2, \end{aligned} \tag{97}$$

where we recognize in general that $f_\infty < 10^{-1}$ N/m, $\rho(f_\infty) < 10^{-5}$ kg/m^2 and $\Omega_{disk} r_{disk} < 2\times10^{-1}$ m/s. Thus, we obtain $\Delta f < 2\times10^{-7}$ N/m, a negligibly small value. The estimated Δf allows us to approximate ρ as a constant and neglect the dependence of σ on ρ, even when we consider a finite monolayer and take into account nonlinearity in the dependence of σ on g_{rd}.

Here, it should be emphasized that η_{L} does not appear in Eq. (95), and thus its magnitude is irrelevant to the generated flow except for an unphysical case of $\eta_{\mathrm{L}}=0$. In other words, a non-zero η_{L} is prerequisite for the present disk-induced flow to occur, although a very small η_{L} is sufficient for transmitting the torque from the inner to the outer part of the monolayer. In the case of deposition-induced flow, on the contrary, η_{L} can be neglected if it is small, because the flow is in this case driven by surface-pressure difference. The apparently different treatments of η_{L} between the deposition-induced and the disk-induced cases do not contradict each other, although some new insights might be obtained if we take a non-zero η_{L} into account also in the former case.

6.2.3. Flow in Finite Monolayer

In the experiment, the actual monolayers were square in shape as we have seen. To model the monolayer in the centrosymmetric condition, let us approximate its shape as a circular one with a radius r_{trough}. When the monolayer is assumed to be of Newtonian as predicted by Eq. (94), a flow like Eq. (95) is derived again as,

$$\Omega(r) = \frac{M}{4\pi\eta_{\mathrm{L}}} + C, \qquad (98)$$

where the constants M and C should be determined so that $\Omega(r_{disk})=\Omega_{disk}$ and $\Omega(r_{trough})=0$. (The former is the constant torque in Eq. (92).) However, experimentally obtained $\Omega(r)$ decreases more rapidly with increasing r than predicted by Eq. (98) when Ω_{disk} is large, although this equation well fits the data for Ω_{disk}=0.63 rad/s (Fig. 28).

This rapid decrease in $\Omega(r)$ has been shown to be explained by a pseudo-plastic behaviour of the monolayer [62]. The pseudo-plastic behaviour of three-dimensional fluids was extensively discussed by Eyring, who introduced a hyperbolic sine function to describe it [65], [66]. The advantage of his hyperbolic-sine model is that higher-order effects of the shear rate are taken into account to a certain extent by only one parameter. This forms a contrast to the usual expansion approximation, in which the introduction of a single parameter corresponds to taking into account only the first order effect of the variable. Eyring's calculation is based on a one-dimensional rate process with a potential energy barrier: the forward and the backward hopping rates of a molecule, identical with each other originally, are differentiated by the shearing. Because his model is essentially one-dimensional, it is also applicable to the two-dimensional fluids. The pseudo-plastic behaviour of the monolayer is then expressed as,

$$\sigma = \eta_L K \operatorname{arcsinh}\left(\frac{g_{rd}}{K}\right). \tag{99}$$

by replacing the original three-dimensional microscopic parameters of Eyring's model by two-dimensional phenomenological parameters η_L and K. Here, η_L and K are, respectively, the surface viscosity of the monolayer in the Newtonian region and the cross-over shear rate between the Newtonian and non-Newtonian regions. Putting Eqs. (90) and (99) into Eq. (92), we finally obtain a model for the monolayer flow as,

$$\Omega(r) = \Omega_{disk} + \int_{r_{disk}}^{r} K\lambda^{-1} \sinh\left(-\frac{M}{2\pi\eta_L K}\lambda^{-2}\right) d\lambda, \tag{100}$$

where the value of M/η_L should be determined by the boundary condition of $\Omega(r_{trough})$=0. It should be mentioned that the torque M must be measured to determine η_L for a given monolayer.

Equation (100) reproduces the experimental data satisfactorily with K=0.4 and 1.0 s^{-1} for the 1 : 2 and the 1 : 4 mixtures of DS and C_{22}, respectively, as shown in Fig. 28. Note that we have only one fitting parameter, K, and we must use the same K-value with different Ω_{disk}-values for describing each monolayer species. The different K-values reflect the difference in the viscoelastic properties of the mixtures: the pseudo-plasticity is more remarkable in the 1 : 2 mixture than in 1 : 4. This difference must be correlated to the lateral interactions within the monolayer.

6.3. Orientation due to Shearing

In the preceding subsection, we have seen that the disk-induced flow can be modeled by taking into account the pseudo-plasticity of the monolayer. Let us model now the process of orientation due to the shearing. The orientation mechanism is essentially the same as that illustrated in Fig. 5(b), but it is quantitatively discussed here.

6.3.1 Angular Distribution Function

Substituting Eq. (89) for Eq. (17), we obtain,

$$\dot{\varphi} = \Omega + g_{rd}\left(\frac{1}{2}\sin 2\psi \sin 2\varphi + \cos^2\psi \cos^2\varphi + \sin^2\psi \sin^2\varphi\right). \tag{101}$$

Let us introduce an angle $\theta \equiv \varphi - \psi$ to discuss the orientation with respect to the radius vector r. Its time derivative is obtained as,

$$\omega_{rd} \equiv \dot{\theta} = \dot{\varphi} - \dot{\psi} = g_{rd} \cos^2 \theta, \tag{102}$$

where ω_{rd} has the same sign for all θ (ω_{rd}<0 when g_{rd}<0), since the mesoscopic domains are in a "simple shear" field. This indicates that there is no mechanically stable point for θ in contrast to the transfer-induced case discussed in §3, where a "pure stretch" field governs the orientation.

The rotatory diffusion equation for the mesoscopic domains is the same as Eq. (25). In the stationary state, the left-hand side of Eq. (25) is zero and we obtain,

$$P_{rd}(\theta) = A\exp\left(\frac{g_{rd}\theta}{2D_{eff}} + \frac{g_{rd}\sin\theta}{4D_{eff}}\right)\left\{\int_0^\theta \exp\left(-\frac{g_{rd}\theta'}{2D_{eff}} - \frac{g_{rd}\sin\theta'}{4D_{eff}}\right)d\theta' + B\right\}, \tag{103}$$

where, A is the normalization factor, and B is determined by the cyclic boundary condition of $P_{rd}(\theta+\pi)=P_{rd}(\theta)$ required for the symmetry of the system:

$$B = \frac{1}{\exp\left(-\frac{\pi g_{rd}}{2D_{eff}}\right) - 1}\int_0^\pi \exp\left(-\frac{g_{rd}\theta'}{2D_{eff}} - \frac{g_{rd}\sin\theta'}{4D_{eff}}\right)d\theta', \tag{104}$$

The present DS-C_{22} mixed system is suggested to behave with a rotatory yield value as is the case with the DS-C_{20} mixture discussed in §3 (cf. Eq. (39)) [63]. Accordingly, g_{rd}/D_{eff} can be written as,

$$\frac{g_{rd}}{D_{eff}} = \frac{\zeta' g_{rd} + \frac{g_{rd}}{|g_{rd}|}\tau_0}{k_B T}. \tag{105}$$

Examples of $P_{rd}(\theta)$ are shown in Fig. 29 for given g_{rd}/D_{eff}-values.

The lack of a mechanically stable point results in a characteristic of P_{rd}: the location of the maximum of this function depends on the parameter g_{rd}/D_{eff} and deviates from $\theta=\pi/2$ (direction of the streamlines) largely when g_{rd}/D_{eff} is small. The maximum along the streamline may be one of the most important characteristics of the stretch-induced orientation as we have seen in §3. For LB films, the deviation of the P_{rd}-maximum was not successfully observed because of the predominance of the deposition-induced flow with small g_{rd}/D_{eff} [68]. This deviation was, however, detected by a Brewster-angle microscope study upon a disk-rotated monolayer [69], although it was difficult to carry out the quantitative comparison between the experimental results and the predictions from Eq. (103) due to the instrumental restrictions.

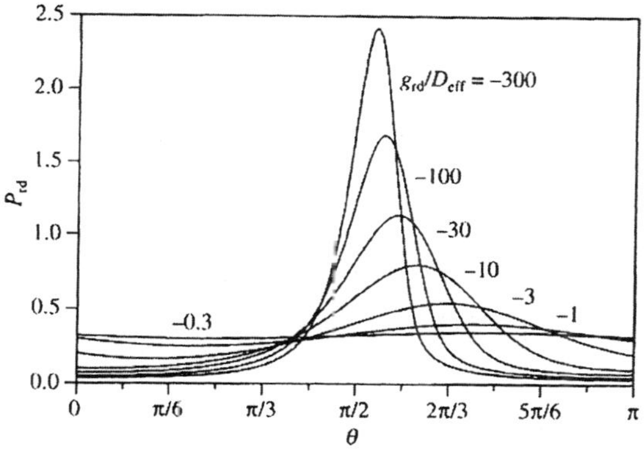

Fig. 29 Examples of the orientation distribution function of mesoscopic rod-like particles in a monolayer under shear given by Eq. (103).

6.3.2. Order Parameter P_2

When we neglect the effect of the dipping process and the disturbance of the monolayer flow due to the substrate immersed across the monolayer, Eq. (103) is basically applicable also to LB films. With the rotating-disk method, a substrate is set parallel to the streamline during the LB deposition process (Fig. 26), and then θ corresponds to the angle between the long axis of a mesoscopic domain and the dipping-raising direction of the substrate during the deposition. A usual Y-type LB film, however, contains four different kinds of monolayers, depending on whether they were deposited during the upward or downward strokes and on the front or rear side of the substrate. The mirror symmetry due to this situation requires to rewrite the distribution function as $\{P_{rd}(\theta)+P_{rd}(\pi-\theta)\}/2$.

For the deposition-induced orientation, the dichroic ratio R is useful to characterize the degree of the in-plane anisotropy of LB films. In the rotating-disk method, however, the in-plane order parameter P_2, a more symmetric quantity than R, is preferred to characterize LB films. This parameter is defined as $P_2 \equiv \langle\cos 2\Theta\rangle$, where Θ is the angle between the projection of a transition dipole moment onto the film plane $\boldsymbol{\mu}_{\parallel}$ and the dipping direction of the substrate, and $\langle\ \rangle$ denotes the average over the molecular distribution within the LB film. The P_2 parameter can be experimentally determined by,

$$P_2 = \frac{A(0) - A(\pi/2)}{A(0) + A(\pi/2)}, \tag{106}$$

which varies from +1 to −1, and these extremes correspond to perfect orientation of $\boldsymbol{\mu}_{\parallel}$ along the axis parallel and perpendicular, respectively, to the dipping direction. Generally, $\boldsymbol{\mu}_{\parallel}$ may not be parallel to the long axis of mesoscopic domains, and the angle α between $\boldsymbol{\mu}_{\parallel}$ and the long axis of domains may be statistical. Even in such cases, the P_2 parameter can be written as a simple product of two "partial" order parameters [62], one of which is determined by the distribution of α and the other by that of θ as,

$$P_2 = \langle\cos 2\alpha\rangle\langle\cos 2\theta\rangle, \tag{107}$$

because the symmetry of the system requires $\langle\sin 2\alpha\rangle\langle\sin 2\theta\rangle=0$, and normally $\langle\sin 2\alpha\rangle=0$ and $\langle\sin 2\theta\rangle=0$ as well. In our model, $P_{rd}(\theta)+P_{rd}(\pi - \theta)=P_{rd}(-\theta)+P_{rd}(\pi + \theta)$ and Eq. (107) is valid. The "partial" order parameter $\langle\cos 2\theta\rangle$ is a convolution of a cosine function and the distribution function, and then the P_2 parameter is given as,

$$P_2 = \frac{1}{2}\langle\cos 2\alpha\rangle \int_0^{\pi} \cos 2\theta \{P_{rd}(\theta) + P_{rd}(\pi - \theta)\}\, d\theta. \tag{108}$$

6.3.3. Comparison with Experimental Results

The rotating-disk method was applied to the DS-C_{22} mixtures under the following conditions. The rotating disk used was the same as that for the observation of monolayer flow (r_{disk}=2.5 cm). The distance between the center of disk and the position of substrate d was varied from

2.8 to 15 cm, while the distance between the substrate and the trough wall d' was fixed at 2.4 cm. As a consequence, the monolayer was not square but rectangular in shape. The surface pressure was kept at 25 mN/m during the deposition, and the dipping-raising velocity of the substrate was about 0.03 cm/s. The P_2-data obtained for LB films of the DS-C_{22} mixtures are plotted against d^{-1} in Fig. 30.

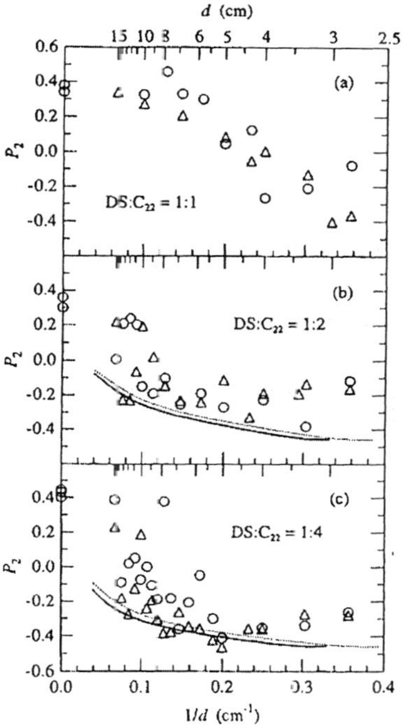

Fig. 30 The P_2 data observed for LB films of DS-C_{22} 1 : x mixtures transferred by the rotating-disk method are plotted against the inverse of the distance between the dipping position and the rotary axis, d^{-1}. The mixing ratios are (a) x=1, (b) x=2 and (c) x=4. Ω_{disk} is set to 6.3 (Δ) or 2.1 rad/s (○). Here we take d^{1} for the abscissa to expand the region near the disk and to include the P_2 data obtained without disk, which is plotted at d^{-1}=0. For the 1 : 2 and 1 : 4 mixtures, simulated curves by Eq. (108) with $\langle\cos 2\alpha\rangle$=0.55, $\zeta'=7\times10^{-19}$ kg m^2 s^{-1} and $\tau_0=6\times10^{-21}$ kg m^2 s^{-2} are also indicated. In this simulation, the pseudo-plastic parameter K is assumed to be 0.4 and 1.0 s^{-1}, respectively, for the 1 : 2 and 1 : 4 mixtures, to evaluate the shear rate by Eq. (100). The solid and dotted curves correspond to Ω_{disk} of 6.3 and 2.1 rad/s, respectively.

The model introduced in this section can be compared with the data obtained for the DS-C_{22} 1 :

4 mixed system, since for this system a regular monolayer flow was observed (Fig. 28). A high efficiency of the rotating-disk method is demonstrated by the fact that the absolute values of P_2 of the LB films deposited with and without a rotating disk are approximately identical. The disk-induced flow of this material is approximated by Eq. (99) with K=1.0 s^{-1} and $r_{trough}=d+d'$, and the calculated shear rate is used in Eq. (108). The simulation based on this model is fairly acceptable (Fig. 28(c)) with $\langle\cos 2\alpha\rangle=0.55$, $\zeta'=7\times10^{-19}$ kg m^2 s^{-1} and $\tau_0=6\times10^{-21}$ kg m^2 s^{-2}. It should be mentioned that the above estimated values give also acceptable P_2-values based on the flow orientation model in §3 for the LB films fabricated without a rotating disk.[5]

The discrepancy between the simulation and the experiment is not negligible in the large d (> 8 cm) and small d (<3.5 cm) regions. The discrepancy in the large d region is probably due to the flow caused by the transfer, since the effect of the disk-induced and the transfer-induced shearing may be comparable in this region. This may also result in the relatively low reproducibility of the P_2 data. The origin of the discrepancy in the small d region is not clear at present, but we speculate that the substrate itself disturbs the monolayer flow [68], reducing the $|P_2|$-value. The difference between the experimental data of P_2 for ω_{disk}=6.3 and 2.1 rad/s is small. The simulation could well predict this small difference, suggesting that the nonlinearity of the monolayer flow and the saturation behaviour of the P_2-value restrict the advantage of large ω_{disk}. This result implies that too large ω_{disk} is useless if the monolayer exhibits a regular flow.

The smaller efficiency of the rotating-disk method in the 1 : 2 mixture case may be related to

---

[5] In these particular cases, τ_0 was not important, because the experiments with and without the rotating disk were performed with large g_{rd} and c, respectively. Accordingly, the value of τ_0 estimated here is unreliable. In addition, the monolayers of DS-C_{22} mixtures used here were metal-free and those of DS-C_{20} used in the previous sections contain divalent cadmium ions. This difference may cause the differences between $\langle\cos 2\alpha\rangle$ and Bingham parameters of these two series of monolayers.

the dipping process. The pseudo-plasticity of this mixture, more prominent than that of the 1 : 4 mixture, may be correlated with this reduction of the efficiency. For the 1 : 1 mixture, we have fairly large $|P_2|$-values especially in case of small d, i.e., small r_{trough}, although the observed flow was not regular for large r_{trough}.

Finally, it should be pointed that the present model provides a guideline for designing a rotating-disk apparatus. The rotation speed is *not* necessarily large (2 rad/s is sufficient), but a larger disk may be useful due to the following reason. The magnitudes of side effects like surface waves may be proportional to r_{disk}, $r_{disk}\Omega_{disk}$, or $r^2_{disk}\Omega_{disk}$.[6]

With the same $r^2_{disk}\Omega_{disk}$, we can produce larger Ωd by increasing r_{disk}, when the distance between the substrate and the edge of the disk ($=d-r_{disk}$) is fixed, since $\Omega(r)$ decreases more rapidly than r^{-2}. Another important point is that small r_{trough} may have an advantage: when r_{trough} is smaller than the size of the macroscopic structure of the monolayer, the regular flow could be generated, even when the monolayer has macroscopic inhomogeneities as the 1 : 1 mixture case. In other words, the homogeneous area required is reduced for a small r_{trough}. Furthermore, a smaller trough enhances $|d\Omega/dr|$ and $|g_{rd}|$ as well.

6.4. Multi-disk Versions

The rotating-disk method is proved effective for producing laterally anisotropic LB films in the preceding subsections. This method is however found to be associated with restrictions. Its efficiency is directly related to the shear rate around the solid substrate, which highly depends on the structure and the viscoelastic properties of the monolayer. Consequently, the

---

[6] When the rotary axis of the disk tilts, or the bottom surface of the disk is not exactly horizontal, the difference in the height at the disk edge is proportional to r_{disk}. The velocity at the disk edge is $r_{disk}\Omega_{disk}$. The magnitudes of side effects may be proportional to the former, the latter or the product of the both values.

pseudo-plastic parameter K should be estimated to predict the degree of the in-plane orientation of LB films. To avoid these problems, two-disk version has been proposed [70], [71]. In the following, the cases with two or more disks will be reviewed.

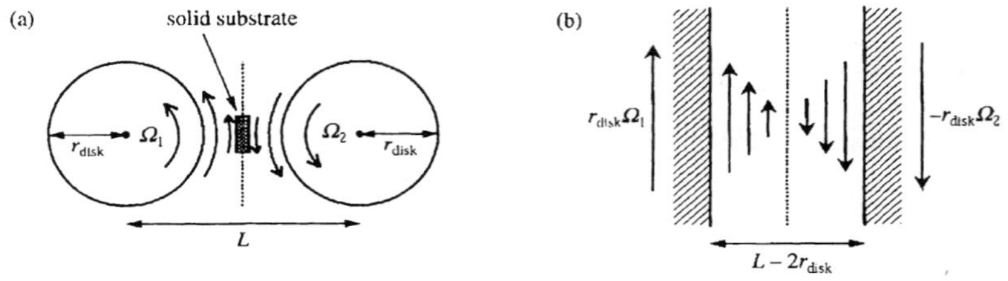

Fig. 31 Schematic illustration of the 2-disk version of the rotating disk method. (a) Two disks are superimposed on a monolayer at an air-water interface. The radii of the disks are the same and denoted by r_{disk}. The distance between the centers of disks L. The solid substrate is immersed at the center of the system. (b) A simplified flow field is used at calculate the shear rate.

6.4.1. Two-disk Version

Figure 31(a) illustrates an apparatus used for the 2-disk version [70], [71]. The disks are rotated with constant rotation speeds, Ω_1 and Ω_2, during the compression and deposition of monolayers. This configuration can provide a simple shear field, like one displayed by Fig. 5(c), in the central region between the disks. Although the 2-disk version has a too complicated geometry to calculate the monolayer flow exactly, we can approximate it as a parallel one (Fig. 31(b)) when r_{disk} is much larger than $L-2r_{disk}$, where L is the distance between the centers of the disks. The shear rate is then easily obtained as,

$$g_{rd} = -\frac{r_{disk}(\Omega_1 + \Omega_2)}{L - 2r_{disk}}. \tag{108}$$

Here, it should be emphasized that this shear field is determined by the geometrical factors. In other words, we can neglect effects of the non-Newtonian properties of monolayers, because the shear rate is constant, allowing to assume a constant surface viscosity. Furthermore, a large homogeneous area of the monolayer is not necessary, since the flow in its central region may be insensitive to the structure of the peripheral part. When $\Omega_l=\Omega_2$, the flow velocity should be zero along the centerline between two disks (dotted line in Fig. 31(a)) because of the symmetry. In the LB deposition process, we can minimize the reduction of shear rate due to the substrate if it is immersed on this line. Although a slight difference between Ω_l and Ω_2 may be occasionally needed to supply the monolayer to the deposition area for maintaining a constant quality of the LB transfer, we can choose a position for the LB deposition appropriate for minimizing the disturbance of the flow [70]. Another important advantage of the 2-disk version is that it can generate the same shear rate with smaller Ω_l compared with the 1-disk version. Both Ω_l and the maximum shear rate are reduced at the same time, resulting in the reduction of the side effects inherent in the rotating disks, such as waves and collapses.

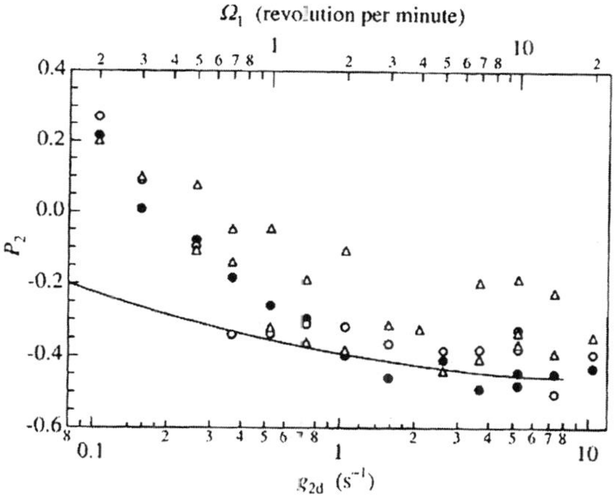

Fig. 32 The P_2 parameter observed for the DS-C_{22} mixed LB films deposited with the 2-disk version of the "rotating disk method" is plotted against the shear rate g_{2d}. The rotation speeds of the disks are identical ($\Omega_l=\Omega_2$) and indicated at the top of the figure. The mixing ratios are 1 : 1 (•), 1 : 2 (○), and 1 : 4 (△). The solid line is calculated according to the model described in the text.

The 2-disk version with r_{disk}=2.5 cm and L=6 cm was applied to the DS-C_{22} mixtures. In Fig.

32, the order parameter P_2 of the LB film is plotted against g_{2d}. The rotation speeds of the disks, $\Omega_1=\Omega_2$, are also indicated in this figure. The P_2–g_{2d} relations for the three mixtures are almost identical with one another. A small Ω_1-value (≈5 revolution per minute) is sufficient to attain $\langle\cos 2\alpha\rangle$ close to −1, i.e., $|P_2|$ close to the estimated value $|\langle\cos 2\theta\rangle|\approx 0.55$. The solid line in Fig. 32 refers to the simulation based on the afore-mentioned model. The curve reasonably reproduces the experimental results obtained for all the three mixtures, indicating the weak dependence of the efficiency on the viscoelastic properties. The deviation between the simulation and the experiment in the small-Ω_1 region is clearly due to the effect of the transfer-induced flow.

6.4.2. Four-disk Version

G. Fuller and collaborators applied the "four-roll mill" method, a well-known technique for characterizing three-dimensional fluids, to monolayers at the air-water interface [72]–[75]. In contrast to the 1-disk and 2-disk versions discussed above, this 4-disk version can provide a pure stretch field around the center of the system, where there is a stagnation point. Accordingly, this method is very useful to observe the dynamic behaviour of mesoscopic domains.

The orientation phenomenon reported by Fuller et al. essentially supports the model described §3. In their analysis, they developed $P(\theta)$ in terms of moments of the orientation distribution, i.e., $\langle\cos 2n\theta\rangle$ and $\langle\sin 2n\theta\rangle$ (n is a positive integer), and truncated the series after the 4-th moment terms ($n\geq 2$) to solve Eq. (25). With this approximation, they succeeded to explain the dynamical process of the orientation phenomenon. Furthermore, with this approximation, they could incorporate the interaction between the domains into their analysis; they added a "nematic potential", which represents the "molecular field", to Eq. (25). Actually, the orientation phenomena are inevitably cooperative ones. In the models described in this article, this aspect is taken into account by introducing the rotatory Bingham parameters.

7. CONCLUDING REMARKS

We have reviewed the present status of the analytic model of flow orientation effect in LB films. The model aims to give a quantitative explanation of the in-plane anisotropy in LB films starting from the results of optical and ESR studies on the mixed LB films of a merocyanine dye and a fatty acid.

The experimental results are correlated with the parameters characteristic of LB deposition process by introducing the following assumptions. The monolayer is regarded as a continuum, and the velocity potential is derived for a system with a two-dimensional ideal fluid and a line sink, each representing the monolayer at the air-water interface and the substrate. The rotatory motion of rod-like particles, such as crystallites and aggregates, in this velocity field is examined assuming a Bingham plasticity.

According to thus constructed model, the flow orientation effect is governed by seven parameters as reviewed in §3: the viscosity and the yield value of monolayer as a Bingham body, the size of particle, the temperature, the velocity of dipping and raising, the width of the substrate, and the relative position on the substrate. The angular distribution function of the particle is evaluated by assuming the local thermal equilibrium, i.e., the distribution immediately before the transfer onto the substrate is conserved in the LB film. The model leads to a satisfactory explanation of the experimental results, as we have reviewed, in the case of a single substrate with a large trough obeying the $4a$-rule, indicating the approximations introduced are relevant to the actual deposition process.

The model is however not versatile in all cases, but associated with limitations. As suggested from the channel-and-iceberg model [41], [42], for instance, the continuum approximation may not hold any more if the monolayers are phase-separated into two or more components having very different viscoelastic properties from one another.

The model has been extended to more complicated cases in §4, which include the batch production of LB films, the deposition in troughs of finite dimensions, and the cases with peeling-off, slipping, stretching and compression of the monolayer. The ideal-fluid approximation is found to be irrelevant to the face-to-face case of batch production for the substrate spacings smaller than six times the substrate width: when this 6a-rule is violated: we have seen that an unsteady flow is observed between the adjacent substrates. The violation of incompressibility have been discussed in §4.3, leading to relevant predictions about the compression orientation effect which has occasionally been regarded to be beyond the scope of the flow orientation model.

In §5, another approach, describing the LB deposition as a Markov process, is reviewed together with the results of numerical calculation. The intermediary region between the water and the substrate surfaces has been explicitly taken into consideration by approximating the meniscus with 1/4 of a cylinder surface. The thermal equilibrium approximation adopted in the analytic approach has been critically examined, leading to a $5a$-rule for a particle to lose the memory of the initial state. This $5a$-rule places limitations on the validity of the approximation, and may also serve as a criterion in discussing the compromise between the deposition- and the compression-induced orientation effects.

The rotating-disk method is reviewed in §6 as an advanced technique aiming at the more active control of the in-plane anisotropy in LB films. The quantitative models are shown in comparison with the experimental results. In this method, the monolayer flow is regulated beyond the scope of ideal fluid approximation, including the cases with vortical flows. The 2-disk and the 4-disk versions are found to be appropriate for simulating two extreme cases, the pure shear and the pure stretch fields, respectively. Of these two versions, the 4-disk may be more effective for obtaining LB films with a large in-plane anisotropy. In that case, however, the solid substrate inevitably disturbs the monolayer flow. For this version, the horizontal-lifting method of Nakahara and Fukuda [52] may be recommended to pick up the monolayer when the relaxation of the induced anisotropy is slow enough.

The present analytic model is found to be effective for a proper understanding of the in-plane anisotropy observed in LB films. The understanding is however still in a primitive stage, and there are indeed many approaches remaining to be adopted. From the theoretical viewpoints, the following are exemplified.

What will be brought about if the monolayer flow is evaluated starting from the Navier-Stokes or any other equations that are more general than the ideal-fluid approximation?

What will be changed if the "nematic field" or any other microscopic model of inter-particle interactions is adopted instead of the Bingham parameters, in both the deposition-induced and the rotating-disk-induced orientation models?

How can the actual meniscus region be better described?

From the experimental viewpoints, the following may be regarded. The measurement of nonlinear dielectric susceptibility allows to estimate the 4th and the higher moments of the angular distribution function which may serve for examining the model. The ESR line shape involve also the information on the higher-order moments of the distribution function, and the precise analysis of it may determine, e.g., whether Eq. (71) or Eq. (74) is appropriate for the compression orientation.

The authors would be happy if the present article stimulates some of the readers to further endeavors of better understanding of various features of LB deposition process to construct more advanced models.

8. ACKNOWLEDGEMENTS

The present authors would like to appreciate Dr. N. Minari and Prof. S. Kuroda for their significant contribution in constructing the framework of the flow orientation model reviewed in this article. They would like to thank Dr. M. Saito and Dr. K. Saito for the fruitful collaboration in elaborating the model. Thanks are due to Prof. H. Nakahara, Prof. K. Iriyama and Prof. T. Ishii for their useful suggestions on the J-aggregate. One of the authors (K. I.) is grateful to Prof. P. Delhaes, Dr. C. Mingotaud, Dr. C. Jego and Ms. B. Agricole for their stimulating discussion about the rotating-disk method.

9. BIBLIOGRAPHY

1. K. B. Blodgett: J. Am. Chem. Soc. **57**, 1007 (1935).
2. See, e.g., H. Kuhn: Naturwissenschaften **54**, 429 (1967) ; H. Kuhn, D. Möbius and H. Bücher: *Techniques of Chemistry*, ed. by A. Weissberger and B. W. Rossiter (Wiley, New York, 1973) Vol. 1, Part IIIB ; M. Sugi: Journal of Molecular Electronics 1, 3 (1985) ; G. G. Roberts: Adv. Phys. **34**, 475 (1985) ; M. Sugi: Thin Solid Films **152**, 305 (1987) ; H. Kuhn: Thin Solid Films **178**, 1 (1989).
3. See, e.g., papers included in: Thin Solid Films **68** (1980); **99** (1983); **132-134** (1985); **159/160** (1988); **178-180** (1989); **210/211** (1992); **242** (1994); **284/285** (1996); **327-329** (1998).

4. See, e.g., J. B. Bateman and E. J. Covington: J. Colloid Sci. **16**, 531 (1961); Y. Koyama, M. Yanagishita, S. Toda and T. Matsuo: J. Colloid Interface Sci. **61**, 438 (1977); M. Fuyuki, H. Tabei and T. Kurihara: Langmuir **4**, 1123 (1988).

5. R. Eirich: *Rheology -Theory and Applications* (Academic Press, New York, 1956).

6. K. Iriyama, M. Yoshiura and F. Mizutani: Thin Solid Films **68**, 47 (1980).

7. M. Sugi and S. Iizima: Thin Solid Films **68**, 199 (1980).

8. M. Sugi, T. Fukui, S. Iizima and K. Iriyama: Mol. Cryst. Liq. Cryst. **62**, 165 (1980).

9. M. Sugi, M. Saito, T. Fukui and S. Iizima: Thin Solid Films **99**, 17 (1983).

10. S. Kuroda, M. Sugi and S. Iizima: Thin Solid Films **99**, 21 (1983).

11. M. Saito, M. Sugi, T. Fukui and S. Iizima: Thin Solid Films **100**, 117 (1983).

12. T. Fukui, M. Saito, M. Sugi and S. Iizima Thin Solid Films **109**, 247 (1983).

13. M. Sugi, M. Saito, T. Fukui and S. Iizima Thin Solid Films **129**, 15 (1985).

14. K. Sakai, M. Saito, M. Sugi and S. Iizima: Jpn. J. Appl. Phys. **24**, 865 (1985).

15. S. Kuroda, K. Ikegami, M. Sugi and S. Iizima: Solid State Commun. **58**, 493 (1986).

16. H. Nakahara, K. Fukuda, D. Möbius and H. Kuhn: J. Phys. Chem. **90**, 6144 (1986).

17. S. Nishikawa, Y. Tokura, T. Koda and K. Iriyama: Jpn. J. Appl. Phys. **25**, L701 (1986).

18. G. Scheibe: Angew. Chem. **49**, 563 (1936).

19. E.E. Jelly: Nature **138**, 1009 (1936).

20. T. Kobayashi (ed.): *J-Aggregates* (World Scientific, Singapore, 1996).

21. N. Minari, K. Ikegami, S. Kuroda, K. Saito, M. Saito and M. Sugi: Solid State Commun. **65**, 1259 (1988).

22. N. Minari, K. Ikegami, S. Kuroda, K. Saito, M. Saito and M. Sugi: J. Phys. Soc. Jpn. **58**, 222 (1989).

23. M. Saito, M. Sugi, K. Ikegami, M. Yoneyama and S. Iizima: Jpn. J. Appl. Phys. **25**,

L478 (1986).

24. S. Kuroda, K. Ikegami, K. Saito, M. Saito and M. Sugi: J. Phys. Soc. Jpn. **56**, 3319 (1987).

25. S. Kuroda, K. Ikegami, K. Saito, M. Saito and M. Sugi: Thin Solid Films **159**, 285 (1988).

26. S. Kuroda, K. Ikegami, K. Saito, M. Saito and M. Sugi: Solid State Commun. **71**, 333 (1989).

27. S. Kuroda, K. Ikegami, K. Saito, M. Saito and M. Sugi: Thin Solid Films **178**, 555 (1989).

28. S. Kuroda, K. Ikegami, K. Saito and M. Saito, M. Sugi and H. Shirakawa: *Proc. Annual Int. Conf. IEEE/EMBS* (1989) p. 1313.

29. S. Kuroda, K. Ikegami and M. Sugi: Mol. Cryst. & Liq. Cryst. **190**, 111 (1990).

30. S. Kuroda, K. Ikegami, Y. Tabe, K. Saito, M. Saito and M. Sugi: Phys. Rev. **B43**, 2531 (1991).

31. K. Murata, H. Shin, S. Kuroda and K. Saito: Thin Solid Films **327-329**, 446 (1998).

32. N. Kato, K. Saito, H. Aida and Y. Uesu: Chem. Phys. Lett. **312**, 115 (1999).

33. K. Ikegami, C. Mingotaud and M. Lan: J. Phys. Chem. **103**, 11261 (1999).

34. Y. Hirano, T. M. Okada, Y. F. Miura, M. Sugi and T. Ishii: J. Appl. Phys. **88**, 5194 (2000).

35. V. L. Streeter (ed.): *Handbook of Fluid Dynamics* (McGraw-Hill, New York, 1961).

36. I. Imai: *Fluid Dynamics* (Syokabo, Tokyo, 1973) Vol. 1.

37. N. Saito: *Introduction to Polymer Physics* (Syokabo, Tokyo, 1967).

38. M. Sugi, N. Minari, K. Ikegami and S. Kuroda: *Proc. Annual Int. Conf. IEEE/EMBS* (1988) p. 1001.

39. M. Sugi, Y. Tabe, K. Ikegami, S. Kuroda, K. Saito and M. Saito: *Proc. 8th Symp. Future Electron Devices* (1989) p. 183.

40. M. Sugi, N. Minari, K. Ikegami, S. Kuroda, K. Saito and M. Saito: Thin Solid Films **178**, 157 (1989).

41. M. Matsumoto, T. Sekiguchi, H. Tanaka, M. Tanaka, T. Nakamura, H. Tachibana, E Manda, Y. Kawabata and M. Sugi: J. Phys. Chem. **93**, 5877 (1989).

42. M. Matsumoto, H. Miyasaka, T. Sekiguchi, H. Tanaka, H. Tachibana, T. Nakamura, E. Manda, Y. Kawabata and M. Sugi: Thin Solid Films **178**, 367 (1989).

43. S. Chandrasekhar: Rev. Mod. Phys. 15, 1 (1943).

44. See, e.g., Papers included in *Bimgham Memorial Symposium*: J. Colloid Sci. **2**, 1-222 (1947).

45. E. C. Bingham: J. Phys. Chem. **29**, 1201 (1925).

46. G. S. Patil, S. S. Katti and A. B. Biswas: J. Colloid & Interface Sci. **25**, 462 (1967).

47. Y. Tabe, K. Ikegami, S. Kuroda, K. Saito, M. Saito and M. Sugi: Appl. Phys. Lett. **57**, 1191 (1990).

48. Y. Tabe, K. Ikegami, S. Kuroda, K. Saito, M. Saito and M. Sugi: J. Appl. Phys. **70**, 1425 (1991).

49. H. Kumehara, T. Kasuga, T. Watanabe and S. Miyata: Thin Solid Films **178**, 175 (1989).

50. S. Schwiegk, T. Vahlenkamp, y. Xu and G. Wegner: Macromolecules **25**, 2513 (1992).

51. K. Ikegami, S. Kuroda, Y. Tabe, K. Saito, M. Sugi, M. Matsumoto, T. Nakamura and Y. Kawabata: Jpn. J. Appl. Phys. **31**, 1206 (1992).

52. H. Nakahara and K. Fukuda: J. Colloid Interface Sci. **69**, 24 (1979).

53. Y. Tabe, K. Ikegami, S. Kuroda, K. Saito, M. Saito and M. Sugi: Thin Solid Films **210**,

32 (1992).

54. Y. Tabe, K. Ikegami and M. Sugi: J. Appl. Phys. **73**, 905 (1993).

55. K. Ogawa, H. Yonehara, T. Shoji, S. Kinoshita, E. Maekawa, H. Nakahara and K. Fukuda: Thin Solid Films **178**, 439 (1989).

56. T. Sorita, S. Miyake, H. Fujioka and H. Nakajima: Jpn. J. Appl. Phys. **30**, 131 (1991).

57. M. Vandevyver, P.-A. Albouy, C. Mingotaud, J. Perez, A. Barraud, O. Karthaus and H. Ringsdorf: Langmuir **9**, 1561 (1993).

58. T. Seki, T. Tamaki, K. Ueno, and Y. Tanaka: Thin Solid Films **243**, 625 (1994).

59. S. Pfeiffer, C. Mingotaud, C. Garrigou-Lagrange, P. Delhaes, A. Sastre and T. Torres: Langmuir **11**, 2705 (1995).

60. C. Jego, N. Leroux, B. Agricole, M. Mauzac and C. Mingotaud: Liq. Cryst. **20**, 691 (1996).

61. C. Mingotaud, B. Agricole and C. Jego: J. Phys. Chem. **99**, 17068 (1995).

62. K. Ikegami, C. Mingotaud and P. Delhaès: Phys. Rev. **E56**, 1987 (1997).

63. G. K. Batchelor: *An Introduction to Fluid Dynamics*, Jpn. ed. (Tokyo Denki Univ. press, 1972).

64. Hiemenz-1P.C. Hiemenz: *Principles of Colloid and Surface Chemistry*, 2nd ed. (Mercel Dekker, New York, 1986) p. 173.

65. H. Eyring: J. Chem. Phys. **4**, 283 (1936).

66. H. Eyring, The *Theory of Rate Processes*, ed. by S.Glasstone, K.J. Laidler and H. Eyring (MacGraw-Hill, New York, 1941) Chapter IX, pp. 480-483.

67. R. Miller, R. Wüstneck, J. Krägel and G. Kretzschmar: Colloid and Surfaces **111**, 75 (1996).

68. K. Ikegami, C. Mingotaud, C. Jego, B. Agricole and P. Delhaès: Supramol. Sci. **4**, 323

(1997).

69. N. Cuvillier, C. Mingotaud and K. Ikegami: J. Chem. Phys. **111**, 6982 (1999).

70. K. Ikegami, C. Jego, C. Mingotaud and P. Delhaès: Jpn. J. Appl. Phys. Part 2 **36**, L883 (1997).

71. K. Ikegami, C. Mingotaud and P. Delhaès: Thin Solid Films **327-329**, 24 (1998).

72. T. Maruyama, G. Fuller, C. Frank and C. Robertson: Science **274**, 233 (1996).

73. T. Maruyama, M. Friedenberg, G. G. Fuller, C. W. Frank, C. R. Robertson, A. Ferencz and G. Wegner: Thin Solid Films **273**, 76 (1996).

74. M. C. Friedenberg, G. G. Fuller, C. W. Frank and C. R. Robertson: Macromolecules **29**, 705 (1996).

75. M. C. Friedenberg, G. G. Fuller, C. W. Frank and C. R. Robertson: Langmuir **12**, 1594 (1996).

CHAPTER 3

EQUILIBRIUM AND DYNAMICS OF 2D AGGREGATING MIXED MONOLAYERS CONSISTING OF SOLUBLE AND INSOLUBLE AMPHIPHILES

V. B. Fainerman[1] and D. Vollhardt[2]

[1] International Medical Physicochemical Center, Donetsk Medical University, 16 Ilych Avenue, Donetsk 83003, Ukraine

[2] Max-Planck-Institut für Kolloid- und Grenzflächenforschung, D-14424 Potsdam/Golm, Germany

1. INTRODUCTION

The equilibrium and dynamic behaviour of mixed monolayers of soluble and insoluble amphiphiles at fluid/liquid interfaces plays an important role in various technologic and biologic processes. The equilibrium and dynamic behaviour of these monolayers was studied in numerous publications [1-19]. However, even for very simple systems, say, gaseous mixed monolayers, the thermodynamic analysis is cumbersome. That explains the fact that only recently few studies were published dealing with the extremely important case of the first-order phase transition in insoluble monolayer experimentally observed by the formation of two-dimensional (2D) domains [17-19]. In the review papers [20, 21] the analysis is given of the main theoretical methods used to describe the mixed monolayers, with special emphasis on the systems where the insoluble component of the monolayer is capable for the two-dimensional condensation.

On the other hand, rapid progress in the understanding of the molecular organisation of condensed monolayer phases has been made in the last decade [22]. The application of sensitive optical microscopy, particularly Brewster angle microscopy (BAM), has revealed a texture variety of condensed phases formed not only in Langmuir monolayers [23] but also in adsorption layers (Gibbs monolayers) [24]. Simultaneously, the introduction of Synchrotron X-ray diffraction at grazing incidence (GIXD) has made possible the determination of the two-dimensional lattice structure. It has been demonstrated that even in the condensed phase of an adsorption layer, the microscopic textural features are related to the two-dimensional lattice structure [25]. First application of these highly sensitive, direct experimental techniques to penetration experiments has provided new interesting information on the effect of the dissolved species on the state of the condensed phase of different penetration systems [19, 26, 27].

In the theoretical analysis of penetration systems, it is important to account not only for the effect of 2D aggregation on the penetration equilibrium and dynamics, but also the effect of adsorbed molecules of a dissolved surfactant on the process of the 2D aggregation of the insoluble monolayer. Some new theoretical approaches enabled the analysis of these systems, in particular, i) the use of Butler's equation to construct the equations of state and adsorption isotherms for multicomponent solutions and monolayers; ii) the application of the generalised Pethica's adsorption equation to derive the adsorption isotherm equation for a soluble surfactant in the presence of a Langmuir monolayer with or without 2D aggregation; and iii) the use of the generalised Volmer equation to derive the equations of state for insoluble and mixed monolayers assuming the 2D aggregation.

To study the dynamic behaviour of mixed monolayer of soluble and insoluble amphiphile, two methods are of common use. In the first one, namely the penetration, first the insoluble monolayer with some pre-defined coverage is formed, and then the dynamics of the adsorption (penetration) of the soluble amphiphile into this monolayer is studied. It was found, in particular, that the penetration of the soluble amphiphile results in the fact that the insoluble component starts to undergo the condensation at lower monolayer coverage as compared with its individual monolayer [17, 19, 20]. Another experimental method, in which the diluted monolayer of the insoluble amphiphile is brought to the equilibrium with the solution of the soluble amphiphile, and then the mixed monolayer is compressed, is a yet more difficult task. The attempts to analyse such a system for the case of two-dimensional condensation of the insoluble component were made in [20, 21]. However, it was assumed that no desorption of the soluble amphiphile takes place (the monolayer is conserved).

In the present review on the penetration of soluble surfactants into Langmuir monolayers and on the compression of mixed monolayer, the theoretical problems are systematically described; the experimental techniques are presented and some results of experimental model systems are reported.

2. GENERAL PRINCIPLES OF PENETRATION THERMODYNAMICS

The main problem in the thermodynamic theory of penetration is to determine the dependence of the adsorption of soluble surfactant on its concentration for any given (constant) adsorption (surface concentration) of the insoluble surfactant, and the onset of the surface pressure jump in mixed monolayers, caused by the adsorption of a soluble surfactant in the presence of the insoluble component of the Langmuir monolayer.

There exist several main theoretical approaches to the description of the penetration thermodynamics [20, 21]. One is based on the Gibbs' adsorption equation for multicomponent monolayers [5-8,14]. Another approach, initially proposed by Pethica [1,2], employs either the original Pethica equation [17-19], or more general expressions of the same type [4] to derive the adsorption isotherm for a soluble surfactant in presence of an insoluble monolayer at the interface. Another method is the application of the Butler's equation [28] to construct equations of state and adsorption isotherm for multicomponent monolayers [17].

For constant pressure and temperature, the Gibbs' adsorption equation for the multicomponent monolayer is:

$$d\Pi = \sum_{i=1}^{n} \Gamma_i d\mu_i \qquad (1)$$

where $\Pi=\gamma_0-\gamma$ is the surface pressure, γ_0 and γ are the surface tension of solvent and solution respectively, Γ and μ are the adsorbed amount and chemical potential for the ith soluble component. The value of μ for the case of soluble components can refer both to the bulk and to the surface layer (as equilibrium exists), and for the case of insoluble components to the surface layer only. It is seen from Eq. (1) that the Gibbs' dividing surface is used here, for which $\Gamma_{i=0}$. For the case when one insoluble and one soluble component exist, denoted by subscripts 1 and 2, respectively and noting that the area per one mole of the insoluble component 1 is $A_1 = 1/\Gamma_1$, Eq. (1) can be rewritten as:

$$d(\mu_1 - A_1\Pi) = \Pi dA_1 - A_1\Gamma_2 d\mu_2 \qquad (2)$$

which leads to the Maxwell equation:

$$\left(\frac{\partial\Pi}{\partial\mu_2}\right)_{A_1} = \left(\frac{\partial A_1\Gamma_2}{\partial A_1}\right)_{\mu_2} \qquad (3)$$

This equation, being integrated for constant μ_2 between the limits corresponding to two states of the monolayer (I and II, respectively) with different Γ values for the soluble or insoluble component, yields the difference of the product $(A_1\Gamma_2)$ in these states [5,8]:

$$(A_1\Gamma_2)_{II} - (A_1\Gamma_2)_I = \int_I^{II} \left(\frac{\partial \Pi}{\partial \mu_2}\right)_{A_1} dA_1 \tag{4}$$

where Γ_2 the adsorption of the dissolved surfactant. Thus, for any given μ_2, the change in the adsorption of the soluble surfactant can be calculated from the experimentally measured dependence of the surface pressure on A_1 and μ_2. .

The Gibbs' equation which expresses the total differential of the difference between the surface pressure of mixed monolayer and that of the monolayer which does not contain the insoluble surfactant (denoted by subscript 0) can be derived from Eq. (1):

$$d(\Pi - \Pi_0) = d\mu_1 / A_1 + (\Gamma_2 - \Gamma_2^0) d\mu_2 \tag{5}$$

which leads to a relation similar to Eq. (4):

$$(A_1(\Gamma_2 - \Gamma_2^0))_{II} - (A_1(\Gamma_2 - \Gamma_2^0))_I = \int_I^{II} \left(\frac{\partial(\Pi - \Pi_0)}{\partial \mu_2}\right)_{A_1} dA_1 \tag{6}$$

which was argued by Hall [8] to be more correct than Eq. (4) because, for infinitesimal adsorption of the two components, the products in the left hand side of this equation converge to a certain limiting value.

To summarise, the rigorous thermodynamic analysis of the penetration equilibrium, based on the Gibbs' equation, cannot provide the equation of state of the monolayer, nor the soluble component adsorption isotherm equation. This analysis only enables one to formulate the conditions for the penetration experiment, which are, however, very difficult to implement. Therefore, to elaborate the description of the thermodynamic behaviour for the case of actual mixed monolayer, one should use some approximate theoretical models. For example, in [14, 29] Eq. (3) was solved simultaneously with the model equations which describe the adsorption isotherm for the soluble component 2 in the presence of the insoluble component. These model equations, in turn, were constructed by a analysis of the formal kinetics of the adsorption process. In this way, various equations of state for mixed monolayers were derived.

It will be shown below that Pethica's model [1] also relies on the condition $\mu_1 = \text{const}$; in this case, however, the additional model equation used is the equation of state of the surface layer, which either follows rigorously from the analysis of the chemical potentials of the monolayer components (Butler's equation), or is postulated on a semi-empirical basis to obey the experimental trends, e.g., the generalised Volmer equation. The simultaneous solution of Pethica's equation and the equation of state for the monolayer yields the adsorption isotherm equation for the soluble component. It will be shown below that these equations, except the simple case of the ideal monolayer comprised of homologues, are quite different from the empirical equations used in [14, 29].

Assuming fixed coverage or adsorbed amounts of all the components except the jth one, the expression for the partial derivative $\partial\Pi/\partial\mu_j$ can be calculated from Eq. (1) as:

$$\left(\frac{\partial\Pi}{\partial\mu_i}\right)_{\Gamma_{i\neq j}} = \Gamma_j + \sum_{i\neq j}^{n}\Gamma_i\left(\frac{\partial\mu_i}{\partial\mu_j}\right)_{\Gamma_{i\neq j}} \tag{7}$$

The sum in the right hand side of Eq. (7) can be expressed as:

$$\sum_{i\neq j}^{n}\Gamma_i\left(\frac{\partial\mu_i}{\partial\mu_j}\right)_{\Gamma_{i\neq j}} = \left(\frac{\partial\Pi}{\partial\mu_j}\right)_{\Gamma_{i\neq j}} \sum_{i\neq j}^{n}\Gamma_i\left(\frac{\partial\mu_i}{\partial\Pi}\right)_{\Gamma_{i\neq j}} \tag{8}$$

Then, introducing expression (8) into Eq. (9), one obtains:

$$\left(\frac{\partial\Pi}{\partial\mu_j}\right)_{\Gamma_{i\neq j}} = \Gamma_j\left[1 - \sum_{i\neq j}^{n}\Gamma_i\left(\frac{\partial\mu_i}{\partial\Pi}\right)_{\Gamma_{i\neq j}}\right]^{-1} \tag{9}$$

Next, the well-known expression for the variation of the free energy of a Gibbs' dividing surface at constant pressure and temperature should be used:

$$dG = -A^s d\gamma + \sum \mu_i^s dm_i^s, \tag{10}$$

where A^s is the surface area, m_i is the excess of moles of the ith component in the monolayer and the subscript 's' refers to the dividing surface (interface). As dG is a total differential, each component at the surface obeys the following Maxwell relationship:

$$\left(\frac{\partial\mu_j^s}{\partial\gamma}\right)_{m_j^s} = -\left(\frac{\partial A^s}{\partial m_j^s}\right)_{\gamma} \tag{11}$$

where the derivative in the left hand side is calculated at constant pressure, temperature and numbers of moles of all the components present in the monolayer. The derivative on the right hand side of Eq. (12) is by definition the partial molar area ω_j of the jth component:

$$\left(\frac{\partial A^s}{\partial m_j^s}\right)_\gamma = \omega_j \tag{12}$$

at constant pressure, temperature and numbers of molecules differently from j.

Note that in Eqs. (9), Γ (adsorption, surface concentration) of all components except the jth Γ are kept fixed, while in Eqs. (11) the number of moles of all components are fixed. Neglecting this difference, taking into account that $d\gamma = -d\Pi$, and using Eqs. (11) and (12) one obtains instead of Eq. (9):

$$\left(\frac{\partial \Pi}{\partial \mu_j}\right)_{\Gamma_{i\neq j}} = \Gamma_j\left[1-\sum_{i\neq j}^{n}\theta_i\right]^{-1} \tag{13}$$

where $\theta_i = \Gamma_i/\Gamma_{\infty i}$ is the monolayer coverage. As for the soluble components, the value of the chemical potential in the bulk is equal to that within the surface layer, provided equilibrium conditions exist. Then recalling that $d\mu_i = RT d\ln a_i$, where a_i is the activity of the ith component in the solution bulk, R is the gas law constant, T is the temperature, one can express Eq. (13) for these components in the form of the generalised Pethica equation [30-32]:

$$\left(\frac{\partial \Pi}{\partial \ln a_j}\right)_{\Gamma_{i\neq j}} = RT\Gamma_j\left[1-\sum_{i\neq j}^{n}\theta_i\right]^{-1} \tag{14}$$

In the case of one insoluble and one soluble component, denoted by 1 and 2, respectively, and an ideal bulk solution, the ordinary Pethica equation [1] follows from Eq. (14):

$$\left(\frac{\partial \Pi}{\partial \ln c_2}\right)_{\theta_1} = \frac{RT\Gamma_2}{(1-\theta_1)} \tag{15}$$

where c is the bulk concentration. Equation (14) is the most general form of Pethica's equation. A possible error introduced into Eqs. (14) and (15) by keeping fixed the Γ values (adsorption or surface concentration) of all the components except the jth component in Eqs. (9) and fixing the number of moles of all the adsorbed components in Eqs. (11) was considered in [4], where a more general expression was obtained for the derivative entering the left hand side of Eq. (15). It follows from the analysis performed in [4] that Eqs. (14) and (15) can be used, when

$$\left(\frac{\partial \mu_1}{\partial \Gamma_2}\right)_{\Pi,\Gamma_1} \left(\frac{\partial \Gamma_2}{\partial \Pi}\right)_{\Gamma_1} = 0 \tag{16}$$

It follows from this relation that the chemical potential of the insoluble component does not depend on the adsorption of the soluble component provided that both the surface pressure and the adsorption of the insoluble component are kept fixed. In turn, as the surface concentration of the insoluble component is fixed, the requirement for the activity of this component to be constant implies the independence of the activity coefficient of the insoluble component on the adsorption of the soluble component. Clearly, this requirement is satisfied not only for the trivial case of the ideal monolayer, but also for the non-ideal monolayer, provided that the activity cross-coefficients of the monolayer components (or intermolecular interaction parameters) are equal to zero [20, 21].

Clearly, the condition of Eq.(16) imposes certain restrictions on the applicability of Pethica's model. On the other hand, an attempt to verify Eq. (14) experimentally was undertaken in [32], which also had confirm its validity for mixed monolayers comprised of two non-ionic surfactants, or the mixture of non-ionic and ionic surfactant, or two ionic surfactants. Moreover, in the review [20] was demonstrated that using the Pethica equation (14) and generalised Szyszkowski-Langmuir equation [33-35]:

$$\Pi = -\frac{RT}{\omega}\ln(1-\theta_1-\theta_2) \tag{17}$$

one obtains the adsorption isotherm equation for a soluble surfactant in presence of an insoluble monolayer:

$$b_2c_2 = \frac{\theta_2}{1-\theta_1-\theta_2} \tag{18}$$

where ω is the mean molar area, and b_2 is the adsorption equilibrium constant. It can be seen that Eq. (18) is just an ordinary Langmuir equation in the generalised form [33-35] which to follow rigorously from the analysis of chemical potentials of the components of mixed monolayer [20, 36]. Also in papers [14, 37], Eq. (18) was derived starting from Eq. (17).

The chemical potentials of components within surface layer μ_i^s depend on the composition of the surface layer and its surface tension γ. The dependence of μ_i^s on the composition of the surface layer is given by the well-known Butler equation [28], often used in the derivation of surface equations of state and adsorption isotherms

$$\mu_i^s = \mu_i^{0s} + RT \ln f_i^s x_i^s - \gamma \omega_i \tag{19}$$

where $\mu_i^{0s}(T,P)$ is the standard chemical potential of component i and depends on temperature T and pressure P; f_i are the activity coefficients. The main guidelines for the application of this equation and numerous examples for the mixtures of soluble surfactants were presented in [36]. It should be noted that, as the solubility or insolubility of the ith component does not affect Eq. (19), then Eq. (19) can be used for the penetration analysis in the way quite similar to how it was employed in the case when the mixtures comprised of soluble components only. However, one remark should be made as follows. The expression for the chemical potentials of the components in the bulk solution, similar to Eq. (19),

$$\mu_i^\alpha = \mu_i^{0\alpha} + RT \ln f_i^\alpha x_i^\alpha, \tag{20}$$

where the superscript 'α' refers to the bulk solution, and the standard chemical potentials $\mu_i^{0\alpha}$ depend on pressure and temperature, is true for soluble components only. Therefore, the adsorption isotherm equations can be derived only for soluble components of the monolayer.

The application of Butler's equation has been illustrated for the case of the equation of state and adsorption isotherm equation for the non-ideal monolayer (with respect to both the enthalpy and the entropy) of an insoluble (1) and a soluble (2) component [36, 37]:

$$\Pi = -\frac{RT}{\omega}\left[\ln(1-\theta_1-\theta_2)+\theta_1\left(1-\frac{1}{n_1}\right)+\theta_2\left(1-\frac{1}{n_2}\right)+a_1\theta_1^2+a_2\theta_2^2+a_{12}\theta_1\theta_2\right], \tag{21}$$

$$b_2c_2 = \frac{\theta_2}{(1-\theta_1-\theta_2)^{n_2}}\exp(-2a_2\theta_2-2a_{12}\theta_1)\cdot\exp\left[(1-n_2)\left(a_1\theta_1^2+a_2\theta_2^2+a_{12}\theta_1\theta_2\right)\right], \tag{22}$$

where $n_i = \omega_i/\omega$.

It can be concluded that Butler's equation, Eq. (19) always leads to the logarithmic form of the equation of state for mixed monolayers; which often disagrees with the experimental results obtained for insoluble monolayers. For these systems, Volmer's or Van der Waals' equations of state are more appropriate. Therefore the method based on the Pethica equation is advantageous, enabling one to apply model semi-empirical equation of state for mixed monolayer.

Prior to considering the equilibrium and dynamics of the penetration of soluble component into insoluble monolayer, the properties of a single amphiphilic Gibbs and Langmuir monolayers should be discussed.

3. DESCRIPTION OF THE GIBBS MONOLAYERS WITH 2D AGGREGATION

There exist some models which describe the equilibrium adsorption behaviour of soluble amphiphiles [36, 37-41]. The equation of state and adsorption isotherm for the Frumkin model (which becomes the Langmuir model for a = 0) are [36, 38]

$$-\frac{\Pi\omega}{RT} = \ln(1-\theta) + a\theta^2, \tag{23}$$

$$bc = \frac{\theta}{1-\theta}\exp(-2a\theta). \tag{24}$$

where θ is the surface coverage, $\theta = \Gamma\omega$, and a is the intermolecular interaction constant. The critical phenomena in an adsorption layer (condensation) have been explained qualitatively using an equation of the Frumkin equation of state [42-44]. If the interaction parameter $a>2$, then at constant temperature Eq. (23) has three real roots: three different values the surface coverage θ (or area per mole $A=1/\Gamma$) correspond to the same surface pressure Π. As the monolayer should be mechanically stable, $d\Pi/dA \leq 0$, so that the system undergoes a jump through the instability region, from a value of θ determined by the smallest root of Eq. (23) to that corresponding to the largest root. This jump is known as two-dimensional condensation of the monolayer. The surface pressure that corresponds to the coexistence of a condensed and gaseous state can be determined by using the so-called Maxwell construction [42-44]. In spite of the obvious capability of the models described by Eq. (23) to provide explanation of the phenomenon of surface condensation due to strong intermolecular interaction, there is evidence that this model is incompatible with the actual physical process governing the two-dimensional condensation of surfactants in interfacial layers. In both regions, precritical and transcritical, Eq. (23) involves the same surface layer entities, namely the monomers and no distinction is made between free monomers and molecules involved into aggregates. Clearly, the intermolecular interaction between the molecules in the aggregates (condensed phase) cannot affect the surface pressure in the same manner as the interaction between non-aggregated monomers does. Therefore the theory which assumes the formation of the aggregates in the monolayer should be most appropriate for such systems [45, 46].

The quasi-chemical model ($\mu_n^s = n\mu_1^s$) for aggregation in the adsorption layer with the arbitrary aggregation number n is described by the following equations of state and adsorption isotherm [45, 46]

$$-\frac{\Pi\omega}{RT} = \ln\left\{1 - \Gamma_1\omega\left[1 + (\Gamma_1/\Gamma_c)^{n-1}\right]\right\} \tag{25}$$

$$bc = \frac{\Gamma_1\omega}{\left\{1 - \Gamma_1\omega\left[1 + (\Gamma_1/\Gamma_c)^{n-1}\right]\right\}^{\omega_1/\omega}} . \tag{26}$$

Here Γ_1 and Γ_c are the partial and critical adsorption of monomers, respectively, and the average molar area ω should be expressed via the equation

$$\frac{\omega}{\omega_1} = \frac{1 + n(\Gamma_1/\Gamma_c)^{n-1}}{1 + (\Gamma_1/\Gamma_c)^{n-1}} . \tag{27}$$

where $\omega 1$ is the molar area of monomers.

In the limiting case of large aggregates (n>>1) equations (25) and (26) expressed via the molar area of monomers $\omega 1$ and the critical adsorption (surface concentration) of aggregation Γ_c, can be written in the following form [47]

$$\Pi = -\frac{RT\Gamma_c}{\omega_1\Gamma}\ln\left[1 - \omega_1\Gamma\right], \tag{28}$$

$$bc = \frac{\Gamma\omega_1}{\left[1 - \omega_1\Gamma\right]^{\Gamma_c/\Gamma}} , \tag{29}$$

In this equations the adsorption Γ is expressed via the observable total adsorption of monomers Γ_1 and aggregates Γ_n recalculated as monomers: $\Gamma = \Gamma_1 + n\Gamma_n$. Eq. (28) predicts that above the critical adsorption Γ_c, or above the corresponding critical bulk concentration determined from Eq. (29), the increase in equilibrium surface pressure is much slower as compared to the case where molecules do not form aggregates. The position of the inflection point in the surface pressure isotherm for large clusters depends on Γ_c: the higher Γ_c is, the higher is the surfactant concentration at which the cluster formation sets in, the less pronounced is the inflection, and the less steep is the isotherm beyond the inflection point [48].

4. EQUATION OF STATE FOR LANGMUIR MONOLAYERS WITH 2D AGGREGATION

Under corresponding conditions, the behaviour of Langmuir monolayers formed by insoluble amphiphiles is characterised by the existence of a sharp break in the surface pressure (Π) vs area per one mole or molecule (A) isotherm in the region where the main phase transition from a gaseous monolayer (GM) (or so-called liquid-expanded (LE) monolayer) to the condensed (LC) state takes place. In these cases, a sharp decrease of the isotherm slope is observed in the two phase coexistence region during the transition from the GM or LE to the LC state, while the area per one molecule which corresponds to the commencement of the transition from (GM)LE to LC monolayer state becomes lower with increasing temperature.

It was suggested by Langmuir [49] that in the LE-LC transition region, the formation of 2D domains surrounded by non-aggregated molecules occurs, while in the transition region the number and size of the condensed phase domains increase. The van der Waals equation for the non-ideal 2D gas

$$\Pi = \frac{RT}{A-\omega} - \frac{a}{A^2} \qquad (30)$$

similar to the Frumkin equation of state (Eq. 23) predicts the existence of metastable states for high values of the intermolecular constant a [50]. These states are characterised by the stepwise decrease of the area A during the 2D condensation, while the surface pressure value remains unchanged. However, contrary to the experimental data, it predicts the horizontal shape of Π-A isotherm in the GM-LC transition region.

Several attempts were made to give an explanation for the non-horizontal shape of Π-A isotherms in 2D transition region, also the effect of trace impurities was discussed [51-59]. In some studies, GM-LC transitions were treated in the framework of a quasi-chemical approach, where the mass action law was employed for the description of the monomer/aggregate equilibrium. In the studies performed by Ruckenstein and Li [54-56], the change of standard chemical potential related to the transition from the monomers to the clusters (n-mers) is considered. For calculating the free energy of the 2D transition, five constituents of standard free energy were taken into account in [56], in particular, the energy associated with the transition of hydrocarbon chains from the gaseous phase to the liquid phase, the conformational energy of these chains in the monolayer and their interaction with the aqueous phase, steric and electrostatic interactions between the polar groups. A still more detailed analysis of the free energy constituents was presented by Ruckenstein and Li in [59]. To our knowledge, in the studies performed by these authors the most rigorous and detailed thermodynamic description is provided for the monomer/aggregate equilibrium within the insoluble monolayer. The theories laid out in [56], and especially in [59], agree well with the experimental data in the prediction of non-horizontal shape of the isotherm for the case when the size of aggregates is reasonable. At the same time, the implementation of the theory for the description of actual monolayers is hampered by the fact that the resulting equations are extremely cumbersome, being of an implicit differential [56] or integro-differential [59] form, and involve 5 to 7 rather unknown parameters (domain radius, dipole-dipole, electrostatic and van der Waals interaction constants, parameters which define the conformational free energy and the entropy of mixing etc.).

The combination of a quasi-chemical approach with model considerations was proposed in [53, 57], resulting in rather simple equations which contain a few parameters only. Satisfactory agreement with the experimental Π - A isotherms has been obtained in the following way. In paper [36], the generalised Volmer equation which describes the multicomponent insoluble monolayer of monomers or aggregates was derived theoretically on the basis of Gibbs' (1) and Butler's (19) equations:

$$\Pi = RT\frac{\sum_i \Gamma_i}{1-\sum_i \Gamma_i \omega_i} - \Pi_{coh} \tag{31}$$

where Γ_i is the adsorption (surface concentration) value of the ith component or state, and Π_{coh} is cohesion pressure. The approach based on the combination of the generalised Volmer equation (31) and the quasi-chemical equilibrium model was developed further in a recent study [61]. It was assumed that the area per one molecule in the aggregate can differ from the area per free monomer molecule. Also, the condition which describes the equilibrium between the monomers and aggregates was formulated considering the free surface existing in the monolayer. If the monolayer consists only of monomers and large clusters (bimodal distribution), then the equation of state for the monolayer at $A \leq Ac$ is [61]:

$$\Pi = \frac{RT(A/A_{c\Pi})^2}{A-\omega_1\left[1+\varepsilon\left((A/A_{c\Pi})^2-1\right)\right]} - \Pi_{coh}, \tag{32}$$

where

$$A_{c\Pi} = A_c \exp\left(\frac{(\Pi-\Pi_c)\varepsilon\omega_1}{RT}\right), \tag{33}$$

Here ω_1 is the partial molar area for monomers; $A = 1/\Gamma$, A is the area per mole; $\theta = \omega_1/A = \Gamma\omega_1$ the monolayer coverage; ε the correcting factor which accounts for the decrease of the area per one molecule during the clusterisation; $A_c = 1/\Gamma_c$ is the molar area which corresponds to the point of the commencement of the phase transition, i.e., at $\Pi = \Pi_c$. For $A \geq A_c$, Eq. (32) becomes the Volmer equation for monomers [61]

$$\Pi = \frac{RT}{A-\omega_1} - \Pi_{coh} = \frac{RT\Gamma}{1-\theta} - \Pi_{coh} \tag{34}$$

Therefore, the equation of state (33) involves only four constants (Π_{coh}, A_c, ω and ε) and describes the entire region of surface pressure, from the gaseous monolayer to the condensed state. Two constants (Π_{coh} and ω) are applicable to the gaseous monolayer region. The value A_c is the co-ordinate of the break point of the Π-A isotherm located at the beginning of the main phase transition region. Only the small parameter ε refers explicitly to the 2D phase transition region, and additionally affects the isotherm shape at $A < A_c$.

It was shown in [19] that the results obtained from the quasi-chemical model for large aggregates are identical to those which follow from the phase model of the aggregation in the monolayer. This model assumes that the chemical potential of the monomers which exist outside of the aggregates, μ_1, is equal to that characteristic for the molecules comprised into the aggregates, $\mu_{1(n)}$, that is, $\mu_1 = \mu_{1(n)}$.

The model described by Eq. (40) was extended in [62] for the case of quasi-bimodal distribution which includes small aggregates (dimers and trimers) in the range $A \geq A_c$ and large aggregates (clusters) in the range $A \leq A_c$. The phospholipids are usually ionised in the monolayers (e.g. Dipalmitoyl-phosphatidyl glycerol, DPPG) or, existing as zwitter-ions (e.g., Dipalmitoyl-phosphatidyl choline, DPPC), are capable of forming monolayers possessing spatially separated charges. Therefore, the equation of state for such monolayer should include the contributions from Coulomb interaction, and account for the variation of the molar area of ionised molecules and a possible increase in the number of kinetic entities caused by the dissociation. Equation (34) was generalised for the case of ionised monolayers in [63]. For phospholipids these corrections to Eq. (34), however, were found to be of minor importance [64].

5. PENETRATION THERMODYNAMICS FOR HOMOLOGUES

The theoretical description of the equilibrium penetration of soluble surfactant into the different phases of Langmuir monolayers becomes less complicated if the partial molar areas of the two components are approximately equal to each other ($\omega_1 \cong \omega_2$). This case includes the penetration of the soluble homologue into the monolayer of the insoluble homologue; obviously, the range of the systems which can be thus described covers other mixtures composed of molecules with similar geometry.

So far, the equation of state which, for monolayers formed by the mixture of the insoluble and soluble components, expresses both the dependence of the surface tension on the adsorption of the soluble surfactant, similarly to the Szyszkowski-Langmuir equation Eq. (17), and the surface tension dependence on the adsorption of the insoluble component, similarly to the Volmer equation Eq. (31) is unfortunately not available. Therefore, for low adsorption of the soluble component, the generalised Volmer equation Eq. (31) is used to describe the two components of the monolayer, while for low adsorption of the insoluble component, the generalised Szyszkowski-Langmuir equation Eq. (17) is employed. In the following sections, some examples of such systems are considered.

5.1. Generalised Szyszkowski-Langmuir equation

The equation of state Eq. (17) and adsorption isotherm equation Eq. (18) correspond to the mixture of two homologues, the insoluble 1 and soluble 2 ones, for the case when 2D aggregation does not occur. These equations can be transformed into (see [18]):

$$\Pi = \frac{RT}{\omega_2} \ln \frac{(1 + b_2 c_2)}{(1 - \theta_1)} \tag{35}$$

$$\Gamma_2 = \frac{1 - \theta_1}{\omega_2} \frac{b_2 c_2}{1 + b_2 c_2} \tag{36}$$

which lead to interesting conclusions First, it follows from Eq. (35) that the variation (jump) of monolayer surface pressure which arises due to the adsorption of soluble surfactant, that is, the difference between the pressures of the mixed monolayer and the pure insoluble monolayer $\Delta\Pi$, does not depend on θ_1, but depends on the value of b_2c_2 only, $\Delta\Pi = (RT/\omega_2)\ln(1 + b_2 c_2)$. Second, according to the isotherm equation Eq. (36), the monolayer coverage by the soluble component $\theta_2 = \omega_2\Gamma_2$ in presence of the insoluble component decreases proportionally to $(1 - \theta_1)$. Therefore, the θ_2 value, which corresponds to some $\Delta\Pi$ values for the mixed monolayer, is lower than the θ_2 value that corresponds to the same $\Delta\Pi$ for the case when the monolayer does not contain the insoluble amphiphile. It will be shown below that this behaviour can result in more rapid achievement of the adsorption equilibrium in the case that an insoluble amphiphile is present.

For an ideal (both in the bulk and at the surface) mixture of two amphiphiles (1 and 2), of which only the first one is capable of the formation of large two-dimensional aggregates in the adsorption layer, is derived the equations [65]

$$\Pi = -\frac{RT(\theta_c + \theta_2)}{\omega(\theta_1 + \theta_2)} \ln(1 - \theta_1 - \theta_2) \tag{37}$$

$$b_1 c_1 = \frac{\theta_1}{(1 - \theta_1 - \theta_2)^{(\theta_c + \theta_2)/(\theta_1 + \theta_2)}} \tag{38}$$

$$b_2 c_2 = \frac{\theta_2}{(1 - \theta_1 - \theta_2)^{(\theta_c + \theta_2)/(\theta_1 + \theta_2)}} \tag{39}$$

where $\theta_1 = \omega_1\Gamma_1 = \omega_1(\Gamma_{11} + n\Gamma_{1n})$, $\theta_2 = \omega_2\Gamma_2$ and $\theta_c = \omega_1\Gamma_{1c}$ are the degrees of monolayer coverage; Γ_{11} and $n\Gamma_{1n}$ are the surface concentrations of the component 1 in the monomeric and aggregated (recalculated in the terms of monomers) form, respectively; Γ_{1c} is the critical adsorption which corresponds to the aggregation onset for the component 1; and $\omega \cong \omega_1 \cong \omega_2$. If the component 1 is insoluble, than the isotherm equation (38) is excluded, and $\theta_1 = \omega_1/A$ and $\theta_c = \omega_1/A_c$.

It appears that the simplifications used in [65] to derive Eqs. (37)-(39) are too rough to provide any satisfactory agreement between the experimental data and theoretical predictions. This concerns not only the assumption of ideal behaviour of the monolayer and the equality $\omega_1 = \omega_2$, but also in what regards the influence of the second component on the critical adsorption value (aggregation onset) for the component 1 (see below), the possible variation of the ω_i values in mixed monolayer etc. In paper [65] new procedure has been applied to calculate the theoretical dependence of surface pressure on concentration, which to be less dependent on the deficiencies of the model. To derive this procedure, one transforms Eqs. (37)-(39) to the form

$$\exp(\overline{\Pi}/\alpha) = b_1 c_1 (1 - \theta_1 - \theta_2)^{\nu - 1} + b_2 c_2 (1 - \theta_1 - \theta_2)^{\nu - 1} + 1 \tag{40}$$

where $\overline{\Pi} = \Pi\omega_1/RT$ is a dimensionless surface pressure, and $\nu = (\theta_c + \theta_2)/(\theta_1 + \theta_2)$ is the fraction of monomers in the total adsorption of the two surfactants. In this equation the $b_i c_i$ values can be expressed via the corresponding monolayer pressures for single solutions, using Eqs. (28) and (29) for aggregating surfactant and the Langmuir equation for the solution of surfactant 2 (Eq. 22 and 23 with a=0). This procedure leads to the expression

$$\exp(\overline{\Pi}/\nu) = \exp(\overline{\Pi}_1/\beta)\frac{(1-\theta_1-\theta_2)^{\nu-1}}{(1-\theta_1^*)^{\beta-1}} + \exp(\overline{\Pi}_2/\nu)(1-\theta_1-\theta_2)^{\nu-1} - 1 \quad (41)$$

where $\theta_1^* = \omega_1\Gamma_1^*$ is the degree of monolayer coverage by component 1 in the single solution at any given $\overline{\Pi}_1 = \Pi_1\omega_1/RT$; here $\beta = \Gamma_{1c}^*/\Gamma^*$ and $\overline{\Pi}_2 = \Pi_2\omega_2/RT$ is the dimensionless surface pressure for the single solution of component 2.

In the precritical adsorption region of the component 1, the relation $\nu = \beta = 1$ is valid, and Eq. (41) becomes

$$\exp\overline{\Pi} = \exp\overline{\Pi}_1 + \exp\overline{\Pi}_2 - 1 \quad (42)$$

To analyze Eq. (41) in the transcritical region of the adsorption of component 1, one transforms it into the form

$$\exp\overline{\Pi} = k_1\exp\overline{\Pi}_1 + k_2\exp\overline{\Pi}_2 - k_3 \quad (43)$$

where $k_1 = \exp(1/\beta - 1/\nu)\frac{(1-\theta_1-\theta_2)^{\nu-1}}{(1-\theta_1^*)^{\beta-1}}$, $k_2 = \exp(-1/\nu)(1-\theta_1-\theta_2)^{\nu-1}$, and $k_3 = \exp(-1/\nu)$.

As $\nu > \beta$, and also $\nu < 1$ and $\beta < 1$, then $k_1 > 1$, $k_2 < 1$ and $k_3 < 1$. As approximation, it can be therefore assumed that $k_1 = k_2 = 1$, while the value of k_3 can vary from $k_3 = 1$ at the onset of component 1 phase transition to $k_3 \cong 0$ in the extremely saturated monolayer for relatively low Γ_{1c} and Γ_2 values. It follows then that the equation

$$\exp\overline{\Pi} = \exp\overline{\Pi}_1 + \exp\overline{\Pi}_2 - k_3 \quad (44)$$

in the range k_3 = 1 - 0 can be used to calculate approximate values of surface pressure for mixed solutions of two surfactants in the transcritical adsorption region. It should be noted that the account for non-ideality effects which take place both in the bulk and in the surface layer of single solutions leads to rather insignificant changes in the form of Eq. (44), because these effects are almost entirely incorporated into the $\overline{\Pi}_1$ and $\overline{\Pi}_2$ values.

In the case that the formation of large aggregates of insoluble surfactant is stimulated by the adsorption of soluble surfactant, the Szyszkowski-Langmuir equation of state were derived in [19, 20]:

$$\Pi = -\frac{RT(A/A_c)}{\omega(1+\Gamma_2 A)}\ln[1-\theta_1-\theta_2] \tag{45}$$

As aggregation of the insoluble component occurs only when its surface concentration is high enough, the description of the two components based on Volmer's equation seems to be more appropriate than that based on the Szyszkowski-Langmuir equation. The results following from this approximation are now considered in more detail.

5.2. Generalised Volmer equation

If a first-order phase transition does not occur in the monolayer ($A \geq A_c$) than Volmer's equation Eq. (31) for two components 1 and 2 can be transformed into:

$$\Pi = RT\frac{\Gamma_2 + 1/A}{1-\omega_2\Gamma_2 - \omega_1/A} - \Pi_{coh} \tag{46}$$

Using Eq. (36), one can express the actual value of the adsorption of soluble amphiphile 2 in Eq. (46) via its equilibrium value in the individual solution $\Gamma_2^* = (1/\omega_2)b_2c_2/(1+b_2c_2)$. Finally one obtains from Eq. (46):

$$\Pi = RT\frac{\Gamma_2^*(1-\omega_1/A)+1/A}{1-\omega_2\Gamma_2^*(1-\omega_1/A)-\omega_1/A} - \Pi_{coh} \tag{47}$$

Note that Eq. (46) covers the cases of both equilibrium and non-equilibrium adsorption of the component 2, while Eq. (47) can be used only if the equilibrium adsorption takes place.

In rigorous development, the application of the generalised Volmer equation for the two components results in the modified adsorption isotherm equation for the soluble component 2. It was shown in [20] that the simultaneous solution of Eq. (46) and Pethica's equation (15) results in the adsorption isotherm equation for the soluble component 2:

$$b_2c_2 = \left(\frac{\theta_2}{1-\theta_1-\theta_2}\right)^{\frac{1}{1-\theta_1}} \cdot \exp\left[\frac{\theta_2}{(1-\theta_1)(1-\theta_1-\theta_2)}\right] \tag{48}$$

For small surface coverages of the two components, Eq. (48) is seen to coincide with Eq. (18). However, the presence of the insoluble monolayer affects the adsorption of the soluble surfactant for a fixed bulk concentration. For $\theta 1 = 0$ the isotherm equation which follows from (48):

$$b_2c_2 = \frac{\theta_2}{1-\theta_2}\exp\left(\frac{\theta_2}{1-\theta_2}\right), \tag{49}$$

which differs from Eq. (24) for a = 0 by an exponential factor.

Considering now the area range $A \le A_c$ or $A \le A_{cm}$ (where the subscript "m" at the critical area value is introduced to denote that this value in the mixed monolayer can be different from that for the individual monolayer; this will be explained below in more detail) [17-20]. Assuming that the formation of mixed aggregates can take place, one can transform the generalised Volmer equation Eq. (31) into:

$$\Pi = RT\frac{\Gamma_{11} + \Gamma_{21} + \Gamma_a}{1-\omega(\Gamma_{11} + \Gamma_{21} + n\Gamma_a)} - \Pi_{coh} \tag{50}$$

where the second suffix 1 refers to monomers, Γ_a is the adsorption of aggregates, and $n = n1 + n2$ is the total aggregation number for both insoluble and soluble component. It is seen from Eq. (50) that the behaviour of the surface pressure of the mixed monolayer depends on the type of the formed aggregates [17-20]. The denominator of Eq. (50) is the fraction of the free surface of the monolayer; therefore, with our assumptions concerning the molar areas of the components, it does not depend on the aggregation conditions. However, the numerator of Eq. (50) depends on these conditions Further we will consider some basic opportunities [20].

If the soluble component 2 does not participate in the formation of the mixed aggregates, that is, assuming $n2 = 0$ and $n1 >> 1$, Eq. (50) can be transformed into the expression (see [18]):

$$\Pi = RT\frac{\Gamma_2 + \Gamma_{1cm}}{1-\omega_1\Gamma_1 - \omega_2\Gamma_2} - \Pi_{coh} \tag{51}$$

If one takes into account the effect of the variations of the aggregate/monomer equilibrium conditions during the monolayer compression ($\Gamma_{1cm} \approx A/A_{cm}$), similarly to the procedure used to derive Eq. (32), then one can transform Eq. (51) into:

$$\Pi = RT\frac{\Gamma_2 + (1/A_{cm})(A/A_{cm})}{1-\omega_2\Gamma_2 - \omega_1/A} - \Pi_{coh} \quad (52)$$

On the other hand, if the adsorption of component 2 takes place in the equilibrium with the solution adjacent to the monolayer, with this equilibrium governed by Eq. (36) then:

$$\Pi = RT\frac{\Gamma_2^*(1-\omega_1/A) + (1/A_{cm})(A/A_{cm})}{1-\omega_2\Gamma_2^*(1-\omega_1/A) - \omega_1/A} - \Pi_{coh} \quad (53)$$

In the case that mixed aggregates are formed, and the conditions n1 >> 1 and n2 >> 1 are held, then it follows from Eq. (50)

$$\Pi = RT\frac{\Gamma_{2c} + \Gamma_{1cm}}{1-\omega_1\Gamma_1 - \omega_2\Gamma_2} - \Pi_{coh} \quad (54)$$

where Γ_{2c} is the adsorption of component 2 corresponding to the inset of the 2D transition of component 2.

Equations (51)-(54) predict different shapes of the Π - A isotherms in the region $A \le A_{cm}$, with the most pressure increase corresponding to Eq. (51).

Differentiating the equations of state (51)-(54) with respect to Γ_2, and solving the resulting differential equations simultaneously with Pethica's equation, one obtains the adsorption isotherm equations for the soluble component:

a) for Eq. (51):

$$b_2c_2 = \left(\frac{\theta_2}{1-\theta_1-\theta_2}\right)^{1+\theta_{cm}/(1-\theta_1)} \cdot \exp\left[\frac{\theta_2\theta_{cm} + (1-\theta_1)}{(1-\theta_1)(1-\theta_1-\theta_2)}\right] \quad (55)$$

b) for Eq. (52)

$$b_2c_2 = \left(\frac{\theta_2}{1-\theta_1-\theta_2}\right)^{(1-\theta_1+\theta_{cm})/(1-\theta_1)} \cdot \exp\left[\frac{\theta_2(1-\theta_1+\theta_{cm})}{(1-\theta_1)(1-\theta_1-\theta_2)}\right] \quad (56)$$

c) for Eq. (54):

$$b_2 c_2 = \left(\frac{\theta_2}{1-\theta_1-\theta_2} \right)^{(\theta_{cm}+\theta_{2c})/(1-\theta_1)} \cdot \exp\left[\frac{\theta_2(\theta_{cm}+\theta_{2c})}{(1-\theta_1)(1-\theta_1-\theta_2)} \right] \quad (57)$$

It is seen that even for low adsorption of the two components (including critical values) the shape of the adsorption isotherm of soluble component is determined by the aggregation conditions in the monolayer.

5.3. Influence of the soluble amphiphile adsorption on the aggregation conditions

For the formation of aggregates in the mixed monolayer, both components affect the value of the critical adsorption of the aggregate formation (or A_c) [17-20]. In most cases known so far, the formation of mixed aggregates does not take place, and therefore in the mixed monolayer the domains consist on the component 1 only. In this case, the presence of component 2 results only in the decrease of the fraction of free surface of the monolayer. Calculating the sum of the adsorptions (surface concentrations) of the two components, one obtains simple expression for A_{cm} (i.e., the critical area per mole of insoluble amphiphile in the mixed monolayer), see [19, 20]:

$$\frac{1}{A_{cm}} = \frac{1}{A_c} - \Gamma_2 \quad (58)$$

or, if the homologues are considered, this can be expressed via the monolayer coverages:

$$\theta_{cm} = \theta_c - \theta_2 \quad (59)$$

For some systems, Eqs. (58)-(59) were shown in [17-20] to give quite satisfactory correspondence with experimental results.

It is possible to derive more general expression for the critical area A_{cm} in the mixed monolayer. Possibly, the most correct procedure is to compare and to add the activities of the components in the individual and mixed monolayer, rather than the adsorptions (or coverages) of the components. The activity of the insoluble component in the monolayer is given by the $x_1\gamma_1$ value, where x_1 is the molar fraction of the component, and the activity coefficient γ_1, arising from mixing enthalpy non-ideality, depends on the composition of the monolayer [66, 67]:

for the individual monolayer:

$$RT\ln\gamma_1^* = H_{01}(1-x_1^*)^2, \tag{60}$$

for the mixed monolayer (if the component 2 is present):

$$\mathrm{RT}\ln\gamma_1 = \mathrm{H}_{01}(1-\mathrm{x}_1-\mathrm{x}_2)^2 + \mathrm{H}_{12}\mathrm{x}_2^2 + (\mathrm{H}_{01}+\mathrm{H}_{12}-\mathrm{H}_{02})(1-\mathrm{x}_1-\mathrm{x}_2)\mathrm{x}_2\,. \tag{61}$$

Usually for close homologues the condition $H_{12} = 0$ can be assumed as a good approximation [38]. Then it follows approximately from Eq. (61) that

$$\mathrm{RT}\ln\gamma_1 = \mathrm{H}_{01}(1-\mathrm{x}_1-\mathrm{x}_2)^2 + (\mathrm{H}_{01}-\mathrm{H}_{02})(1-\mathrm{x}_1-\mathrm{x}_2)\mathrm{x}_2 \tag{62}$$

Introducing Frumkin's constants $a_1 = H_{01}/RT$ and $a_2 = H_{02}/RT$, one obtains the expressions for the activity coefficients of component 1 in the individual and mixed monolayers:

$$\gamma_1^* = \exp\left(\mathrm{a}_1(1-\mathrm{x}_1^*)^2\right) \tag{63}$$

$$\gamma_1 = \exp\left(\mathrm{a}_1(1-\mathrm{x}_1-\mathrm{x}_2)^2 + (\mathrm{a}_1-\mathrm{a}_2)(1-\mathrm{x}_1-\mathrm{x}_2)\mathrm{x}_2\right) \cong \exp\left(\mathrm{a}_1(1-\mathrm{x}_1-\mathrm{x}_2)^2\right) \tag{64}$$

The approximate equality in Eq. (30) corresponds to the case when the values of a_1 and a_2 are roughly the same; for more general case $a_1 > a_2$. Similarly, for component 2 in the mixed monolayer one obtains:

$$\gamma_2 = \exp\left(\mathrm{a}_2(1-\mathrm{x}_1-\mathrm{x}_2)^2 + (\mathrm{a}_2-\mathrm{a}_1)(1-\mathrm{x}_1-\mathrm{x}_2)\mathrm{x}_1\right) \cong \exp\left(\mathrm{a}_2(1-\mathrm{x}_1-\mathrm{x}_2)^2\right) \tag{65}$$

The commencement of the aggregation in the monolayer corresponds to certain activity of the insoluble component in the individual monolayer, or the sum of the activities of insoluble and soluble components – for mixed monolayer. Therefore, instead of Eq. (58) or (59), one obtains:

$$\mathrm{x}_1^*\gamma_1^* = \mathrm{x}_1\gamma_1 + \mathrm{x}_2\gamma_2 \tag{66}$$

Taking into account Eqs. (63)-(65), one obtains

$$x_1 = x_1^* \exp k_1 - x_2 \exp k_2 \cong x_1^* \exp k_3 - x_2 \tag{67}$$

where

$$\mathrm{k}_1 = \mathrm{a}_1(1-\mathrm{x}_1^*)^2 - \mathrm{a}_1(1-\mathrm{x}_1-\mathrm{x}_2)^2 - (\mathrm{a}_1-\mathrm{a}_2)(1-\mathrm{x}_1-\mathrm{x}_2)\mathrm{x}_2,$$

$$\mathrm{k}_2 = (\mathrm{a}_2-\mathrm{a}_1)(1-\mathrm{x}_1-\mathrm{x}_2)^2 + (\mathrm{a}_2-\mathrm{a}_1)(1-\mathrm{x}_1-\mathrm{x}_2)(\mathrm{x}_1+\mathrm{x}_2),$$

and $k_3 = a(1-x_1^*)^2 - a(1-x_1-x_2)^2$; here the approximate relations a ≅ a_1 ≅ a_2 are assumed to be valid. As for the case of homologues the molar fractions are equal to the coverage values [36], then instead Eq. (59) one can obtain from Eq. (67) a more general equation:

$$\theta_{cm} = \theta_c \exp k_1 - \theta_2 \exp k_2 \cong \theta_c \exp k_3 - \theta_2 \qquad (68)$$

where $k_1 = a_1(1-\theta_c)^2 - a_1(1-\theta_{cm}-\theta_2)^2 - (a_1-a_2)(1-\theta_{cm}-\theta_2)\theta_2$,

$k_2 = (a_2-a_1)(1-\theta_{cm}-\theta_2)^2 + (a_2-a_1)(1-\theta_{cm}-\theta_2)(\theta_{cm}+\theta_2)$,

and $k_3 = a(1-\theta_c)^2 - a(1-\theta_{cm}-\theta_2)^2$. It is seen that, in contrast to Eq. (59), the enthalpy non-ideality gives rise to the exponential factors at two (or one for $a_1 \cong a_2$) terms in the right hand side of Eq. (68). It should be noted that Eq. (68) also is the approximate relation, because it was assumed that $H_{12} = 0$, and the entropy non-ideality of mixed monolayer was disregarded.

6. PENETRATION DYNAMICS

Information on the penetration dynamics, i.e., the time dependence of the surface pressure of a penetration system during the adsorption of the soluble surfactant, is still far from comprehensive, and related publications are rather scarce [11, 16-20].

We proceed with the discussion of the penetration kinetics of the component 2 into the insoluble monolayer, and the variation in the surface pressure jump (ΔΠ) rate determined by this penetration process. The well-known equation derived by Ward and Tordai [68] is the most general relationship between the dynamic adsorption Γ(t) and the subsurface concentration $c_2(0,t)$

$$\Gamma = 2\sqrt{\frac{D}{\pi}}\left[c_2\sqrt{t} - \int_0^{\sqrt{t}} c_2(0,t-\lambda)\,d\sqrt{\lambda}\right] \qquad (59)$$

where D is the diffusion coefficient, t is time, λ is a dummy integration variable. Eq. (69) is valid for non-diffusion, mixed and pure diffusion-controlled adsorption mechanisms. In the latter case the adsorption isotherm is an additional equation between the adsorption and the subsurface concentration. Equation (69) can be clearly applied also to describe the penetration. Provided that 2D aggregation does not occur, the relation between the subsurface concentration and the adsorption is given by the isotherm equation Eq. (38) for the diffusion mechanism of the penetration process. The simultaneous numerical solution of Eqs. (69) and (36) was performed in Ref. [19] using the collocation method reported previously in Ref. [69]. The variation of the dynamic surface pressure for mixed monolayers caused by penetration of the soluble surfactant was calculated from equation

$$\Delta\Pi = \frac{RT}{\omega}\ln\left(1 + b_2 c_2(0,t)\right) \tag{70}$$

which follows from Eq. (35). The results of the calculations performed for a single value of bulk concentration of soluble surfactant and various values of monolayer coverage by insoluble surfactant are presented in Fig. 1 [18]. It can be seen that the increase of θ_2 leads to a decrease in time necessary to achieve the equilibrium state, and to a significant increase of $\Delta\Pi$. As was mentioned above, this phenomenon is caused by the decrease of the equilibrium adsorption value for the soluble surfactant in presence of an insoluble monolayer.

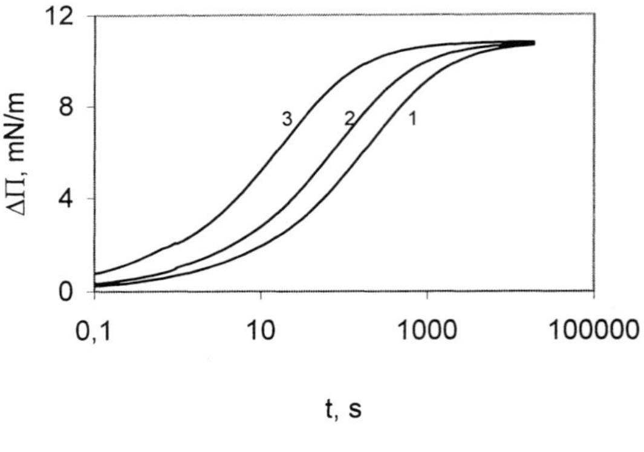

Fig. 1. Dynamic surface tension of monolayers for θ = 0 (curve 1), 0.35 (curve 2) and 0.7 (curve 3) due to the adsorption of the soluble surfactant [18]. The curves are calculated from Eqs. (69) and (70) for c2 = 10-5 mol/l, D = 5·10-6 cm2/s, ω = 2.28·109 cm2/mol and b2 = 1.69·105 l/mol.

A simple theoretical model Eq. (44) for the description of the coadsorption of surfactant mixtures enables one to estimate the dynamic surface pressure for the diffusion adsorption mechanism. The situation becomes most simply when the mixture of two surfactants is considered, of which one (say, component 1) possesses significantly higher activity than the other component, but the bulk concentration of this component 1 is much lower than that of component 2. This results in a situation that the adsorption of component 2 takes first place. Therefore, the dynamic surface pressure of component 2 in Eq. (44) is just the equilibrium surface pressure of this component.

Comparing the dynamic adsorption of component 1 in the single solution (Γ_1^*) with the dynamic adsorption in the mixed solution (Γ_1), from Eq. (69) one obtains the approximate expression for the adsorption time in the mixed solution

$$t = t^* \left(\Gamma_1 / \Gamma_1^*\right)^2 \tag{71}$$

where t^* is the adsorption time of component 1 in the single solution. Using Eqs. (28), (29), (37) and (38) one can transform Eq. (71) into the form [65]

$$t = t^* \exp\left(-\frac{2\omega_1(\Pi - \Pi_1)}{RT}\right) \tag{72}$$

In this expression the Π value is calculated from Eq. (44), while the Π_1 and t^* values can be taken from the experiment for the single component 1. As $\Pi > \Pi_1$, it follows from Eq. (72) that $t < t^*$, that is, the time necessary to achieve equilibrium in the mixed solution is lower than that in the single solution.

7. ADSORPTION OF A SOLUBLE COMPONENT IN THE COMPRESSED MIXED MONOLAYER

In some cases, the initial phase of the penetration experiment is performed in a usual way (the equilibrium pressure of the mixed monolayer is measured for fixed adsorption or area per molecule of the insoluble component), and subsequently the behaviour of the mixed monolayer compression isotherm is analysed [18]. Consider the constant rate dilation deformation of a mixed monolayer, which was first brought in equilibrium with the solution of component 2. The equation describing the balance of the soluble amphiphile in the monolayer is:

$$\frac{d\Gamma_2}{dt} = -\Gamma_2 \frac{d\ln A}{dt} - D\frac{c_2(0,t) - c_2^*}{\delta} \tag{73}$$

where D is the diffusion coefficient, t is time, $c_2(0,t)$ and c_2* are the surface and bulk concentrations of component 2, respectively, and δ is the thickness of the diffusion boundary layer. It is convenient to express the concentrations in Eq. (73) via corresponding adsorptions. The value of $c_2(0,t)$ is determined by arbitrary current values of θ_1 and θ_2, and can be expressed by the adsorption isotherm equation, as long as the diffusion adsorption/desorption mechanism is assumed. Therefore, it holds for the generalised Szyszkowski-Langmuir model, throughout the whole range of the area per one molecule of the insoluble homologue, equation:

$$c_2(0,t) = \frac{\theta_2}{b_2(1-\theta_1-\theta_2)}, \tag{74}$$

and for the generalised Volmer – Pethica model in the pre-critical range ($A \geq A_{cm}$) equation

$$c_2(0,t) = \frac{1}{b_2}\left(\frac{\theta_2}{1-\theta_1-\theta_2}\right)^{\frac{1}{1-\theta_1}} \cdot \exp\left[\frac{\theta_2}{(1-\theta_1)(1-\theta_1-\theta_2)}\right] \tag{75}$$

which follows from Eq. (48). In the range $A \leq A_{cm}$, Eq. (57) with $c_2 = c_2(0,t)$ is valid for this model.

The concentration in the bulk solution c_2* can be expressed via the adsorption in the single solution using the adsorption isotherm equation Eq.(24) for the Szyszkowski-Langmuir model

$$c_2^* = \frac{\theta_2^*}{b_2(1-\theta_2^*)}, \tag{76}$$

or Eq. (49) for the Volmer - Pethica model

$$c_2^* = \frac{1}{b_2}\frac{\theta_2^*}{(1-\theta_2^*)}\exp\left(\frac{\theta_2^*}{1-\theta_2^*}\right) \tag{77}$$

The δ value can be approximately expressed as $\delta = \sqrt{D\pi t}$ [70]. Noting that $dA/dt = \alpha$, $t = \alpha(A - A_0)$, and making corresponding substitutions, one obtains for the Szyszkowski-Langmuir model the differential equation which is valid throughout the whole A range:

$$d\theta_2 = -\frac{\theta_2}{A}dA + \frac{\omega_2}{b_2}\left(\frac{D}{\pi\alpha(A-A_0)}\right)^{1/2}\left(\frac{\theta_2}{1-\theta_1-\theta_2} - \frac{\theta_2^*}{1-\theta_2^*}\right)dA \tag{78}$$

For the Volmer – Pethica model, different equations for $d\theta_2/dA$ in the pre-critical and trans-critical range should be used,

for $A \geq A_{cm}$

$$d\theta_2 = -\frac{\theta_2}{A}dA + \frac{\omega_2}{b_2}\left(\frac{D}{\pi\alpha(A-A_0)}\right)^{1/2} \cdot \left(\left(\frac{\theta_2}{1-\theta_1-\theta_2}\right)^{\frac{1}{1-\theta_1}} \cdot \exp\left[\frac{\theta_2}{(1-\theta_1)(1-\theta_1-\theta_2)}\right] - \frac{\theta_2^*}{1-\theta_2^*}\exp\left(\frac{\theta_2^*}{1-\theta_2^*}\right)\right)dA \tag{79}$$

and for $A \leq A_{cm}$

$$d\theta_2 = -\frac{\theta_2}{A}dA + \frac{\omega_2}{b_2}\left(\frac{D}{\pi\alpha(A-A_0)}\right)^{1/2} \cdot \left(\left(\frac{\theta_2}{1-\theta_1-\theta_2}\right)^{(1-\theta_1+\theta_{1c})/(1-\theta_1)} \cdot \exp\left[\frac{\theta_2(1-\theta_1+\theta_{1c})}{(1-\theta_1)(1-\theta_1-\theta_2)}\right] - \frac{\theta_2^*}{1-\theta_2^*}\exp\left(\frac{\theta_2^*}{1-\theta_2^*}\right)\right)dA \tag{80}$$

Equations (79)-(80) can be integrated numerically for the compressible monolayer ($\alpha < 0$, $dA < 0$, $A - A_0 < 0$) throughout the range of area per one molecule of insoluble amphiphile: from the initial value $A = A_0$ to the final value $A = A$ provided that $\theta_2 = \theta_2^*(1-\omega_1/A_0)$ for $A = A_0$. Each of Eqs. (78)-(80) has two exact solutions for the case of infinitely large compression rate and infinitely small compression rate of the monolayer. Consider here Eq. (78) as an example. For $\alpha \to 0$ the equilibrium is established between the subsurface layer and the bulk. It follows, therefore, from Eq. (80) that

$$\theta_2 = \theta_2^*(1-\omega_1/A)\,. \tag{81}$$

On the contrary, for very rapid monolayer compression (it is assumed that at $t = 0$ and $A = A_0$ the monolayer exists in the equilibrium state) the mass of component 2 in the monolayer remains constant (desorption does not take place). Therefore, assuming $\theta_2 + \omega_1/A < 1$ one obtains from Eq. (78)

$$\theta_2 = \theta_2^* \frac{(1-\omega_1/A_0)^2}{(1-\omega_1/A)} . \tag{82}$$

It can be seen that, while according to Eq. (81) adsorption of the soluble amphiphile decreases during the slow compression of the monolayer, according to Eq. (82), the rapid compression results in an increase of the adsorption, and this adsorption can even exceed the θ_2^* value. The dependencies shown in Fig. 2 [71] were calculated from Eq. (78). All these curves correspond to the concentration of the soluble amphiphile in the solution equal to 0.01 mmol/l.

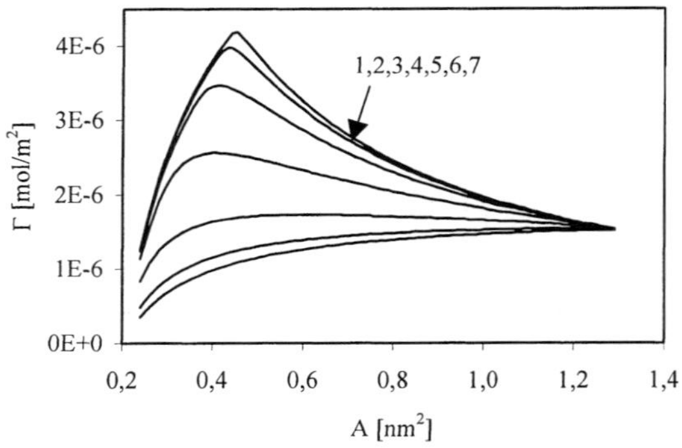

Fig. 2. The dependence of soluble surfactant adsorption in mixed monolayer on A for A0 =1.3 nm2/molecule, and the concentration c2 = 0.01 mmol/l and –dA/dt = 10 (curve 1), 1.0 (2), 0.1 (3), 0.01 (4), 0.001 (5), 0.0001 (6) и 0.00001 nm2/molecule·s (7).

For very low compression rates, the decrease in the area per insoluble molecule leads to a decrease of the adsorption of the soluble amphiphile. Note that curve 7 in Fig. 2 is almost coincident with the dependence calculated from Eq. (81), that is, for the conditions of adsorption equilibrium. With increase in the monolayer compression rate, the Γ_2 adsorption becomes higher as compared with the equilibrium value. Here the extreme behaviour of Γ_2 for very high compression rates takes place. Note that in the θ_2 and A range which satisfy the condition $\theta_2 + \omega_1/A < 1$, curve 1 in Fig. 2 coincides with the values calculated from Eq. (82). The variations in the concentration of amphiphile 2 affect significantly the shape of the Γ_2 vs A curves. For fixed α value at concentrations c_2 < 0.005 mmol/l an increase of Γ_2 in the compressed monolayer was observed. On the contrary, for concentrations c_2 > 0.02 mmol/l the adsorption decreases as compared with the initial value corresponding to A = 1.3 nm^2. Similar behaviour of the dependencies on α and c_2 was found for the Γ_2 values calculated from Eq. (79).

8. EXPERIMENTAL SECTION

To obtain quantitative information on penetrated monolayers under equilibrium, but also on their dynamic behaviour, large attention has to be focused on the experimental technique. The penetration experiments were performed by using the sweeping technique [12, 13]. The injection method also often used for penetration experiments has the disadvantage that incomplete distribution of the dissolved material in the subphase after injection can be the reason for poor reproducibility or artifacts. The sweeping technique is not only superior to other penetration techniques because disturbances of the system can be largely avoided but also because the initial conditions can be better defined. The basic idea of the penetration technique is that in a multicompartment trough a Langmuir monolayer, enclosed between two movable barriers and kept at a selected surface pressure, can be swept over subphases of different composition. Thus the two components of the penetration system to be investigated can be characterized independently of each other.

The scheme of the circular penetration apparatus and the principles of the experimental procedure are presented in Fig. 3. Generally the trough can be subdivided into several compartments. For penetration experiments of a dissolved species into a Langmuir monolayer, it can be useful to apply three compartments. In compartment I the monolayer is spread; compartment II filled with pure water can be used for removing possible bulk components of compartment I; and finally compartment III contains the dissolved component. In most of our experiments the trough was partitioned into two semicircular regions only, as the subphase of compartment I consisted of water. To optimize the accuracy of the measurements, a circular Teflon trough has been used with a large total area of 158 000 mm^2 and a depth of 7 mm. A special barrier coupling allows the movement of both barriers separately as well as simultaneously in the range between 0.05 and 15 angular degrees per minute. The small distance between the segment walls and the moving barriers is crucial for the functionality of the sweeping mechanism.

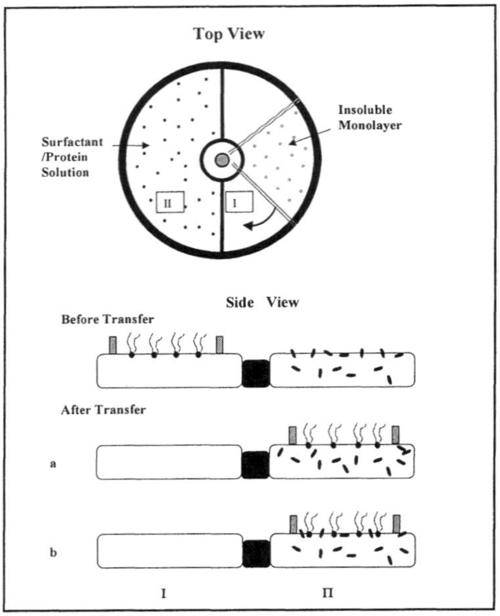

Fig. 3. Schematic of the monolayer penetration trough based on the sweeping technique. The Langmuir monolayer is spread between the movable barriers on region I containing pure water or a buffer solution. After compression to the desired surface pressure the monolayer is swept to region II filled with aqueous surfactant solution.

For the presented penetration experiments the trough was partitioned into two semicircular regions. The insoluble amphiphile of the Langmuir monolayer was spread between the movable barriers at the surface of one region filled with pure buffer solution, whereas the other region was filled by the solution of the dissolved surfactant. Afterward, the monolayer was brought to the desired state, e.g. surface pressure, molecular area, and swept onto the region containing the dissolved surfactant. Then the penetration kinetics experiments coupled with the BAM imaging were performed and the state of the penetrated monolayer in equilibrium was characterized.

The surface tension of the computer-interfaced film balance was measured by the Wilhelmy plate technique. Imaging of the monolayers was performed with a Brewster angle microscope BAM (NFT Göttingen, Germany) [72, 73] mounted on the penetration trough apparatus [23, 24]. The BAM images were treated with an image processing software to correct the distortion resulting from the observation at the Brewster angle. The 80 μm bar in the respective figures indicates the size of all the other BAM images (see later Figs. 15, 16 and 20).

The lattice structure of the condensed monolayer phases can be characterised by Synchrotron grazing incidence X-ray diffraction (GIXD). To obtain direct information on the effect of the dissolved component after penetrating the monolayer, GIXD experiments were performed at the liquid-surface diffractometer on the undulator beamline BW1 at HASYLAB, DESY, Hamburg. A monochromatic X-ray beam strikes the air/water interface at an angle of about 0.12°. This limits the penetration depth to approximately 70 Å. The diffracted intensity is detected by a linear position sensitive detector (PSD) as a function of the vertical scattering angle α_f. A Soller collimator, located in front of the PSD, provides the resolution (1.5 mrad) for the horizontal scattering angle $2\theta_{xy}$ [74-76]. The horizontal component of the scattering vector is given by $Q_{xy} \cong (4\pi/\lambda)\sin(2\theta_{xy}/2)$ and the vertical component is given by $Q_z \cong (2\pi/\lambda)\sin(\alpha_f)$ where λ is the X-ray wavelength. The counts for the Q_z resolved scans were accumulated in a multichannel analyser. The intensities as a function of Q_{xy} and Q_z were corrected for polarization, effective area and Lorentz factor. Model peaks taken to be Lorentzian in the in-plane direction and Gaussian in the out-of-plane direction were fitted to the corrected intensities.

For the GIXD measurements, the multi-compartment apparatus could not be used. Here the effect of protein penetration into DPPC monolayers has been studied, so that due to the slow adsorption kinetics of low protein concentration the DPPC monolayer can be directly spread onto the protein solution.

9. EXPERIMENTAL STUDIES OF PENETRATION AND 2D AGGREGATION

9.1. Mixed monolayer of soluble and insoluble dimethyl phosphine oxides

First some results are discussed which were obtained in penetration studies of soluble dodecyl dimethyl phosphine oxide (C_{12}DMPO) into the monolayer of the insoluble eicosyl dimethyl phosphine oxide (C_{20}DMPO) [18]. Figure 4 shows the surface pressure – area per one C_{20}DMPO molecule isotherm for 20 °C and monolayer compression rate 0.001 nm^2/molecule·s. The existence of a break point at $A \cong 0.4$ nm^2/molecule and the steep portion of the isotherm at low area values indicates that 2D aggregates of the C_{20}DMPO are formed. It is seen that an excellent agreement exists between the theory and the experimental data for both the gaseous and the condensed state of C_{20}DMPO monolayer, with quite realistic parameter values which enter Eqs. (32)-(34), estimated from the fitting program.

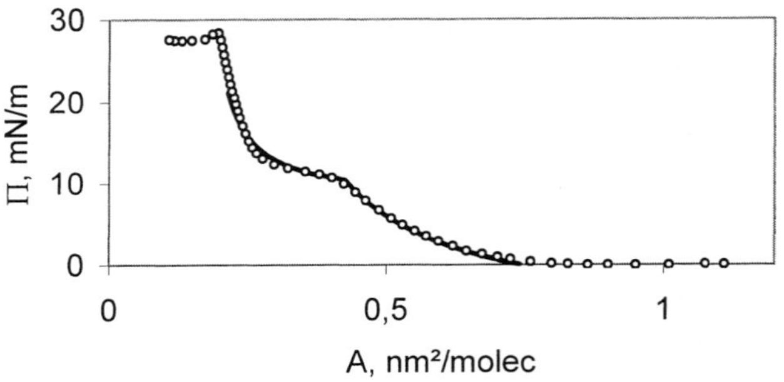

Fig. 4. Pressure vs area per one C_{20}DMPO molecule isotherm. Curve calculated from Eq. (32) for $\omega = 0.2$ nm2, $A_c = 0.42$ nm2/molecule, $\Pi_{coh} = 7.4$ mN/m and $\varepsilon = 0.15$ [18].

The studies of the penetration process of soluble C_{12}DMPO molecules into the C_{20}DMPO monolayer, and the subsequent compression of the mixed monolayer were performed for some fixed initial values of area per one C_{20}DMPO molecule: $A_0 = 0.6$ nm^2/molecule (gaseous monolayer), $A_0 = 0.35$ nm^2/molecule (2D aggregation region), $A_0 = 0.22$ nm^2/molecule (condensed monolayer near the collapse point). The results obtained for $A_0 = 0.6$ nm^2/molecule and C_{20}DMPO concentration $2 \cdot 10^{-5}$ mol/l are presented in Figs. 5 and 6. Figure 5 shows the isotherm for initial compression of the monolayer to $A_0 = 0.6$ nm^2/molecule (curve 1), and the mixed monolayer compression isotherm after the adsorption equilibrium was established (curve 2). Figure 5 illustrates the C_{12}DMPO penetration dynamics for $A_0 = 0.6$ nm^2/molecule.

Consider dynamics and equilibrium of the penetration into a gaseous monolayer at various concentration values of C_{12}DMPO. Figure 7 shows surface pressure isotherms for a single C_{12}DMPO solution, and for the case when C_{20}DMPO is present, at $A_0 = 0.6$ nm^2/molecule, i.e. for $\theta_1 = 0.35$. The theoretical curve for the single isotherm was calculated from the Szyszkowski equation $\Pi = (RT/\omega)\ln(1 + bc)$, while for the mixed monolayer Eq. (35) was used; in both cases the same parameter values ($\omega = 2.28 \cdot 10^9$ cm^2/mol and $b = 1.69 \cdot 10^5$ l/mol) were assumed. The surface pressure isotherm for the single C_{12}DMPO solutions corresponds well to the Szyszkowski equation. It is seen that for the mixed monolayer, a satisfactory agreement between theory and experiment exists at $c \leq 10^{-5}$ mol/l. For higher C_{12}DMPO concentrations, however, the experimental Π values are significantly lower than the theoretical values, because Eq. (35) does not account for the aggregation processes within the monolayer. Therefore, the discrepancy between the theory and experiment for $\Pi > 15$ mN/m proves implicitly the existence of 2D aggregation in the mixed monolayer within this surface pressure range. This conclusion is also supported by the penetration dynamics for various C_{12}DMPO concentrations [18]. For $c \leq 10^{-5}$ mol/l, the theoretical and experimental curves of dynamic surface pressure coincide with each another, while for higher C_{12}DMPO concentration the experimental Π values are significantly lower than the theoretical values, what may be explained by the possibility that a certain portion of the adsorbed C_{12}DMPO molecules are incorporated within the mixed aggregates. The conclusion that condensed phase aggregates are formed by the penetration of the dissolved C_{12}DMPO molecules into the insoluble C_{20}DMPO monolayer, can be also supported by the analysis of mixed monolayer compression isotherms [18]. It was shown that the theoretical model which assumes the formation of mixed aggregates, leads to a better agreement with experimental data.

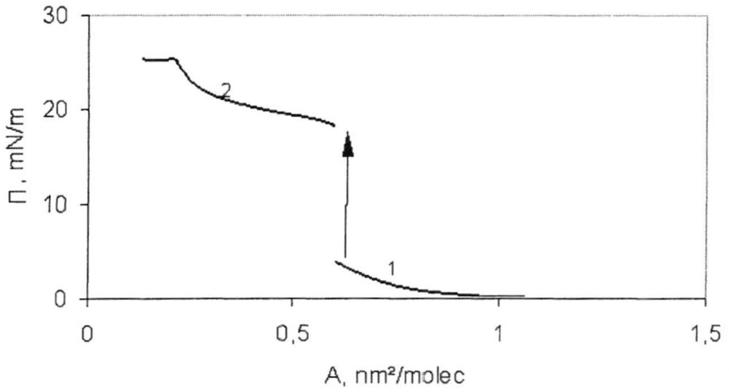

Fig. 5. Pressure – area per one C_{20}DMPO molecule isotherm. Curve 1 - initial compression of the monolayer to $A_0 = 0.6$ nm2/molecule, curve 2 - mixed monolayer compression isotherm [18].

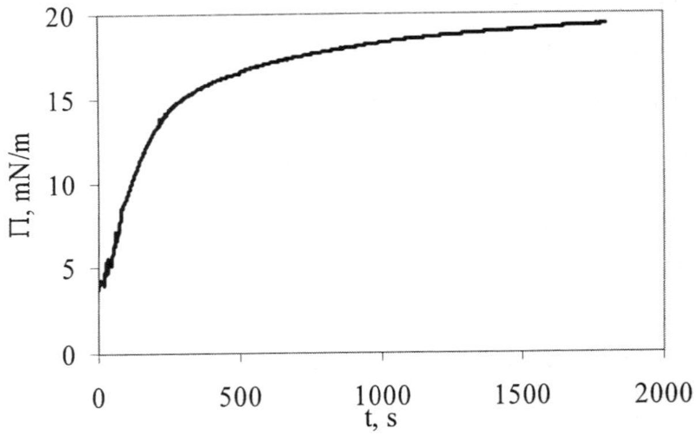

Fig. 6. C_{12}DMPO penetration dynamics for c = 10^{-5} mol/l and A_0 = 0.6 nm^2/molecule [18].

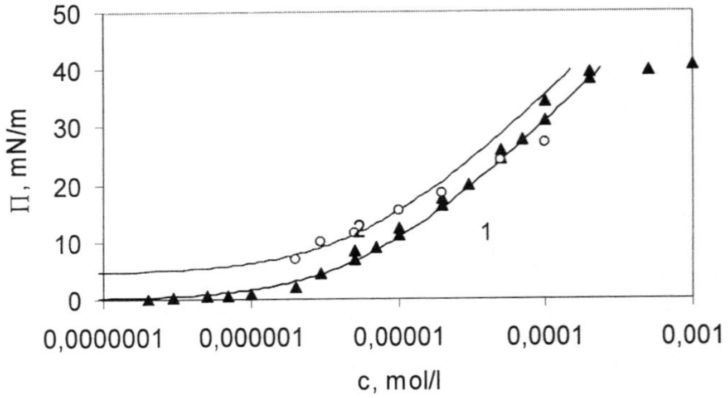

Fig. 7. Surface pressure isotherm for C_{12}DMPO solutions at $\theta_1 = 0$ (curve 1 - theory, ▲ - experiment) and $\theta_1 = 0.35$ (curve 2 - theory, O - experiment). Theoretical curves calculated with ω = 2.28·109 cm2/mol and b = 1.69·105 l/mol [18].

9.2. Mixed monolayer of soluble and insoluble homologues of a tailored acid amide amphiphile

Figure 8 illustrates the equilibrium surface tension of aqueous solution of $C_{10}H_{21}$-NH-CO-CH_2-CH_2-CH_2OH (C_{10}NH) at 20 °C. The theoretical curves were calculated for the Frumkin, Langmuir and aggregation models (model parameters are listed in the legend to Fig. 8). Note that, of the three models used, the best agreement with the experimental data was obtained for the aggregation model. The Frumkin model gives somewhat worse correspondence, while the Langmuir model yields only approximate agreement with the experiment (the discrepancy is 10 times higher as compared with the aggregation model). However, to analyse the behaviour of mixed monolayers we employ here the Langmuir model, because the theoretical model of mixed monolayer presented in the theoretical section does not account for the effect of non-ideality or aggregation of $C_{10}NH$ on the surface pressure.

The dynamic surface tension of aqueous solutions of $C_{10}NH$ for certain concentration values is illustrated in Fig. 9. The theoretical curves shown in this Fig. 9 were calculated from the Ward-Tordai equation Eq. (69). It is seen that a good correspondence exists between the theory and experiment for reasonable value of the diffusion coefficient $D = 4 \cdot 10^{-10}$ m^2/s. This fact is of a paramount importance for the subsequent analysis, because the diffusion mechanism of adsorption for component 2 was assumed for the derivation of Eqs. (78)-(80).

Π vs A isotherms of ($C_{14}H_{29}$-NH-CO-CH_2-CH_2-CH_2OH) $C_{14}NH$ monolayers for various temperatures are shown in Fig. 10. Here an excellent correspondence exists between the theoretical curves calculated from Eqs. (32)-(34) with the parameters listed in Table 1, and the experimental data. It is seen that, as usual, A_c increases with the temperature with simultaneous decrease of Π_{coh}.

Figure 11 illustrates the experimental compression isotherms of mixed monolayers at 20°C for various $C_{10}NH$ concentrations in the adjacent monolayer. Note that in all experiments the initial value of area per one $C_{14}NH$ molecule in the monolayer was 1.3 nm^2/molecule, and before the compression of the monolayer with constant rate $-dA/dt = -\alpha = 0.0013$ nm^2/molecule·s the monolayer was allowed to stand during 5000 s after the insoluble monolayer was transferred onto the $C_{10}NH$ solution. Therefore, equilibrium of the soluble component 2 between the solution and surface layer existed before starting the compression process. It is seen from Fig. 11 that in the pre-critical range the compression of mixed monolayer is accompanied by an increase of surface pressure. This implies that the adsorption equilibrium does not exist for $C_{10}NH$, because for the equilibrium monolayer the curve for $A > 0.7$ nm^2/molecule should be the straight line parallel to the abscissa axis, cf. Eq. (35). Another important experimental result is the shift of the critical area per molecule corresponding to the commencement of the $C_{14}NH$ condensation towards higher values as compared to that characteristic for the single monolayer, as illustrated by the dotted line in Fig. 11. It can be seen that this shift becomes more significant with the increase of the $C_{10}NH$ concentration in the solution adjacent to the monolayer. This is indicative of the influence of

$C_{10}NH$ on the $C_{14}NH$ aggregation equilibrium. The experimental conditions of Fig. 11 almost correspond to curve 5 in Fig. 2. Due to the quite high compression rate of the monolayer, the $C_{10}NH$ adsorption value was higher than the equilibrium value. This explains the increase in the surface pressure of the compressed monolayer in the pre-critical range, shown in Fig. 11.

In Fig. 12, the experimental isotherms at compression of the mixed monolayers for the concentrations of the soluble component of 0.002 and 0.02 mmol/l, respectively, are compared with the theoretical curves. It results that the following options exist for the theoretical description of mixed monolayers at non-equilibrium in the pre-critical ($A \geq A_{cm}$) or trans-critical ($A \leq A_{cm}$) range:

I) the generalised Szyszkowski-Langmuir model: Eqs. (35), (78) and (37), (78) for the pre-critical and trans-critical range, respectively;

II) the combined Langmuir and Volmer models: Eqs. (46), (78) and (52), (78), respectively;

III) the generalised Volmer - Pethica model: Eqs. (46), (79) and (52), (80), respectively.

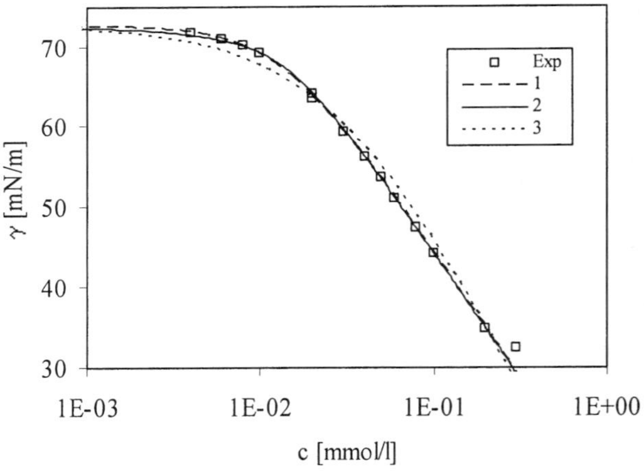

Fig. 8. Dependence of surface tension for $C_{10}NH$ vs c (□); Theoretical isotherms were calculated from the aggregation model (curve 1) with ω = 1.70·105 m^2/mol, Γ_c =3·10-10 mol/m^2, n=2 and b=1.02 l/mmol, Frumkin's model (curve 2) with ω = 1.74·105 m2/mol, a=1.28 and b=20.7 l/mmol, Langmuir model (curve 3) with ω = 1.26·105 m^2/mol and b=30.6 l/mmol.

It is easily seen from the comparison between the theoretical and experimental results for a $C_{10}NH$ concentration of 0.002 mmol/l (equilibrium adsorption Γ_2^* is ca. $4.8 \cdot 10^{-7}$ mol/m^2) that model I is, quite expectably, irrelevant for low concentrations (adsorptions) of amphiphile 2. At the same time, the agreement with the combined models II and III is quite satisfactory in many details. On the contrary, for high $C_{10}NH$ concentration (0.02 mmol/l, equilibrium adsorption for the single solution Γ_2^* is ca. $4.7 \cdot 10^{-6}$ mol/m^2) all three models are in a satisfactory agreement with each other but the best correspondence with the experiment is observed for model III. For all other concentration values illustrated in Fig. 11, the best agreement with the experiment is also obtained by the models II and III.

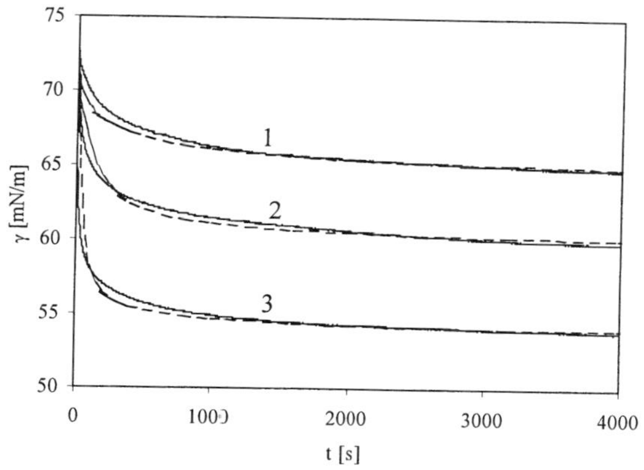

Fig. 9. Dynamic surface tension for $C_{10}NH$ solutions at concentration 0.02 (curves 1), 0.03 (curves 2) and 0.05 mmol/l (curves 3). Solid lines, experiment; dotted lines, calculations from the diffusion model.

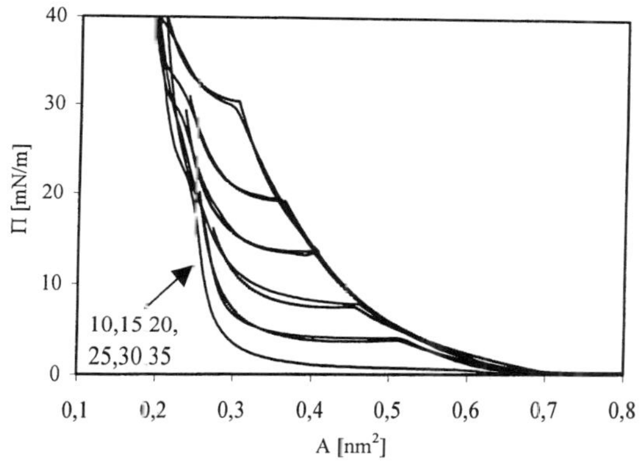

Fig: 10. Experimental and theoretical Π-A isotherms for $C_{14}NH$ at various temperatures (shown at the curves). x x x - experiment; solid lines - calculations. The values of the parameters in Eqs. (32)-(34) are shown in the Table 1.

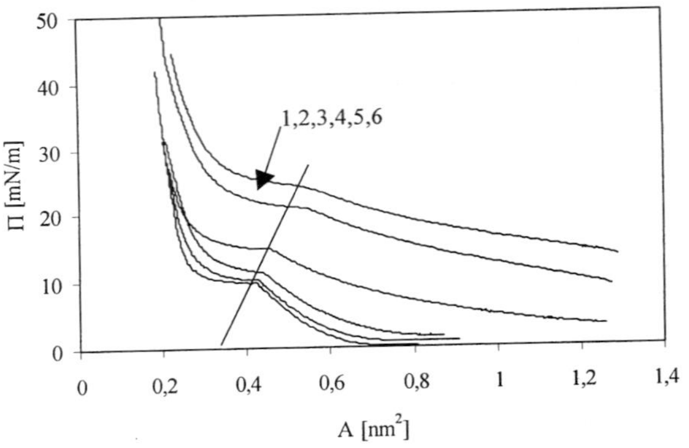

Fig. 11. Experimental Π-A isotherms for the compression (–dA/dt = 0.0013 nm2/molecule·s) of mixed monolayer of amino acids C10NH and C14NH at 20 °C and the C10NH concentrations 0.05 (curve 1), 0.02 (2), 0.01 (3), 0.002 (4), 0.001 (5) and 0.0 mmol/l (6).

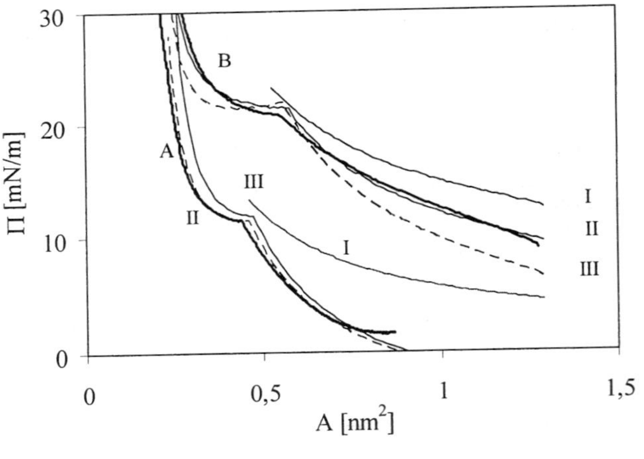

Fig. 12. The experimental (bold solid line) Π-A isotherms of the compression of mixed monolayer for dA/dt = 0.0013 nm2/molecule·s and C10NH concentration c = 0.002 mmol/l (curve A) and c = 0.02 mmol/l (curve B) compared with theoretical dependencies calculated for models I, II and II.

The $C_{10}NH$ adsorption values in the mixed monolayer at $A = A_{cm}$ were calculated from Eq. (78). The corresponding dependence of $\Gamma_{2(cm)}$ on the concentration c_2 in the solution is shown in Table 2. Here also the Γ_2^* values obtained from the Langmuir isotherm, Eq. (24) with a=0, are listed. The $\Gamma_{2(cm)}$ values were then used to estimate θ_{cm} from the exact equation Eq. (68), where $a_1 = 2.0\pm0.2$ and $a_2 = 1.3$ was assumed, cf. Fig. 8. It is known that the value $a_1 = 2.0$

corresponds to the critical value for Frumkin's equation of state and isotherm [42-46].

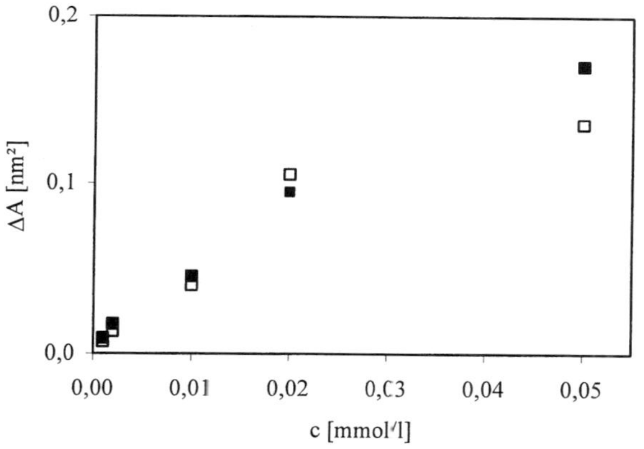

Fig. 13. The theoretical dependence of $\Delta A = A_{cm} - A_c$ on the $C_{10}NH$ concentration (■) compared with the experimental data (□).

Table 1 Values of the equilibrium parameters of Eqs. (32)-(34) for $C_{14}NH$ monolayers

Temperature, °C	A_c, nm^2	ω_1, nm^2/molecule	Π_{coh}, mN/m	ε
15	0.520	0.22	9.20	0.00
20	0.460	0.21	8.50	0.00
25	0.404	0.21	8.40	0.04
30	0.358	0.20	8.14	0.08
35	0.306	0.19	6.76	0.15

The agreement between the experimental data (shown in Fig. 11) and the theoretical values of the critical area A_{cm} per one $C_{14}NH$ molecule in the mixed monolayer is rather good, see Fig. 13. The A_{cm} values calculated from the approximate equation Eq. (68) for $a = 2\pm0.1$ are close to the results obtained from the exact calculations. However, disregarding of the activities in the mixed monolayer, that is, employing of Eq. (59) instead of Eq. (68) leads to an approximately 2-3 times higher increase of the critical area value than that observed in the experiment.

Table 2 The values of the equilibrium adsorption Γ_2^*, and adsorption $\Gamma_{2(cm)}$ at $A = A_{cm}$ calculated from Eq. (78).

c_2, mmol/l	0.001	0.002	0.01	0.02	0.05
$\Gamma_{2(cm)}$, mol/m^2	$3.0 \cdot 10^{-7}$	$5.7 \cdot 10^{-7}$	$1.8 \cdot 10^{-6}$	$2.56 \cdot 10^{-6}$	$3.4 \cdot 10^{-6}$
Γ_2^*, mol/m^2	$2.4 \cdot 10^{-7}$	$4.8 \cdot 10^{-7}$	$1.8 \cdot 10^{-6}$	$3.0 \cdot 10^{-6}$	$4.7 \cdot 10^{-6}$

9.3. Protein penetration into DPPC phospholipid monolayers

Model monolayers of various phospholipids at the air-water interface have been intensively studied for obtaining information on specific processes in biological systems. In recent years considerable progress has been obtained in the understanding of the fundamental monolayer properties of the phospholipids DPPC (dipalmitoyl-phosphatidyl choline) and DPPG (dipalmitoyl-phosphatidyl glycerol) [77-82] which are the major components of the lung surfactant [83-89].

Fig. 14 illustrates the experimental isotherm and a set of theoretical Π-A isotherms for DPPC. Images obtained by Brewster angle microscopy (BAM) (Fig. 15) indicate clearly that a first order phase transition occurs at Ac and condensed phase domains are formed at $A < Ac$. BAM images of the designated points a – e of the Π - A isotherm demonstrate evolution, growth and coalescence of the condensed phase domains of DPPC.

The results of the studies of the β-lactoglobulin penetration dynamics into DPPC monolayer [19] for a bulk concentration of 5·10-7 mol/l and the values of $A > Ac$ (e.g. for $A = 1.20$ nm2/molecule in Fig. 16) show unambiguously that due to the penetration of β-lactoglobulin, a first order phase transition and thus the development of a 2D condensed phase takes place, although at this high area values the pure DPPC monolayer exists only in the fluid-like state (see Fig. 14) and does not form condensed phase domains. The first-order phase transition is seen from the fact that the Π(t) adsorption kinetics curves show a characteristic break point. After this phase transition point, evolution and growth of condensed phase domains can be observed, the shape of which is quite similar to those of pure DPPC domains. This is convincingly demonstrated by a sequence of BAM images (Fig. 16) taken at different time values of the adsorption process according to the letter of the Π(t) adsorption kinetics curve. These results obviously demonstrate that the aggregation of DPPC into condensed phase domains is induced even if the surface concentration of DPPC amounts to 50% of the critical adsorption value Γc only.

The theoretical calculations of the surface pressure *vs* β-lactoglobulin adsorption isotherms according to the Volmer model for mixed monolayer, Eqs. (46) and (52), also show a break point, characteristic for the phase transition [19-21]. The decrease in the slope indicates that the β-lactoglobulin molecules newly adsorbed (penetrated) into the monolayer from the solution compensate the incorporation of free DPPC molecules into the condensed phase domains. The time at which the DPPC aggregation commences was estimated for different A values using the simplest diffusion model of the adsorption kinetics [21]. This estimate agrees satisfactorily with the experimental data for the critical time of 2D aggregation, as shown in Fig. 17.

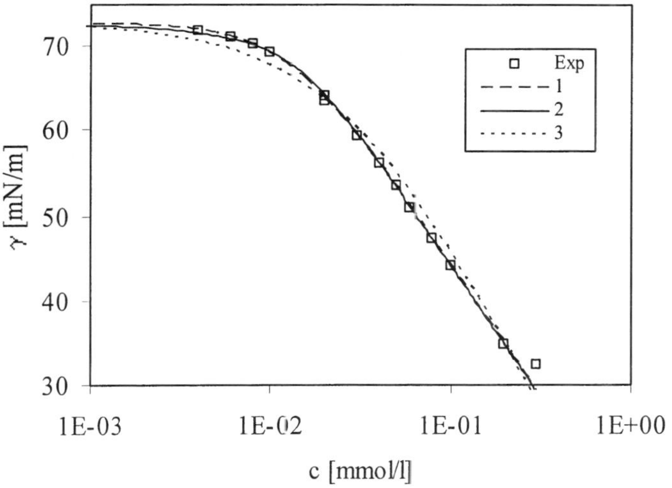

Fig. 14. Experimental and theoretical Π-A isotherms for DPPC at temperature 22°. The values of the parameters in Eq. (32)-(34) are: Π_{coh} =7.8 mN/m, ω=0.445 nm2/molecule, A_c =0.695 nm2/molecule, ε=0.06 [82].

Direct information on phase transition and phase properties of DPPC monolayers penetrated by β-lactoglobulin has been obtained by coupling of the surface pressure transients with BAM and GIXD measurements [20]. GIXD (Grazing Incidence X-ray Diffraction) studies have provided experimental evidence that indeed the condensed phase consists only of DPPC (Fig.18). After reaching an equilibrium pressure, the lattice structure of the penetrated monolayer is very similar to that of DPPC on pure water (Fig.18). At additional compression of the penetrated monolayer in equilibrium, the tilt angles of the DPPC molecules deviate from that of the pure DPPC monolayers (Fig.19). These differences can be understood as initial conformational changes of the protein in the fluid-like monolayer and succeeding squeezing out of the protein at high surface pressures.

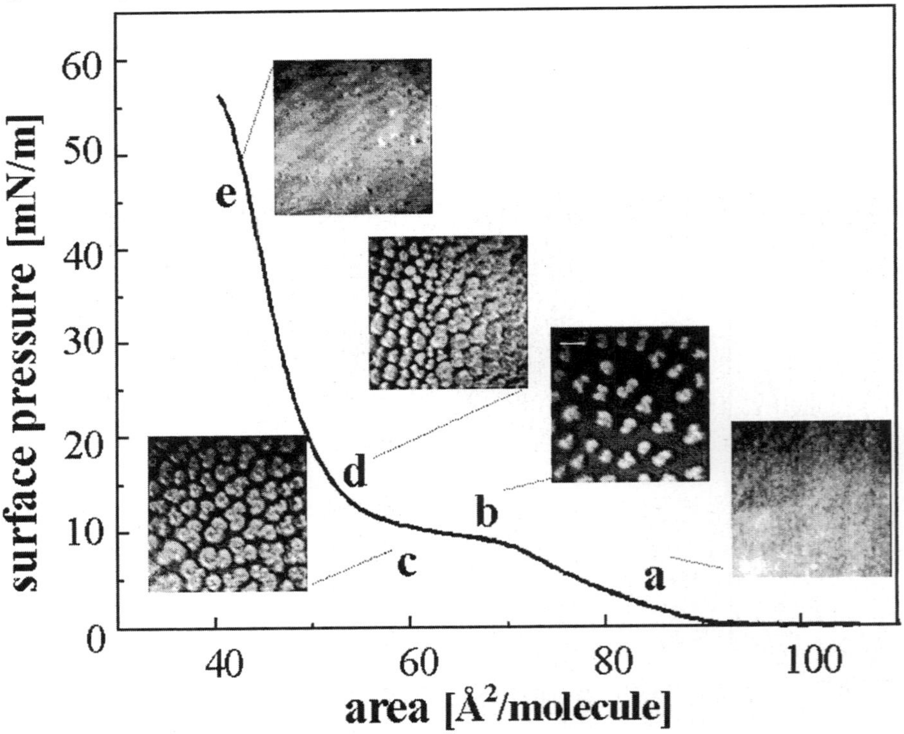

Fig. 15. Formation and growth of condensed phase domains in pure DPPC monolayers corresponding with the designated points (a) – (e) of the Π-A isotherm at T = 22 °C. (a) At A > Ac no condensed phase domains are not formed. (b) – (c) Formation and growth of condensed phase domains in the two-phase coexistence (plateau) region. (d) Deformation and first coalescence with increasing surface pressure. (e) Continuous condensed phase at high surface pressures [19].

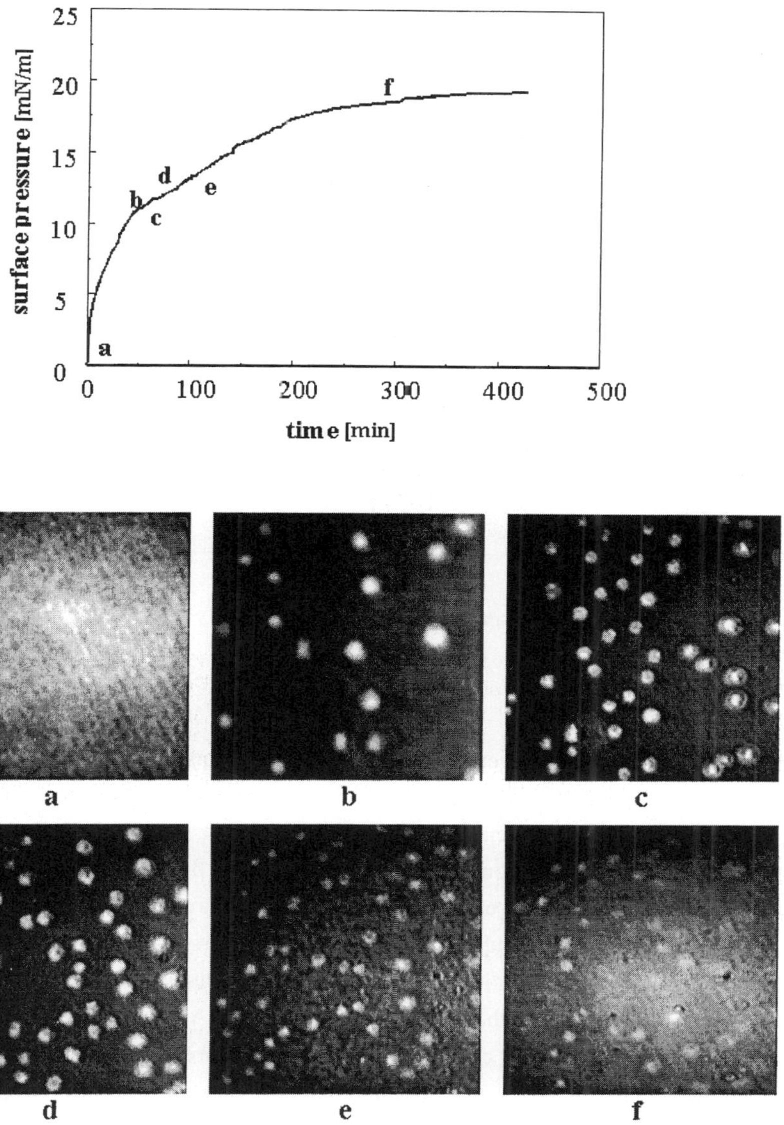

Fig. 16. Penetration dynamics for β-lactoglobulin penetration at A = 1.20 nm2 per one DPPC molecule. The Π(t) penetration kinetics curve indicates the first order phase transition point after 29 min. A sequence of BAM images (a) – (f) is shown according to the letters of the Π(t) transients [19].

Further results were obtained in the studies of the lysozyme penetration into DPPC monolayers [17]. Similarly to the β-lactoglobulin, the lysozyme stimulates the DPPC aggregation. For example, the lysozyme penetration for concentrations above 10^{-4}% leads to 2D aggregation in the DPPC monolayer at the coverage value $\theta \cong 0.45$, while the critical point of the individual monolayer corresponds to the coverage value $\theta \cong 0.64$. It was shown in [17] that the equation of state similar to Eq. (51) agrees satisfactorily with the experimental data. Similarly to the penetration of a homologue into the insoluble monolayer, illustrated by Fig. 7, the 2D aggregation of the DPPC in the monolayer results in a decrease of the surface pressure jump for the mixed monolayer of the lysozyme, as compared to the individual lysozyme solution of the same concentration.

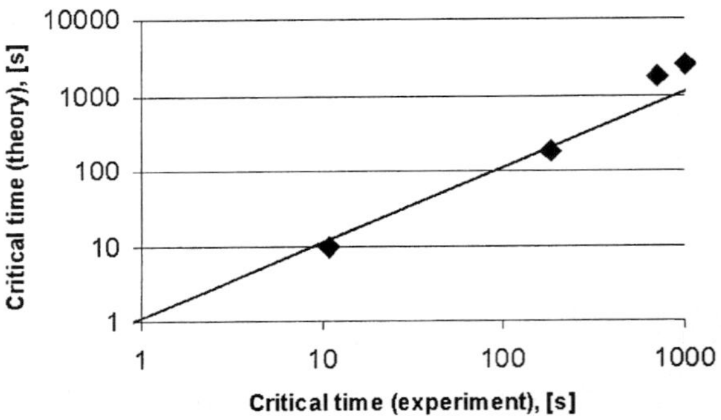

Fig. 17. Experimental and theoretical value of the critical time at which the DPPC aggregation commences as result of β-lactoglobulin penetration.

The penetration of flexible β-casein into the DPPC monolayer also stimulates its 2D aggregation [20]. Figure 20 illustrates the β-casein penetration dynamics for the solution concentration 10^{-7} mol/l. Comparing this with Fig. 12, one can see the behaviour quite similar to that found for the penetration of the globular protein β-lactoglobulin. However, some particular features exist. The equilibrium surface pressure of a mixed monolayer for $A > 0.6$ nm^2/molecule was found to exceed that characteristic for the individual protein solution with the same concentration, while for the mixed monolayer with β-lactoglobulin this pressure was the same as for the individual protein. For $A \leq 0.6$ nm^2/molecule, the initial portion of the dynamic pressure curves Π *vs* t for the mixtures containing β-casein exhibit a quite long period during which the pressure remains almost constant. These effects can possibly be ascribed to the interaction between the protein and the phospholipid in the monolayer, and to the conformation transitions of β-casein molecule between possible states of the molecule in the monolayer. The existing theoretical models are incapable even of qualitative estimating the effect of these processes.

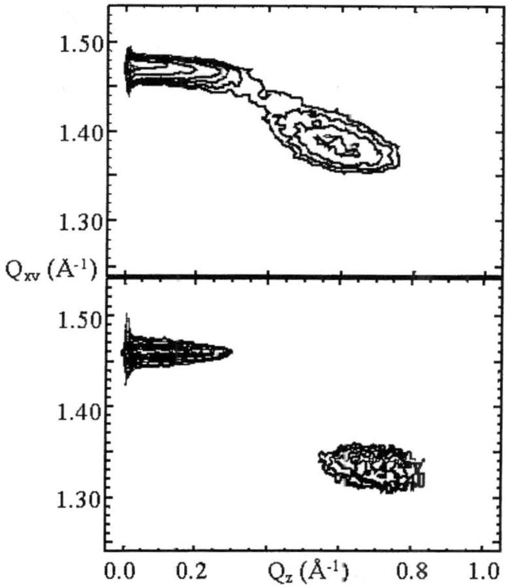

Fig. 18. Grazing incidence X-ray diffraction contour plots as a function of the in-plane and out-of-plane scattering vector components Qxy and Qz, respectively of a DPPC monolayer taken after 3 hours of β-lactoglobulin penetration (at an area of 90 Å^2/DPPC molecule with a protein subphase concentration of 5×10-7 mol/l) penetration followed by compression of the film to a desired lateral pressure at 20 °C; top: Π = 41 mN/m, bottom: Π = 18 mN/m [20].

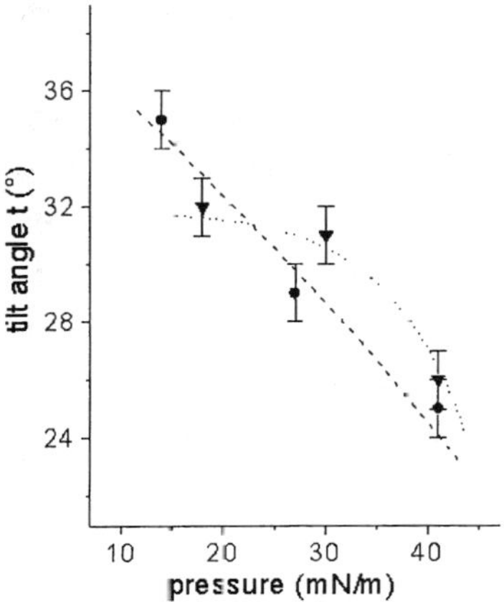

Fig. 19. Tilt angle t as a function of the surface pressure for a pure DPPC monolayer on water (●) and a DPPC monolayer penetrated by β-lactoglobulin which was dissolved in the subphase (▼). Penetration was started at a area of 90 Å2/DPPC molecule with a β-lactoglobulin subphase concentration of 5×10-7 mol/l [20].

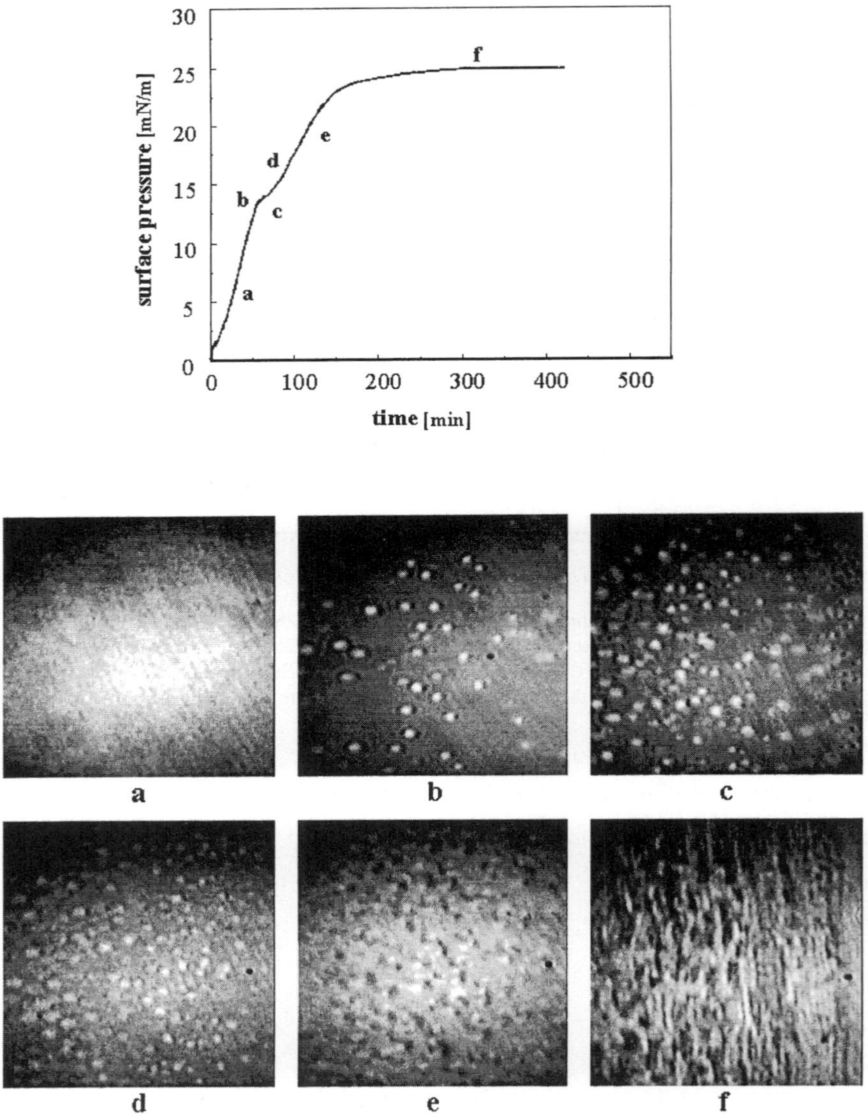

Fig. 20. Penetration dynamics for β-casein penetration at solution concentration 10^{-7} mol/l at A = 0.92 nm^2 per one DPPC molecule. The Π(t) penetration kinetics curve indicates the first order phase transition point after 3100 s. A sequence of BAM images (a) – (f) is shown according to the letters of the Π(t) transients [20].

9.4. Coadsorption of SDS/dodecanol mixtures

Due to the dominant influence of the dodecanol adsorption, coadsorption of sodium dodecyl sulphate (SDS)/dodecanol mixtures from aqueous solution is mainly investigated for trace amounts of n-dodecanol [90-101]. Two limiting cases are of interest for such mixture ratios, namely where a phase transition occurs or cannot occur [65, 101, 102].

Figure 21 shows the dynamic surface pressures for the single solutions of SDS (concentration 3 mM) and n-dodecanol (concentration 15 μM), and the mixed solution of these components with the same concentrations at 15°C. The measurements were performed by the maximum bubble pressure method and by the Wilhelmy plate method so that the range of the lifetimes studied extends over 7 decimal orders from 10^{-3} to 10^4 s. Already at time values of approximately 100 s the surface pressure of the single SDS solution becomes constant being equal to 19 mN/m at 15°C. Comparing the behaviour of the mixed solution with the behaviour of the single dodecanol solution one can see that for the mixture the effect of the dodecanol adsorption on the surface pressure becomes evident at a more early stage of the adsorption process (at ~10 s) than for the pure dodecanol solution (at ~50 s). The $\Pi(t)$ curves for pure dodecanol (curve 2) and also for the mixture show a noticeable break point which indicates a main phase transition of first order in the adsorption layer. In the time interval after the break point, the formation of condensed phase domains surrounded by a fluid phase is expected. The first-order phase transition and thus the two-dimensional condensation (aggregation) commences for the mixed monolayer at much lower time (140 s) as compared with a pure dodecanol solution (930 s). BAM measurements can provide direct evidence for a first order phase transition. A sequence of representative BAM images obtained at different times during the adsorption of 3 mM SDS/12 μM dodecanol solutions at 5 °C are shown in Fig. 22 As expected, condensed phase domains can be observed only after the break point in the dynamic surface pressure curves so that indeed the break point corresponds the main phase transition of first-order. It can be clearly seen that after the phase transition point condensed phase patterns, surrounded by a homogeneous fluid phase, are formed and grow rather rapidly in an approximately shape-preserving way. The fingered domains are irregularly shaped. In the following stages they become more compact, coalesce and approach, near the equilibrium, to an approximately homogeneous condensed phase. The morphological behaviour resembles to that observed for pure dodecanol monolayers [47] but the growth kinetics is more rapid. This similarity indicates that the condensed phase is formed only by dodecanol.

Figures 23 and 24 illustrate the dynamic surface pressures for single dodecanol solutions at a concentration of 12 μM and temperatures of 15°C and 5°C, respectively, and those for mixed dodecanol/SDS solutions with a SDS concentration 3 mM for the same dodecanol concentration and temperatures [65]. Again it can be seen that the dynamic surface pressure curves of all mixed solutions have a break point characteristic for a first-order phase transition, the position of which is located at significantly lower time values. Figure 24 also shows the theoretical curves for the solution of pure dodecanol reproduced from Ref. [47]. The curves were calculated using Eqs. (28)-(29), assuming equilibrium between the average dimers and large aggregates in the dodecanol monolayer.

The results of the calculations using Eqs. (44) and (72) for $k_3 = 1$ and for k_3 varying within the range of 0.5 (for low times) to 0 (at large time limit) are shown in Figs. 23 and 24. It can be seen that in the time range below 10 min the experimental dependencies are located between these theoretical curves. Moreover, the critical time of the dodecanol aggregation onset, which corresponds to the break point of the theoretical curve, is in perfect agreement with the experimental data. Figures 23 and 24 also illustrate the calculations performed using Eqs. (44) and (72) with $\omega_1 = 10^5$ m^2/mol for mixed monolayer and with $\omega_1 = 1.2 \cdot 10^5$ m^2/mol for the single monolayers. At high time values (≥30 min) the calculated surface pressure values are in excellent agreement with the experimental data.

The time for the onset of the dodecanol aggregation in the mixed solution is 4 - 6 times lower as compared to that for the single dodecanol solutions, see Figs. 23 and 24. According to Eq. (71), this fact indicates the 2 - 2.5 times decrease of the critical adsorption value characteristic of the onset of aggregation in the mixed solution. Similar results were also obtained for insoluble monolayers, where the penetration of a soluble surfactant into the monolayer results in the fact that the insoluble surfactant aggregates at lower surface concentrations as compared with the phase transition point of the insoluble monolayer without penetration of dissolved components. This can explain the existence of 2D condensation in form of pure dodecanol in the mixed solution (3 mM SDS and 10 μM dodecanol) at 15°C, see Fig. 25. However, a single dodecanol solution does not aggregate at this concentration and temperature.

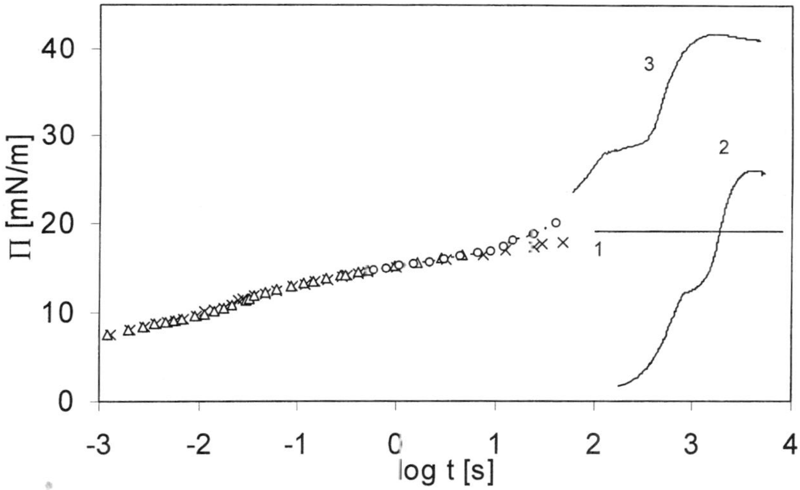

Fig. 21. Time dependence of the dynamic surface pressure for the single 3 mM SDS solution (1), single 15 μM dodecanol solution (2), and the mixed 3 mM SDS/15 μM dodecanol solution (3) at 15°C. Measured with the Wilhelmy method: solid lines; measured with the maximum bubble pressure method: × - (1); ○, △ - (3).

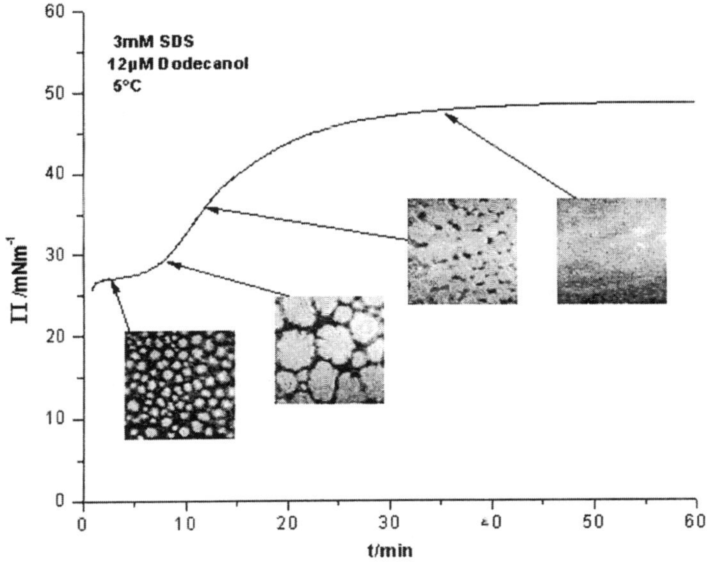

Fig. 22. BAM images obtained after the break point in the Π(t) curve of a aqueous 3mM SDS/12μM dodecanol solution at different times and at 5 °C.

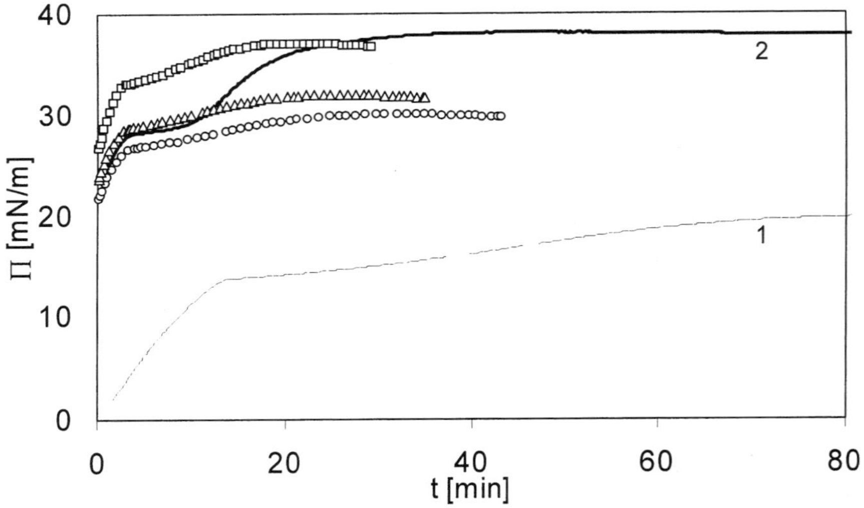

Fig. 23. Dependence of the dynamic surface pressure for the single 12 μM dodecanol solution (lower thin curve 1) and for the mixed 3 mM SDS/15 μM dodecanol solution (upper solid curve 2) at 15°C measured using the Wilhelmy method. The theoretical curves were calculated from Eqs. (44) and (72): ○ for ω_1 = 1.2·105 m2/mol, k_3 = 1; △ for k_3 = 0.5÷0; □ for ω_1 = 1·105 m2/mol, k_3 = 1.

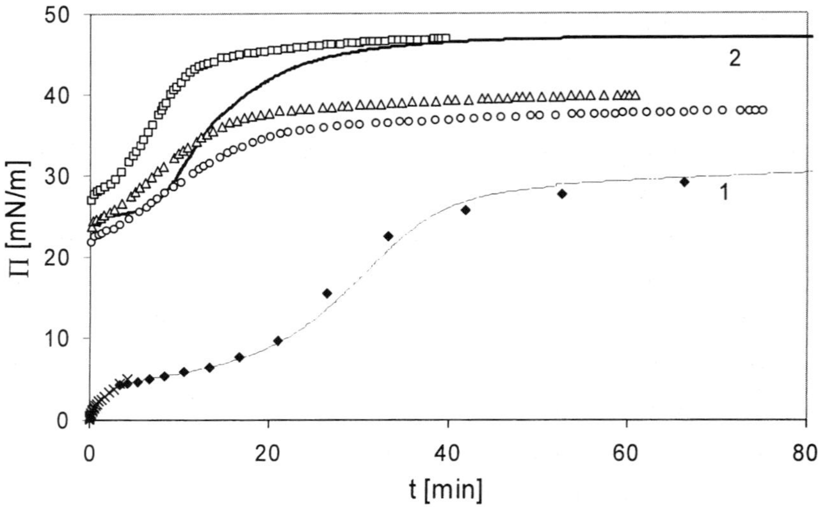

Fig. 24. The same as Fig. 23, but for 5°C. ×, ◆ for the theoretical curves of single dodecanol solution, data presented in Ref. [47].

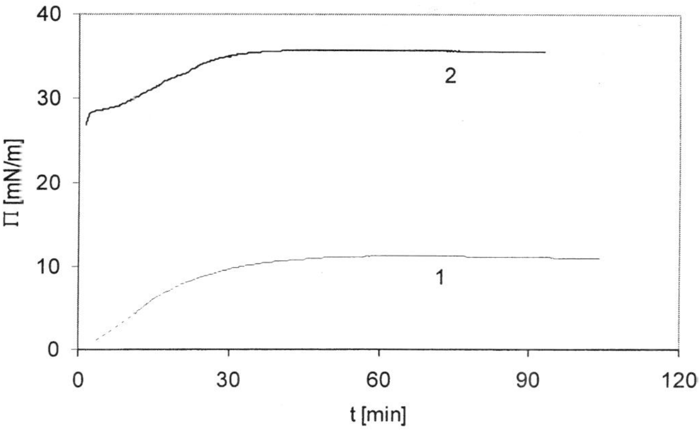

Fig. 25. Dynamic surface pressure for 10 μM dodecanol solution (lower thin line 1) and mixed 3 mM SDS/10 μM dodecanol solution (upper solid line 2) at 15°C, measured by the Wilhelmy method.

10. CONCLUSIONS.

Penetration systems at the air-water interface in which a dissolved amphiphile (surfactant, protein) penetrates into a Langmuir monolayer are interesting models for a better understanding of various technological and biologic processes. A critical review of the existing theoretical models which describe the thermodynamics of the penetration of a soluble surfactant into the insoluble monolayer is presented. The theories which describe the penetration of a soluble surfactant into the Langmuir monolayers were presented for the case of the mixture of the molecules possessing equal partial molar areas (the mixture of homologues). In the theoretical models, various mechanisms for the effect of the soluble surfactant on the aggregation of the insoluble component were considered: (i) no effect on the aggregate formation process; (ii) formation of mixed aggregates; and (iii) the influence on the aggregating process via the change of aggregation constant. Assuming diffusion kinetics for the desorption of the soluble component during the compression of mixed monolayer, an equation is derived for the adsorption of the soluble amphiphile, and the dependence of this value on the compression conditions of the monolayer, the concentration and other parameters is analysed. The analysis of the influence of the non-ideality of mixture on the shift of the value of area per one molecule of insoluble amphiphile, which corresponds to the commencement of the two-dimensional condensation in the monolayer, is also performed. A simple theoretical model has been introduced for the description of the coadsorption of surfactant mixtures, that enables one to calculate the surface pressure isotherm of the mixture from the corresponding dependencies of the single components, including also the case of the aggregation of one surfactant in the adsorbed monolayer.

New information on phase transition and phase properties of Langmuir monolayers penetrated by soluble amphiphiles can be obtained by coupling of the $\Pi(t)$ penetration kinetics curves with BAM and GIXD measurements. The experimental data obtained for the mixture of soluble and insoluble phosphine oxide amphiphiles as well as for the single and mixed monolayers of homologues of a tailored acid amide amphiphile were found to agree well with the theoretical predictions based on the characteristics of these amphiphiles as single-component solutions and monolayers. It is shown that the account for the non-ideality in the mixture of amphiphiles is important for the qualitative and quantitative analysis of the domain formation in the mixed monolayer. The experimental data for both the pure dodecanol and SDS/dodecanol mixtures show two-dimensional phase transition and subsequent formation of condensed dodecanol phase though the conditions of their development are different. Adsorption of SDS accelerates the phase transition in the adsorbed monolayer of the mixture.

Experimental evidence has been provided that the penetration of β-lactoglobulin, β-casein and lysozyme into fluid-like DPPC monolayers induces a first order main phase transition of DPPC even if the area per one DPPC molecule in the mixed monolayer is two times larger than the area of the main phase transition point in the Π - A isotherm of the pure DPPC monolayer. The formation of condensed phase domains caused by the penetration of proteins can be visualised by BAM. It is seen from the experimental results that mixed aggregates of protein with phospholipid are not formed, but the presence of proteins in the monolayer induces the formation of the DPPC aggregates, due to the increase in the total monolayer coverage and the change in the lipid aggregation constant. Penetration studies of β-lactoglobulin of different concentrations into DPPC monolayers shows that protein penetration occurs without any specific interactions with the DPPC molecules and the condensed phase consists only of DPPC. The behaviour of mixed monolayers agree satisfactorily with the predictions based on the theoretical models.

Acknowledgements

The work was financially supported by the Max Planck Society and by the Ukraine SFFR (03.07/00227).

11. References

1. B.A. Pethica, Trans. Faraday Soc., 51(1955)1402.
2. P.J. Anderson, B.A. Pethica, Trans. Faraday Soc., 52(1956)1080.
3. M.A. McGregor, G.T. Barnes, J. Colloid Interface Sci., 54(1976)439; 60(1977)408.
4. D.M. Alexander, G.T. Barnes, J. Chem. Soc. Faraday Trans., 1 76(1980)118.
5. K. Motomura, I. Hayami, M. Aratono, R. Matuura, J. Colloid Interface Sci., 87(1982)333.
6. I. Panaiotov, L. Ter-Minassian-Saraga, G. Albrecht, Langmuir, 1(1985)395.
7. L. Ter-Minassian-Saraga, Langmuir, 1(1985)391
8. D.G. Hall, Langmuir, 2(1986)809
9. K. Tajima, M. Koshinuma, A. Nakamura, Langmuir, 7(1991)2764.
10. Q. Jiang, C.J. O'Lenick, J.E. Valentini, I.C. Chiew, Langmuir, 11(1995)1138.
11. K. Asano, K. Miyano, H. Ui, M. Shimomura, I. Ohta, Langmuir, 9(1993)3587.
12. D. Vollhardt, M. Wittig, Colloids & Surfaces, 47 (1990) 233.
13. S. Siegel, D. Vollhardt, Colloids & Surfaces A, 76(1993)197.
14. S. Sundaran, K.J. Stebe, Langmuir, 12(1996)2028; 13(1997)1729.
15. N.S. Santos Magalhaes, H.M. de Oliveira, A. Baszkin, Colloids & Surfaces A, 118(1996)63
16. S. Sundaram, J.K. Ferri, D. Vollhardt, K.J. Stebe, Langmuir, 14(1998)1208.
17. V.B. Fainerman, D. Vollhardt, Langmuir, 15(1999)1784.
18. V.B. Fainerman, A.V. Makievski, D. Vollhardt, S. Siegel, R. Miller, J. Phys. Chem. B, 103(1999)330.
19. V.B. Fainerman, J. Zhao, D. Vollhardt, A.V. Makievski, J.B. Li, J. Phys. Chem., 103(1999)8998.
20. D. Vollhardt, V.B. Fainerman, Adv. Colloid Interface Sci., 86(2000)103.
21. V.B. Fainerman, R. Miller, D. Vollhardt, in "Food Colloids 2000 – Fundamentals of Formulation", E. Dickinson and R. Miller (Eds.), Special Publication No. 258, Royal Society of Chemistry 2001, p. 210.
22. H. Möhwald, Rep. Progr. Phys., 56(1993)653.
23. D. Vollhardt, Adv. Colloid Interface Sci., 64(1996)143.

24. D. Vollhardt, Adv. Colloid Interface Sci., 79(1999)19.
25. V. Melzer, D. Vollhardt, G. Weidemann, G. Brezesinski, R. Wagner, H. Möhwald. Phys. Rev. E, 57(1998)901.
26. F.M. Mota, M.A. Busquets, F. Reig, M.A. Alsina, I. Haro, J. Colloid Interface Sci., 188(1997)81.
27. M. Amrein, A. Vonnahmen, M. Sieber, Eur. Biophys. J., 26(1997)349.
28. J.A. Butler, Proc. Roy. Soc. Ser. A, 138(1932)348.
29. S.S. Datwani, K.J. Stebe, Colloids & Surfaces A, 192(2001)109.
30. V.B. Fainerman, S.V. Lylyk, Kolloidn. Zh., 45(1983)500
31. V.V. Krotov, Kolloidn. Zh., 47(1985)1075.
32. V.B. Fainerman, Zh. Fiz. Khim., 62(1988)1003.
33. E.H. Lucassen-Reynders, J. Phys. Chem., 70(1966)1777.
34. B.B. Damaskin, A.N. Frumkin, S.L. Djatkina, Izv. AN SSSR, Ser. Chim., (1967)2171.
35. V.B. Fainerman, R. Miller, R. Wüstneck, A.V. Makievski, J. Phys. Chem., 100(1996)7669.
36. V.B. Fainerman, E.H. Lucassen-Reynders, R. Miller, Colloids & Surfaces A, 143(1998)141
37. B.B. Damaskin, Izv. AN SSSR, Ser. Chim., (1969)346.
38. E.H. Lucassen-Reynders, In: Anionic Surfactants (Physical Chemistry of Surfactant Action), (Ed. E.H. Lucassen-Reynders), Marcel Dekker Inc., New York-Basel, 1981, p.1.
39. M. Mulqueen, D. Blankschtein, Langmuir, 15(1999)8832.
40. Y.J. Nikas, S. Puvvada, D. Blankschtein, Langmuir, 8(1992)2680.
41. J.R. Lu, R.K. Thomas, J. Penford, Adv. Colloid Interface Sci., 84(2000)143.
42. A.N. Frumkin, Z. Phys. Chem. (Leipzig), 116(1925)466.
43. J.K. Ferri, K.J. Stebe, Colloids & Surfaces A, 156(1999)567.
44. J.K. Ferri, K.J. Stebe, J. Colloid Interface Sci., 208(1999)1.
45. V.B. Fainerman, R. Miller, J. Phys. Chem. B, 104 (2000) 8471
46. V.B. Fainerman, R. Miller, J. Colloid Interface Sci., 232 (2000) 254
47. D. Vollhardt, V.B. Fainerman, G. Emrich, J. Phys. Chem. B, 104(2000)8536.
48. V.B. Fainerman, R. Miller, E.V. Aksenenko, J. Phys. Chem. B, 104(2000)5744.
49. I. Langmuir, J. Chem. Phys., 1(1933)756.
50. J.H. De Boer, The Dynamical Character of Adsorption, Oxford University Press, London, 1945.

51. T. Smith, Adv. Colloid Interface Sci., 3(1972)161.

52. E. Ruckenstein, A. Bhakta, Langmuir, 10(1994)2694.

53. J.N. Israelachvili, Langmuir, 10(1994)3774.

54. E. Ruckenstein, B. Li, Langmuir, 11(1995)3510.

55. E. Ruckenstein, B. Li, J. Phys. Chem., 100(1996)3108.

56. E. Ruckenstein, B. Li, Langmuir, 12(1996)2309.

57. V.B. Fainerman, D. Vollhardt, V. Melzer, J. Phys. Chem., 100(1996)15478.

58. V.B. Fainerman, D. Vollhardt, V. Melzer, J. Chem. Phys., 107(1997)242.

59. E. Ruckenstein, B. Li, J. Phys. Chem. B, 102(1998)981.

60. M. Volmer, Z. Phys. Chem. (Leipzig), 115(1925)253.

61. V.B. Fainerman, D. Vollhardt, J. Phys. Chem. B, 103(1999)145.

62. D. Vollhardt, V.B. Fainerman, Colloids & Surfaces A, 176(2000)117.

63. V.B. Fainerman, D. Vollhardt, R. Johann, Langmuir, 16(2000)7731.

64. D. Vollhardt, V.B. Fainerman, S. Siegel, J. Phys. Chem. B, 104(2000)4115.

65. V.B. Fainerman, D. Vollhardt, G. Emrich, J. Phys. Chem. B, 105(2001)4324.

66. J.H. Hildebrand and J.R. Scott, Regular Solutions, Prentice–Hall, Englewood Cliffs, New Jersey, 1962

67. R.C. Read, J.M. Prausnitz, T.K. Sherwood, The Properties of Gases and Liquids, 3rd Ed., McGraw-Hill Inc. , New York, London, Paris, Tokyo, 1977.

68. A.F.H. Ward, L. Tordai, J. Chem. Phys., 14(1946)453.

69. M. Ziller, R. Miller, Colloid Polym. Sci., 164(1986)611.

70. S.S. Dukhin, G. Kretzschmar and R. Miller, Dynamics of Adsorption at Liquid Interfaces: Theory, Experiment, Application, D. Möbius and R. Miller (Eds.), in Studies in Interface Science, Vol. 1, Elsevier, Amsterdam, 1995

71. V.B. Fainerman, D. Vollhardt, J. Phys. Chem B, submitted.

72. S. Henon, J. Meunier, Rev. Sci. Instrum., 62(1991)936.

73. D. Hönig, D. Möbius, J. Phys. Chem., 95(1991)4590.

74. J. Als-Nielsen, K. Kjaer, in T. Riste, D. Sherrington (eds.), Phase Transitions in Soft Condensed Matter, NATO Series B, 211, Plenum Press, New York and London, 1989, p. 113.

75. J. Als-Nielsen, D. Jacquemain, K. Kjaer, M. Lahav, F. Leveiller, L. Leiserowitz, Phys. Rep., 246(1994)251.

76. K. Kjaer, Physica B, 198(1994)100.

77. V.T. Moy, D.J. Keller, H.E. Gaub, H.M. McConnell, J. Phys. Chem., 90(1986)3198.

78. V.T. Moy, D.J. Keller, H.M. McConnell, J. Phys. Chem., 92(1988)5233.

79. G. Weidemann, D. Vollhardt, Colloids Surf. 100(1995)187.

80. G. Weidemann, D. Vollhardt, Biophys. J. 70(1996)2758.

81. G. Brezesinski, A. Dietrich, B. Struth, C. Böhm, W.G. Bouwman, K. Kjaer, H. Möhwald, Chem. Phys. Lipids, 76(1995)145.

82. D. Vollhardt, V.B. Fainerman, S. Siegel, J. Phys. Chem. B, 104(2000)4115.

83. C. Mestras, A. Ortiz, I. Haro, F. Reig, M. Alsina, Langmuir 13(1997)5669.

84. J. Perezgil, C. Casals, D. Marsh, Biochemistry 34(1995)3964.

85. A.D. Williams, J.M. Wilkin, R.A. Dluhy, Colloid Surf. A 102(1995)231.

86. K. Nag, J. Perezgil, A. Cruz, K.M.W. Keough, Biophys. J. 71(1996)246.

87. S.G. Taneva, K.M.W. Keough, Biochim. Biophys. Acta 1236(1995)185.

88. A. Vonnahmen, M. Schenk, M. Sieber, M. Amrein, Biophys. J., 72(1997)463.

89. S. Koppenol, F.H.C. Tsao, H. Yu, G. Zografi, Biochim. Biophys. Acta, 1369(1998)221.

90. D. Vollhardt, G. Czichocki, Colloids & Surfaces, 11(1984)209.

91. D. Vollhardt, G. Czichocki, Langmuir, 6(1990)317.

92. K.J. Mysels, A.T. Florence, J. Colloid Interface Sci., 43(1973)577.

93. G. Czichocki, D. Vollhardt, H. Seibt, Tenside Surfactants Detergents, 18(1981)320.

94. D. Vollhardt, G. Czichocki, Tenside Surfactants Detergents, 30(1993)349.

95. K. Lunkenheimer, J. Colloid Interface Sci., 160(1993) 509.

96. J.D. Hines, J. Colloid Interface Sci., 180(1996)488.

97. K. Lunkenheimer, H Fruhner, F. Theil, Colloids & Surfaces A, 76(1993)289.

98. J.R. Lu, I.P. Purcell, E.M. Lee, E.A. Simister, R.K. Thomas, A.R. Rennie, J. Penfold, J. Colloid Interface Sci., 174(1995)441.

99. C.D. Bain, P.B. Davies, R.N. Ward, Langmuir, 10(1994)2060.

100. R.N. Ward, P.B. Davies, C.D. Bain, J. Phys. Chem. B, 101(1997)1594.

101. D. Vollhardt, G. Emrich, Colloids & Surfaces A, 161(2000)173.

102. D. Vollhardt, G. Brezesinski, S. Siegel, G. Emrich, J. Phys. Chem. B, 105(2001) in press.

CHAPTER 4

ORGANISATION OF PORPHYRINS IN MONOLAYERS AND MONOLAYER ASSEMBLIES

María Teresa Martín Romero[1] and Dietmar Möbius[2*]

[1] Departamento de Química Física y Termodynámica Aplicada, Facultad de Ciencias, Universidad de Córdoba, Campus Universitario de Rabanales, C3, 2ª planta, E-14014,Córdoba, Spain

[2] Max-Planck-Institut für biophysikalische Chemie, Am Fassberg 11, D-37077 Göttingen, Germany

The organisation of both amphiphilic and water soluble porphyrins in monolayers at the air-water interface and in monolayer assemblies made by sequential transfer to solid substrates is determined by the intermolecular interactions and may be controlled externally by selecting optimal parameters like the surface pressure. Molecular orientation and association, photo-physical and photochemical processes as well as chemical equilibria at interfaces are fascinating topics that have attracted the interest of a large community of scientist. Some of the results and insights in the behaviour of porphyrins in organised monolayers and assemblies described in this Chapter may be of particular interest in the view of potential use of this class of molecules in devices of nanometric dimensions.

1. INTRODUCTION

Porphyrins play an important role in biological systems. They are involved in the conversion of light to chemical energy (photosynthesis), in electron transfer processes (metabolism) or in the transport of oxygen (respiration). The particular chemical structure of the porphyrins with a circular π-electron system gives rise to unusual photo-physical properties of single molecules as well as to molecular association phenomena causing efficient energy delocalisation. Therefore, this fascinating class of dyes has attracted much attention, and many attempts have been made to organise porphyrins in monolayers at interfaces in order to investigate model systems for biological structures and to construct components of devices for potential applications. The organisation of molecules in monolayers at the air-water interface provides the possibility of external control by variation of relevant parameters like the area per molecule. Strictly speaking, the systems may not be in thermodynamic equilibrium, however, the activation energy for transition to the energetically lowest state may be too high. The resulting metastable systems offer a large potential of use as components of molecular machines, i.e. functional units on the nanometer scale [1].

In the class of porphyrins, biological compounds having only short chain substituents (like hemin) [2] and the chlorophylls [3, 4] have been a starting point for the organisation in monolayers. Although the stability of these materials of biological origin is much smaller than that of synthetic porphyrins, model experiments in order to elucidate the role of porphyrins in biostructures may require the biological materials. Therefore, a variety of methods has been applied to the studies of chlorophylls in monolayers and monolayer systems elucidating the orientation, association and function of these porphyrins until the recent past [5-11]. Here, however, the organization of synthetic porphyrins either amphiphilic or without extended hydrophobic part and therefore water soluble will be discussed. Further, supermolecules containing porphyrins have been synthesised for the systematic investigation of photoinduced electron transfer processes for mimicking biological photosynthesis, and such supermolecules have also been organised in monolayer assemblies.

2. MONOLAYERS AT THE GAS-WATER INTERFACE

The fundamental techniques of forming monolayers at the gas-water interface have been developed long ago [12] and are described in monographs and reviews [4, 13-15]. The most common way is spreading of a solution of the film-forming material in a volatile organic solvent on the surface of water or an aqueous solution (other subphases like water-glycerol mixtures have also been used [16]). After evaporation of the solvent, the molecules at the gas-water interface are pushed together by moving a barrier. The main characteristic of monolayers is that they are only one layer of molecules thick. However, they may consist of an amphiphile as insoluble anchor and molecules bound to these amphiphile molecules. The techniques of monolayer characterization will not be described here in detail but may be mentioned: measurement of surface pressure-area and surface potential-area isotherms, observation of the morphology by Brewster angle microscopy and scanning force microscopy, spectroscopic techniques like reflection, adsorption, fluorescence and ESR spectroscopy, X-ray reflection and diffraction, electrochemical methods.

In this Section, the formation of monolayers of or with synthetic porphyrins and their organisation will be discussed. One class of porphyrins is amphiphilic consisting of the porphyrin ring and hydrophobic substituents sufficiently large to keep the molecules at the gas-water interface. An example is an Fe(III) porphyrin with 4 octadecyl chains used to investigate the formation of μ-oxo dimers [17]. Substitution of the two protons in the center of the porphyrin ring by metal ions from the aqueous subphase has been studied with a variety of amphiphilic porphyrins [18]. Another class are porphyrins that are water soluble due to hydrophilic groups. Such porphyrins may be part of monolayers by binding them to appropriate amphiphilic anchor molecules via intermolecular interactions. The kinetics of binding a water soluble anionic porphyrin from the aqueous solution to a lipid monolayer at the surface with counter ion head groups has been investigated as one of the first applications of reflection spectroscopy [19, 20]. Porphyrin organisation in monolayers means e.g. the location and orientation of the porphyrin ring, the molecular distribution in two- or multi-component monolayers, and porphyrin association (dimers, larger aggregates).

2.1 Location and Orientation of Porphyrins in Monolayers at the Gas-Water Interface

Amphiphilic porphyrins form stable, insoluble monolayers at the gas-water interface. One way to synthesise amphiphilic porphyrins is the substitution of a meso-tetra-pyridinyl-porphine with 4 long hydrocarbon chains (e.g. octadecyl) yielding a tetracationic porphyrin [19, 21]. It is reasonable to assume that with such a molecule the porphyrin ring with 4 positive charges at the periphery will be lying flat on the water surface with the long chains on top. Since the area occupied by a flat lying meso-tetra-pyridinyl-porphine is around 2.50 nm^2, and the cross-sectional area of a vertically oriented hydrocarbon chain is about 0.2 nm^2, the long chains should be disordered. Dense-packing of the hydrocarbon chains on top of the porphyrin ring, therefore, requires the addition of another amphiphile. Then, the area per amphiphilic porphyrin molecule should be 4 times the cross-sectional area of its hydrocarbon substituents. This behaviour has been observed [19] with the amphiphilic porphyrin PO1 (structure see Fig. 1) as concluded from the surface pressure-area isotherms shown in Fig. 1. The isotherm curve 1 refers to the two-component monolayer of eicosanoic acid (C20) and methyleicosanoate (AME), molar ratio 1:1, and curve 2 to the three-component monolayer of the porphyrin PO1, C20, and AME, molar ratio 1:10:10.

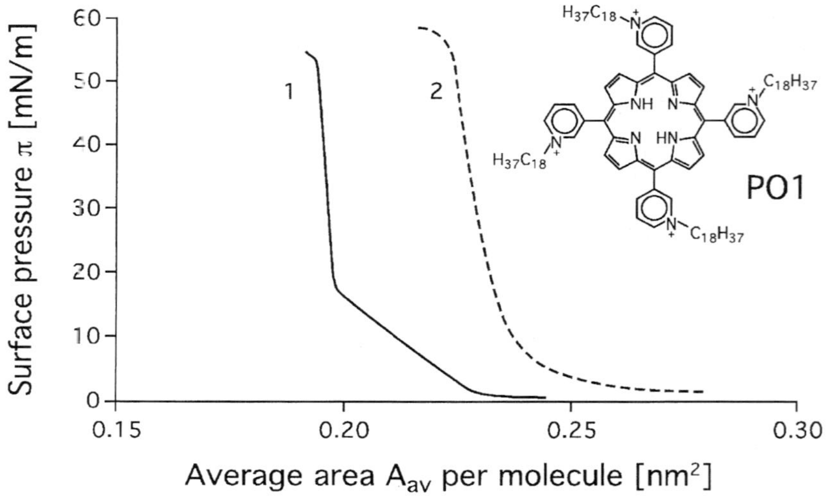

Fig. 1 Surface pressure (π)-average area (A_{av}) isotherms of the two-component monolayer of eicosanoic acid (C20) and methyleicosanoate (AME), molar ratio 1:1 (curve 1), and the three-component monolayer of the porphyrin PO1, C20, and AME, molar ratio 1:10:10 (curve 2) on water, 20°C.

At the surface pressure $\pi = 30$ mN/m, the area is $A_{av}(1) = 0.196$ nm^2 for the two-component monolayer and $A_{av}(2) = 0.228$ nm^2 for the three-component monolayer. From these data, the area per porphyrin PO1 in the three-component monolayer of $A_{PO1} = 0.868$ nm^2 is calculated. This value is close to the area requirement of the 4 hydrocarbon chains, supporting the molecular organisation of the three-component monolayer proposed above. The mixed monolayer of C20 and AME may be considered as a matrix in which the porphyrin is homogeneously distributed.

This organisation of the amphiphilic porphyrin in the three-component monolayer may be verified by reflection spectroscopy. This technique is based on the resonant reflection of transition dipoles arranged at an interface between two media of different refractive index [19].

The light reflection from the interface is modified in the presence of a monolayer of light absorbing molecules. The difference in reflectivity of the monolayer-covered interface and of the bare interface, ΔR, follows the absorption spectrum, since absorption and reflection are strictly related. This reflection spectroscopy should not be confused with a technique, in which a mirror is placed at the bottom of the trough, and the change in the light intensity transmitted twice across the interface with the monolayer is determined. In contrast to the reflection method, where the light reflected from the interface only is detected, the „transmission" method is not applicable when dye solutions are used as aqueous subphase.

Since the reflectivity depends on the angle of incidence and on the average orientation of the transition moments located at the interface, this average orientation may be determined [22]. In the case of the three-component monolayer of PO1:C20:AME = 1:10:10, discussed above, the reflection spectra shown in Fig. 2 have been measured at the air-water interface.

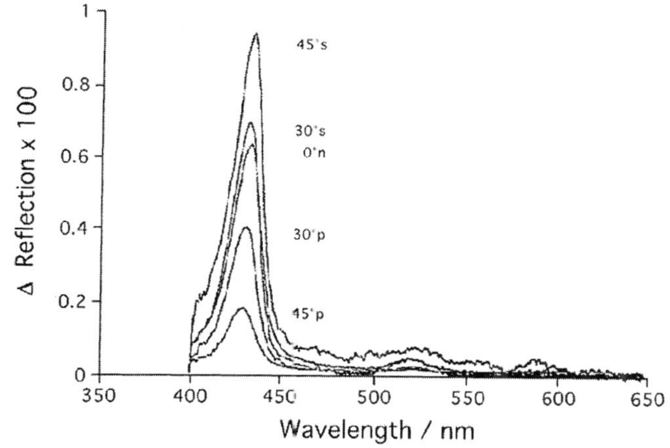

Fig. 2 Reflection spectra of a monolayer of PO1:C20:AME = 1:10:10 at the air-water interface, surface pressure π = 20 mN/m, at different polarisation and angles of incidence of the light with respect to the surface normal; 45° s; 30° s; 3: 0° n; 30° p; 45° p.

The reflection spectra are dominated by the strong Soret band typical for porphyrins with maximum at 430 nm, and at longer waves, the weak Q-bands are clearly seen. With the values of ΔR at the maximum of the Soret band, the ratios of $\Delta R_s/\Delta R_p$ as well as $\Delta R_s/\Delta R_n$ have been calculated for the two angles and compared with the theoretical values for an orientation of the porphyrin ring plane horizontal, i.e. parallel to the water surface. Since the difference between theoretical and experimental values in the case of 45° has been larger than the experimental error, an angle between the surface normal and the porphyrin ring normal of about 10° has been deduced (since the degenerate transitions are in the ring plane) [22]. The reason for this small inclination may be some fluctuation of the molecules around the flat orientation.

A homogeneous distribution of the porphyrin PO1 in the matrix of C20:AME = 1:1 has been assumed and has been confirmed for the case of transferred monolayers on glass plates by the energy transfer test (see Section 3.1). However, fluorescence microscopic studies of mixed monolayers of PO1 and the phospholipid dimyristoyl phosphatidyl ethanolamin (DMPE) clearly show a liquid phase of DMPE with fluorescent PO1 and dark domains of a gel phase upon compression of the monolayer passing a phase transition in the range of area per DMPE of 0.70 $nm^2 > \pi > 0.5\ nm^2$. In this particular case it appears reasonable to conclude that the partitioning of PO1 in the two phases is quite different [23]. Energy transfer from excited PO1 to an amphiphilic phthalocyanine characterized by fluorescence measurements seems to indicate that the acceptor phthalocyanine is present in the gel phase as well as in the fluid phase.

An orientation of the porphyrin ring plane perpendicular to the water surface is desirable for creating strong interactions of the π-electron systems in order to achieve lateral charge transport in monolayers. Such an orientation has been proposed for porphyrin esters without long hydrocarbon chain or extended hydrocarbon portion, however, monolayers at the air-water interface produced with those materials are difficult to handle [24]. Further, the orientation was deduced from isotherm measurements. More convincing are studies with another porphyrin that has a carboxyphenyl group [25], although there is no additional evidence besides isotherm measurements for monolayers at the air-water interface. The absorption of transferred monolayers on glass, however, seems to be in agreement with an orientation of the porphyrin plane perpendicular to the glass surface. A similar orientation has been proposed for an octaoctyl porphyrin (OOP) that undergoes two phase transitions upon compression [26]. At the second transition, starting at an area of about 1.1 nm^2 per OOP, the molecules are forced into a perpendicular orientation. No direct evidence is given for this suggestion. Further, the absorption spectrum of a monolayer transferred at $\pi = 30$ mN/m clearly shows the Soret band shifted to shorter waves with respect to the solution spectrum pointing to the formation of face-to-face dimers.

Porphyrins having one substituent only may adapt an orientation with the plane normal to the substrate surface [27]. In the case of 5,10,15-triphenyl-20-(4dl-α-phenylalanylamino)phenyl porphyrin, according to surface pressure-area isotherms and scanning force microscopy of monolayers transferred to mica the area per porphyrin is about 0.80 nm^2 in agreement with the requirement for the porphyrin ring standing edge-on. Again, no direct evidence for this type of orientation has been provided.

Porphyrins with 4 short fluorocarbon chains (4 to 8 CF_2-groups) form stable monolayers at the air-water interface [28]. The π-A isotherms show the rise of surface pressure upon compression at areas of about 1.6 to 2.0 nm^2. This seems to be too small for flat lying tetra-phenyl-porphin ring, and stacking of the porphyrins may occur like in the case of the amphiphilic Fe(III)-porphyrin mentioned above [17]. According to observations with the Brewster angle microscope (BAM) the monolayer shows a foam-like structure with bright stripes (condensed phase) and dark areas (gas-analogue phase) at an area of 4 nm^2 ($\pi = 0$ mN/m). Those stripes grow slowly and cover a large area fraction after 3h.

Formation and study of monolayer of amphiphilic porphyrins requires the synthesis of derivatives with sufficiently large hydrophobic part. It is, however, possible to use water soluble, easily accessible porphyrins for organisation at the gas-water interface. A compound monolayer can be formed by adsorption of the porphyrin from the aqueous subphase to a preformed monolayer of a simple amphiphile at the air-water interface. This matrix monolayer may have charged hydrophilic head groups that act as counterions of the adsorbent. Based on electrostatic interactions, charged molecules are adsorbed from bulk aqueous solution to oppositely charged head groups of an insoluble monolayer at the air–water interface. This has been demonstrated with different molecules like cyanine dyes [29] or porphyrins [19]. The tetra-anionic porphyrin PO2 (structure see Fig. 3) in the aqueous subphase is adsorbed to a two-component monolayer of eicosylamine (EA) and methyleicosanoate (AME), molar ratio EA:AME = 1:4, as clearly shown by the reflection spectrum in Fig. 3a. The strong Soret band of the porphyrin with maximum at 430 nm and the Q-bands prove the presence of the porphyrin at the interface. This has to be attributed to electrostatic interactions of the anionic solute with the ammonium head groups of the EA, which are protonated under the experimental conditions. If the pH of the aqueous subphase is raised, e.g. to pH > 9, no binding of the porphyrin PO2 takes place [20].

In contrast to this behaviour, a monolayer of eicosanoic acid (C20) prevents the accumulation of the porphyrin PO2 at the interface, see Fig. 3b, even at a concentration 5000 times higher than in the former case. This is also evidence for the fact, that in reflection spectroscopy those molecules only contribute to the signal that are located at the interface and coherently scatter the incident light. This fact has been used to follow the kinetics of adsorption of the porphyrin to the lipid matrix at the surface of the aqueous subphase. In this case, the matrix was a two-component monolayer of eiscolyamine (EA) and octadecanol (ODO), molar ratio EA:ODO = 1:4, and the maximum of the Soret band is at 412 nm. Fig. 4 shows the time dependence of the reflection ΔR (412 nm) for different concentrations of PO2 in the subphase.

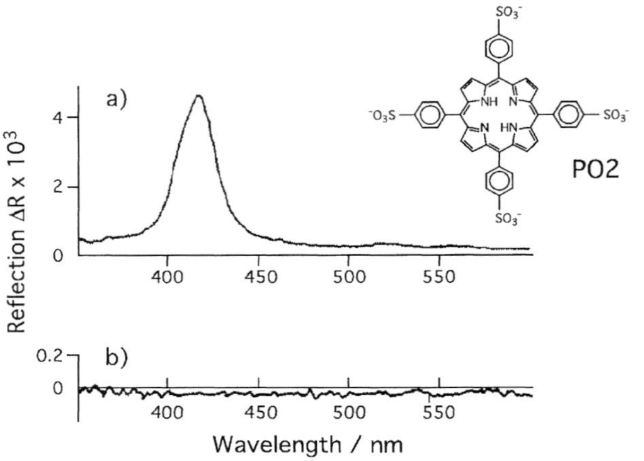

Fig. 3 Interaction of the porphyrin PO2 in the aqueous subphase with two different insoluble monolayers at the solution-air interface; reflection spectra measured in the presence of the two-component monolayer of eicosylamine (EA) and methyleicosanoate (AME), molar ratio EA:AME = 1:4, PO2 concentration $2\cdot10^{-7}$ M (a), and monolayer of eiscosanoic acid, PO2 concentration 10^{-3} M (b); surface pressure 20 mN/m, pH = 5.4, 20°C.

The adsorption technique allows one to form complex monolayers of a lipid matrix and a water soluble porphyrin. However, the presence of the porphyrin in the aqueous subphase may not be desirable during transfer of such monolayers to solid substrates for assembling functional units. The co-spreading technique of spreading a mixed solution of an appropriate lipid anchor and the water soluble component [30] offers many advantages as compared to adsorption procedures such as the use of a minimal amount of the non-amphiphilic water-soluble component, no long adsorption times are required, and the resulting monolayers are quite stable. In this way, non-amphiphilic water-soluble porphyrins in combination with an amphiphilic anchor can be organized into well-defined monolayers at the air-water interface [31]. In fact, the system of complex monolayer and pure water as aqueous subphase is not in thermodynamic equilibrium, since the activation energy for the porphyrin to leave the interface and escape into the bulk is too high at room temperature. The retention of the porphyrin at the interface is obvious from the reflection spectrum. Fig. 5, curve a, shows the reflection spectrum of a monolayer formed by co-spreading of PO2, dimethyl-dioctadecyl-ammonium bromide (DOMA) and methyloctadecanoate (SME), molar ratios PO2:DOMA:SME = 1:4:16, surface pressure 30 mN/m.

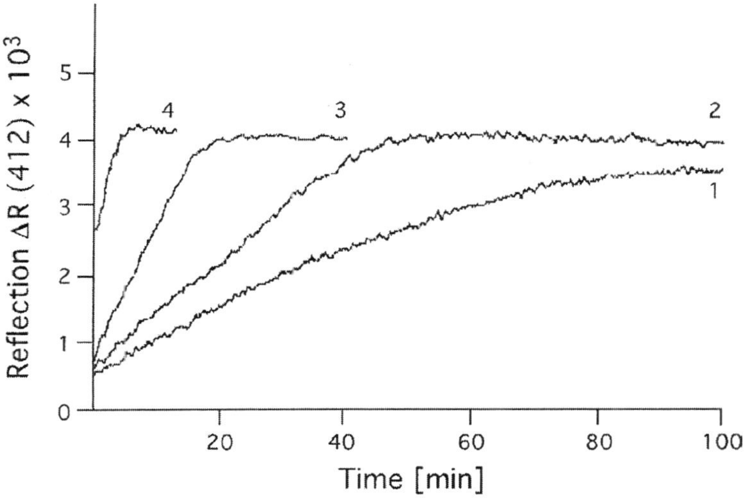

Fig. 4 Time dependence of the reflection signal ΔR (412 nm) at the maximum of the Soret band of a two-component monolayer of eicosylamine (EA) and octadecanol (ODO), molar ratio EA:ODO = 1:4, surface pressure 20 mN/m; the measurement started immediately after compression to this surface pressure, that is kept constant by a feedback system; subphase composition: $2 \cdot 10^{-3}$ M phosphate buffer, pH = 5, with 10^{-7} M PO2 (curve 1), $2 \cdot 10^{-7}$ M PO2 (curve 2), $5 \cdot 10^{-7}$ M PO2 (curve 3), 10^{-6} M PO2 (curve 4).

The particular molar ratios of PO2:DOMA:SME = 1:4:16 have been selected for the following reasons: (a) the stoichiometry requires a ratio of PO2:DOMA of 1:4; (b) for optimal packing of the hydrophobic fraction of the monolayer, at least the area of the flat-lying porphyrin A_{PO} = 3.2 nm^2 should be covered with dense-packed hydrocarbon chains oriented approximately normal to the water surface that have a cross-sectional area of A_{hc} = 0.2 nm^2. Therefore, the minimum number of SME per PO2 is 8 (since 4 molecules of DOMA have 8 chains). The number of 16 was selected in order to reduce possible contacts between PO2 molecules.

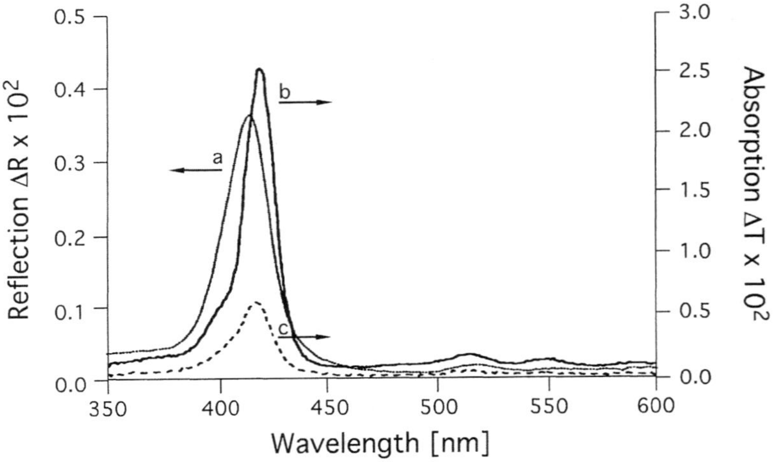

Fig. 5 Lateral organisation of a multi-component monolayer of the porphyrin PO2, dimethyl-dioctadecyl-ammonium bromide (DOMA) and methyloctadecanoate (SME), molar ratios PO2:DOMA:SME = 1:4:16. Reflection spectrum ΔR at the air-water interface, π = 30 mN/m, (curve a) and absorption spectra ΔT of transferred monolayers after adsorption of the porphyrin PO2 from an adjacent $5 \cdot 10^{-8}$ M PO2 aqueous solution to a laterally organised monolayer (curve b) and a reference monolayer (curve c), see text.

It is conceivable that the molecules offering the binding sites for electrostatic interaction with the porphyrin, if statistically distributed in the neutral matrix in the absence of the porphyrin, may adapt optimal places for porphyrin binding in the monolayer formed by co-spreading. Such a type of lateral ordering has been observed after adsorption of a water soluble phthalocyanine from the aqueous subphase to a two-component monolayer with binding sites [32]. The organisation by adsorption requires a certain mobility of the lipid anchor in the matrix. The co-spreading technique may facilitate the coordination of porphyrin and lipid anchor via ion pair formation in the spreading solution. The formation of a template, after immobilization and removal of the porphyrin, should result in enhanced binding of the porphyrin from a dilute solution as compared to a reference monolayer of the same lipid composition that has been formed in the absence of the porphyrin. Immobilisation can be achieved by transfer of the monolayer to a hydrophobic glass plate. Subsequently, the porphyrin has to be removed. The amount of bound porphyrin after adsorption from dilute solution is directly determined by measuring the absorption spectrum of the single monolayer on a glass plate as difference ΔT in transmission of a reference section of the glass plate and the sample section. A typical result [31] is shown in Fig. 5. The adsorption was done from an aqueous $5 \cdot 10^{-8}$ M PO2 solution to the laterally organized monolayer (template), curve b. With the reference monolayer of the same lipid composition of DOMA: SME = 1:4, the absorption spectrum curve c in Fig. 5 was obtained. Obviously, the template binds about 4 times more porphyrin under the same experimental conditions of adsorption as compared to the laterally unorganised surface.

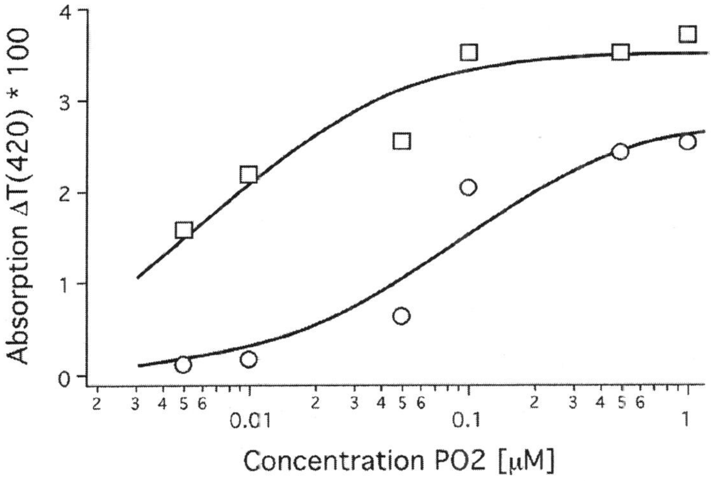

Fig. 6 Adsorption isotherms of PO2 at a template surface formed for the adsorption of the porphyrin PO2 (squares), and at a reference surface of the same statistical composition, however, formed in the absence of the porphyrin (circles); the lines are least square fits of a Langmuir isotherm to the experimental data; 21°C.

This fact becomes even more clear by regarding the adsorption isotherms of both types of surfaces [31] shown in Fig. 6. The values of ΔT at the maximum of the Soret band are plotted vs. the bulk concentration of PO2 for the template (squares) and the reference surface (circles). The full lines in Fig. 6 are least squares fits of a Langmuir isotherm to the experimental data.

The ratio $\Delta T(420)_{temp}/\Delta T(420)_{ref} = 1.25$ of extrapolated (for concentration of PO2 $\rightarrow \infty$) values shows that in the template a larger fraction of binding sites is active than in the reference sample. Further, the ratio of the binding constants is $K_{temp}/K_{ref} = 12.0$. Clearly, binding of the porphyrin is much stronger to the template than to the reference surface. These results show that a charge distribution may be imprinted in an appropriate monolayer on the molecular scale (molecular engineering) leading to improved binding of an analyte.

The porphyrins considered so far were located in the head group region. It has been proposed that a rather hydrophobic nickel porphyrin dimer, whose rings are chemically linked by a diacetylene bridge, in a mixed monolayer with eicosanoic acid, is ultimately squeezed out of the acid monolayer and lies on top of the hydrophobic tails [33]. Direct evidence for this localisation of the porphyrin has not been provided. It should be possible to determine the plane in which the porphyrin is located by energy transfer experiments in monolayer assemblies using an appropriate donor or acceptor, respectively.

2.2 Association Phenomena

Porphyrins may have a strong tendency to form dimers and larger aggregates. This is known from biological systems, e.g. the light harvesting system LH2 of photosynthetic bacteria. Here, according to X-ray data, the porphyrin rings of 18 molecules of bacteriochlorophyll are arranged in a circle with partial overlap of the π-electron systems [34]. This geometry gives rise to rapid energy transport similar to that in J-aggregates of cyanine dyes where the excited state, described as a coherent exciton, moves at a speed of 2 kms^{-1} [35]. The control of molecular association, therefore, is a particular challenge in molecular engineering. The most common way of association of porphyrins is the formation of face-to-face dimers. An example is the μ-oxo-dimer formation of amphiphilic Fe(III)-porphyrins mentioned above [17]. A similar μ-oxo-dimer was used in the investigation of lateral organisation of porphyrins in order to achieve the formation of an ionic network at the air-water interface [36]. With different amphiphilic porphyrins, one with positive charges at the periphery, the other one with carboxyl groups with the possibility to form a salt, heterodimers have been formed in monolayers [37]. However, at the air-water interface, no direct evidence for heterodimer formation has been presented in contrast to the situation in monolayer assemblies on glass. By adsorption, the water soluble tetramethyl pyridinoporphyrazinium bromide (CuTMTPyPz) has been attached to a monolayer of the amphiphilic porphyrin H2P with carboxyl groups as demonstrated by the bands at 350 nm and 680 nm characteristic for CuTMTPyPz in the reflection spectrum, shown in Fig. 7, solid line. For comparison, the reflection spectrum of the amphiphilic porphyrin is shown in Fig 7 , dashed line [37].

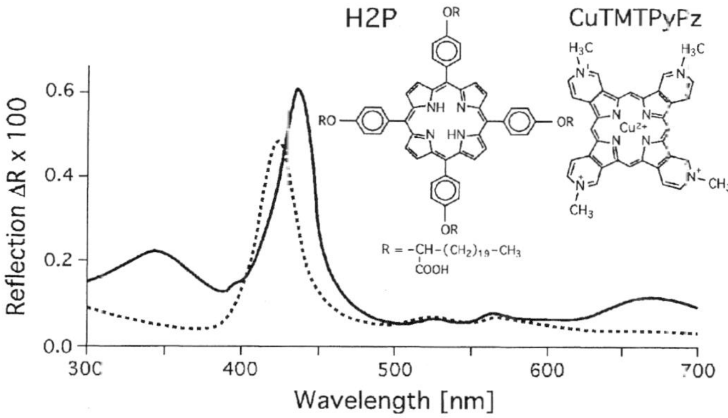

Fig. 7 Formation of heterodimers of the amphiphilic porphyrin H2P and the water soluble CuTMTPyPz by adsorption from the subphase to a monolayer of H2P; aqueous subphase: 10^{-4} M CuTMTPyPz (solid line); 10^{-4} M NaOH (dashed line).

In the co-spreading technique, the tetra-cationic porphyrin PO3 (structure see Fig. 8) has also been used instead of the tetra-anionic porphyrin PO2. Evidently, this requires a lipid anchor with negatively charged head group. The phospholipid dimyristoyl-phosphatidic acid, DMPA, has been used for maintaining the water soluble porphyrin PO3 at the air-water interface. For stoichiometric reason the molar ratio PO3:DMPA = 1:4. In the presence of PO3 the π-A isotherm of DMPA on water is drastically changed as seen in Fig. 8 by comparison of curve 1 (DMPA) and curve 2 (PO3:DMPA = 1:4), where the surface pressure π is plotted vs. area per DMPA. The interaction between the tetra-cationic porphyrin and the lipid anchor causes an expansion. If the molar ratio PO3:DMPA = 1:1 is chosen (data not shown here), the isotherm resembles closely curve 2, i.e. the porphyrin in addition to the stoichiometric ratio may be lost upon compression into the subphase. The isotherm curve 2 has been measured in 2 cycles of compression and decompression. Some hysteresis is observed, however, no loss of PO3 occurred during these two cycles [38].

The observation that no porphyrin is lost is somewhat intriguing. The area per flat lying porphyrin is approximately A_{PO3} = 3.2 nm^2, and the area occupied by DMPA at π = 35 mN/m in the presence of PO3 is A_{DMPA}(PO3,35) = 0.5 nm^2. Therefore, 4 dense-packed molecules of DMPA occupy about 2.0 nm^2. The consequence is either tilting of the porphyrin rings upon compression in order to accommodate all porphyrin molecules underneath the DMPA matrix or partial stacking of the porphyrin molecules, i.e. formation of dimers. A decision which model applies should be possible by measuring the reflection spectra at different surface pressures. A series of reflection spectra of monolayers formed by co-spreading PO3:DMPA = 1:4 on water is shown in Fig. 9 [38]. As typical for porphyrins, the spectra are dominated by the Soret band with maximum at about 430 nm. At $\pi = 5$ mN/m the signal at the maximum is $\Delta R_{max}(5) = 0.476\times10^{-2}$.

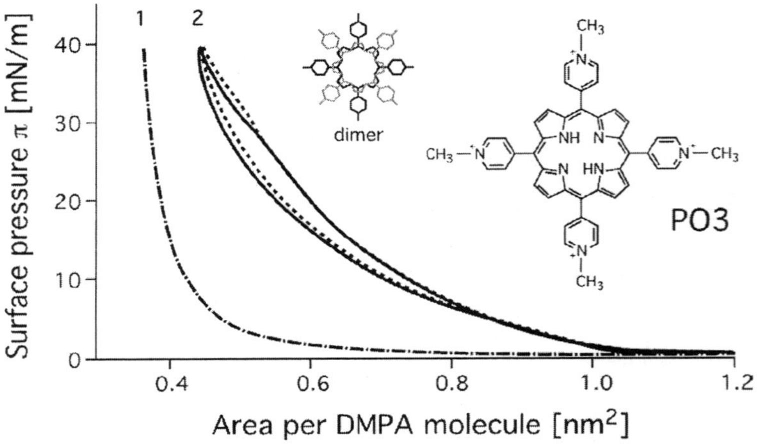

Fig. 8: Surface pressure-area isotherms on water of DMPA (curve 1) and PO3:DMPA = 1:4 (curve 2, two cycles of compression and decompression); isotherms plotted vs. area per DMPA, 21°C; bird's view of a porphyrin dimer with rings twisted by 45°.

The reflection ΔR is strictly related to the absorption. If we assume the same extinction coefficient in the monolayer at the air-water interface as in an aqueous solution ΔR can be approximated in the case of transition moments oriented parallel to the interface with a statistical distribution around the surface normal according to a simplified model [19, 38] by

$$\Delta R = 2.303 \cdot f_{\text{orient}} \left[\frac{\varepsilon}{\text{Lmol}^{-1}\text{cm}^{-1}}\right] \cdot 1.66 \cdot 10^{-7} \cdot \left[\frac{\sigma}{\text{nm}^{-2}}\right] \cdot R_i^{1/2} \qquad (1)$$

Here, f_{orient} = 1.5 is a numerical factor that takes into account the different orientation of PO3 in solution as compared to the interface. The extinction coefficient ε at the maximum of the Soret band is $\varepsilon = 2 \cdot 10^5$ L mol^{-1} cm^{-1}, and the reflectivity R_i at the air-water interface is $R_i \approx 0.02$. From to the π-A isotherm in Figure 8, curve 2 (upon decompression) at 5 mN/m, the surface density $\sigma = 1/A = 0.29$ nm^{-2} is taken. With these parameters, the calculated reflection $\Delta R_{max} = 0.477 \cdot 10^{-2}$ is obtained in excellent agreement with the experimental value of $\Delta R_{max}(5) = 0.476 \cdot 10^{-2}$. This supports the conclusion that no loss of porphyrin to the bulk has occurred.

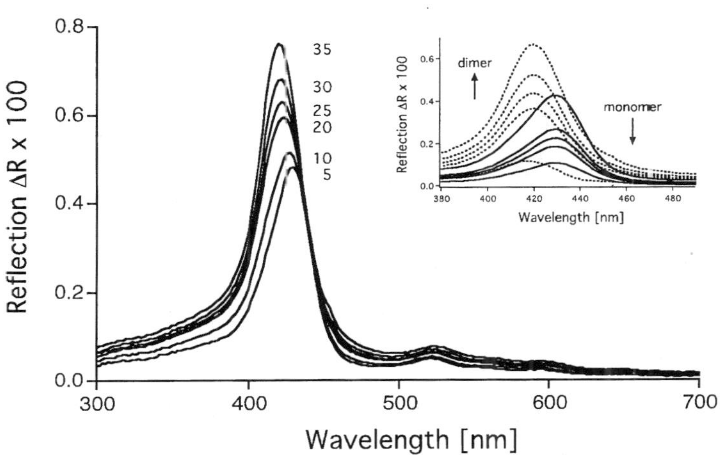

Fig. 9 Reflection spectra of a monolayer formed by co-spreading of PO3:DMPA = 1:4 on water at the different surface pressures π (mN/m): 5; 10; 20; 25; 30; 35; 21°C; inset: monomer and dimer contributions with increasing π to the measured reflection spectra, see text.

The maximum is at λ_{max} = 430 nm for π = 5 mN/m and shifts to shorter waves upon compression of the monolayer with λ_{max} = 420 nm at π = 35 mN/m. Such a shift may indicate the formation of face-to-face dimers. A possible arrangement of the porphyrin rings in a twisted dimer is schematically shown in Figure 8. The twist is necessary in order to provide access of the DMPA head groups to the positively charged methyl pyridinium substituents of the porphyrin PO3. The shift of the Soret band with respect to the monomer band may be calculated for the arrangement shown in Fig. 8. Using the extended dipole model [39], a shift to shorter waves of 9 nm was obtained in good agreement with the observed shift of 10 nm [38].

The area fraction of dimers and monomers, respectively, in the monolayer of PO3:DMPA = 1:4 may be determined by analysing the reflection spectra in combination with the π-A isotherm. The fraction of monomers depends on the average surface density of the porphyrin, σ. For $\sigma \leq 0.31$ nm^{-2}, corresponding to an area per porphyrin of $A_{PO3} = 3.2$ nm^2, this fraction is $f_m = 1$, and the spectrum is that of the porphyrin monomer. The spectrum measured at $\pi = 5$ mN/m (Fig. 9, curve 1) is taken as pure monomer. When the average surface density σ of the porphyrin is larger than $1/A_{PO3}$, porphyrin dimers are formed. The value of σ is taken from the isotherm in Fig. 8, curve 1 after transformation to surface pressure vs. area per porphyrin. The surface density of the porphyrin monomer is $\sigma_I = 2\ \sigma_{PO3} - \sigma$. According to this relation, the contribution of the monomer is determined and subtracted from each spectrum. For example, σ_I(20 mN/m) = [0.62 - 1/(4·0.56)] nm^{-2}, and therefore, the monomer contribution to the spectrum is ΔR_I(20mN/m) = ΔR(5mN/m)·σ_I(20 mN/m)/σ_I(5 mN/m). The remaining $\Delta R_{II} = \Delta R - \Delta R_I$ refers to the dimer. The spectra of the dimer determined in this way for the different surface pressures have exactly the same shape, see Fig. 8, inset, supporting this type of analysis. The area fraction of the dimer $f_d = \sigma_d/\sigma$ is plotted vs. the area per DMPA in Fig. 10 (squares). Evidently, this fraction increases upon compression in a linear way. The dotted line represents the isotherm evaluation model.

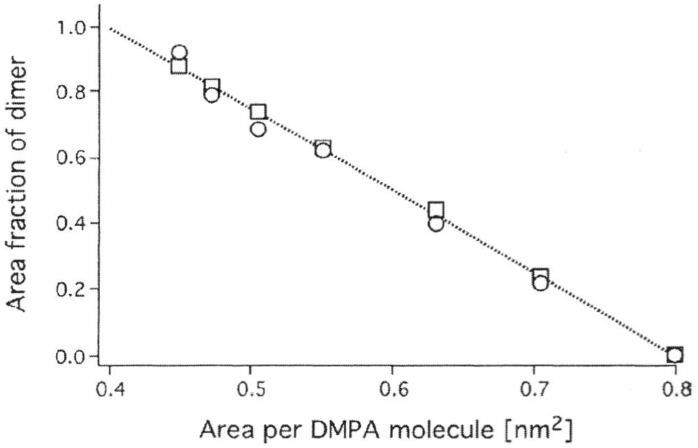

Fig. 10: Area fraction of the monolayer of PO3:DMPA = 1:4 on water with porphyrin dimers as determined by analysing the reflection spectra in combination with the π-A isotherm (squares, data taken from Fig. 8, curve 2 upon decompression), and by evaluating the BAM images shown in Fig. 11 (circles).

The formation of porphyrin dimers may be considered as a phase transition. If domains of DMPA with PO3 dimers underneath of sufficient lateral size are formed, the phase transition should be visible in the Brewster angle microscope. Due to the increased thickness of the monolayer and the larger fraction of the π-electron systems with a refractive index larger than that of the saturated hydrocarbon in the domains of porphyrin dimers, these domains should be brighter than the surrounding phase of DMPA with porphyrin monomers.

A monolayer of DMPA, without porphyrin, shows no defined structures in the BAM images recorded at low and high surface pressures. The behaviour is totally different in the presence of PO3. The BAM images recorded at different surface pressures are shown in Fig. 11. Immediately after spreading the solution, the optical signal due to the presence of the mixed monolayer obtained from the air–water interface was negligible. Upon increasing the surface pressure to approximately 5 mN/m, a continuous, homogeneous film with weak reflectivity was observed (Fig. 11, 5 mN/m). Under further compression, small domains of higher brightness than that of the surrounding area are formed, clearly visible at a surface pressure of $\pi \approx 10$ mN/m (Fig. 11, 10 mN/m). With increasing surface pressure, the density of the bright domains increases (Fig. 11, 25 mN/m) until they cover nearly the whole surface at 35 mN/m (Fig. 11, 35 mN/m).

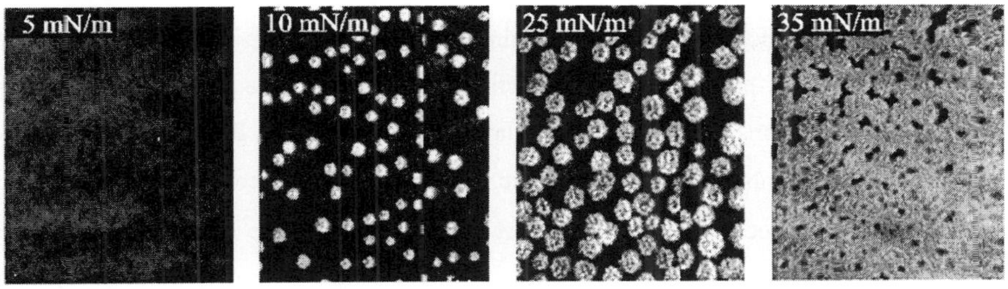

Fig. 11 Brewster angle microscope images of a monolayer of PO3:DMPA = 1:4 at the air-water interface at different surface pressures π (mN/m) as indicated.

The area fraction of the bright phase as determined from the BAM images is plotted in Fig. 10 vs. the area per DMPA (circles). The coincidence with the values of area fraction of dimers from isotherm and reflection measurements is amazing, strongly supporting the assignment of the bright monolayer phase to the dimers of PO3 located underneath the monolayer of the lipid anchor DMPA.

Porphyrins may be metallated, and the central metal ion may have a strong influence on the association behaviour of the porphyrin in organised monolayers at the air-water interface. A particular challenge is the formation of J-aggregate like associates that may exhibit rapid energy transfer as in observed in J-aggregates of cyanine dyes. In order to achieve this, the usually observed formation of face-to-face dimers (see above) has to be suppressed. One way could be binding of a ligand to a central metal ion that prevents the close approach of another porphyrin molecule like in the twisted dimers schematically shown in Fig. 8. An appropriate ligand could force the approaching porphyrin molecule to form a shifted dimer thus producing the smallest subunit of a J-aggregate-like associate. The metal ion Co^{2+} seems to be a good candidate, since it has been demonstrated that ligands bind axially to a Co-porphyrin [40].

However, in first approximation, a behaviour similar to that observed for PO3 in the mixed monolayer of PO3:DMPA = 1:4 is expected for the mixtures of DMPA and the metallated porphyrins NiPO3 and CoPO3, respectively. Therefore, monolayers of MPO3:DMPA (M = Ni, Co), molar ratio 1:4, were prepared at the air–water interface by co-spreading [41]. Surface pressure-area and normalized surface potential (ΔV_n)-area isotherms are shown in Fig. 12.
The influence of the metal complexes on the behaviour of the mixed monolayers formed at the air-water interface is clearly observed in the π–A and ΔV_n–A isotherms. The expansion of the monolayers as compared to the isotherm of DMPA seen at low and medium surface pressures proves that the MPO3 molecules as being retained in the head group region of the DMPA monolayer predominantly by electrostatic interactions, as detected in the case of PO3:DMPA = 1:4. Although the isotherms of the mixed monolayers approach that of DMPA at high surface pressure around $\pi = 35$ mN/m the porphyrins are still present. This is concluded from the fact that the normalised surface potential $\Delta V_n = \Delta V \cdot A$ approaches the value $\Delta V_n = 0.212$ V that is also observed in the case of DMPA on an aqueous subphase of pH = 0.5, when the DMPA monolayer is completely protonated. The same value in the presence of the porphyrins indicates a total compensation of the negative charges of the matrix by interaction with the porphyrins.

The two metallated porphyrins differ in the surface pressure at which a phase transition is detected: $\pi \approx 8$ mN/m for NiPO3:DMPA = 1:4 and $\pi \approx 22$ mN/m for CoPO3:DMPA = 1:4. Therefore, the presence of the metal ion in the central ring of the porphyrin seems to cause an organisation of the monolayer different from that observed with metal-free porphyrin.

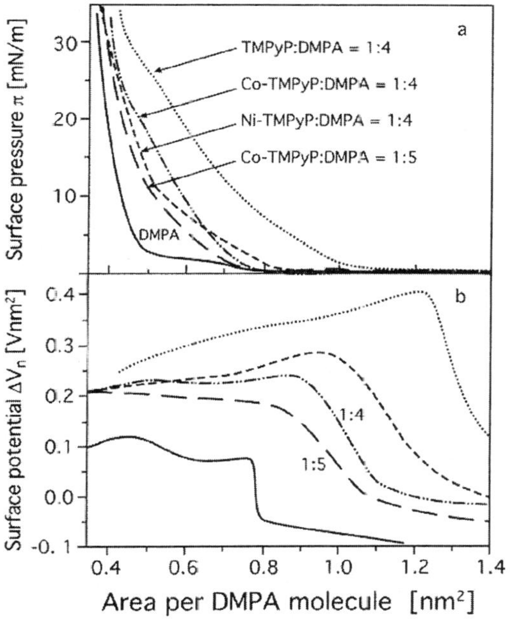

Fig. 12 Surface pressure-area (a) and normalized surface potential (ΔV_n)-area (b) isotherms of monolayers formed by co-spreading metallated porphyrins with DMPA; dashed lines: NiPO3:DMPA = 1:4; dashed-dotted lines: CoPO3:DMPA = 1:4; long-dashed lines: CoPO3:DMPA = 1:5; for comparison the isotherms of DMPA (solid lines) and PO3:DMPA = 1:4 (dotted lines) are also displayed.

Direct evidence of the presence of NiPO3 and CoPO3 in the mixed monolayer with DMPA is obtained by reflection spectroscopy under normal incidence, and a quite different behaviour of the metallated porphyrins was observed, see Fig 13. The reflection spectra of the mixed monolayer of NiTMPyP/DMPA = 1:4 at the air–water interface increase during compression with increasing surface pressure and show a blue shift of the Soret band from λ_{max} = 426 nm (at 2 mN/m) which is close to that measured in solution (λ_{max} = 424 nm), to λ_{max} = 420 nm (at 15 mN/m), i.e. $\Delta\lambda_{max}$ = −6 nm, with increasing surface pressure. These observations, and an analysis of the shift using the extended dipole model suggest an organisation of the monolayer of NiPO3:DMPA = 1:4, similar to that of the metal-free porphyrin PO3:DMPA = 1:4. Consequently, the NiPO3:DMPA =1:4 undergoes a phase transition from NiPO3 monomers to particular NiPO3 dimers with increasing surface pressure.

As in the isotherm measurements, monolayers of CoPO3:DMPA = 1:4 show a behaviour quite different from that of NiPO3:DMPA = 1:4 also in the reflection spectra. Now, a red shift of the Soret band from λ_{max} = 442 nm (at 1 mN/m, as well as in solution) to λ_{max} = 445 nm (at high surface pressure) is observed during monolayer compression, see Fig 14. The normalized reflection spectra $\Delta R_n = \Delta R \cdot A$ (Fig. 14, inset) show a decrease with increasing surface pressure.

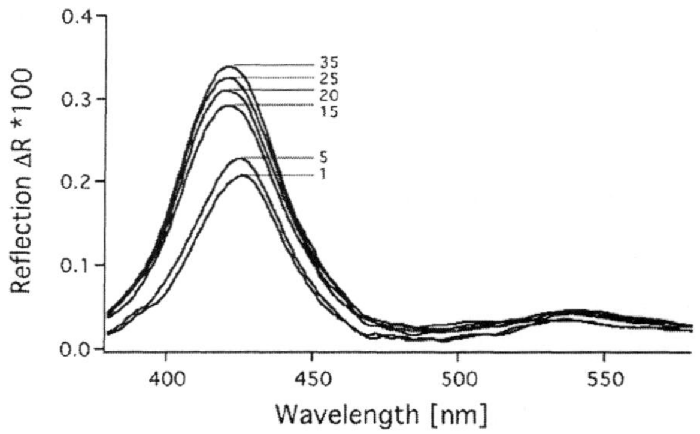

Fig. 13 Reflection spectra ΔR of a monolayer of NiPO3:DMPA = 1:4 formed by co-spreading at the air-water interface at different surface pressures (mN/m) from bottom to top: 2, 5, 15, 20, 25, 30; 21°C.

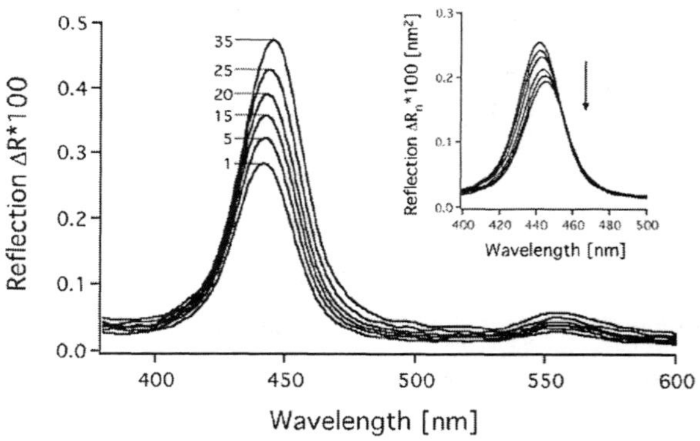

Fig. 14 Reflection spectra ΔR of a monolayer of CoPO3:DMPA = 1:4 formed by co-spreading at the air-water interface at different surface pressures (mN/m) from bottom to top: 1, 5, 15, 20, 25, 30, 35; inset: corresponding normalised reflection spectra ΔR_n, now surface pressure increasing from top to bottom; 21°C.

A constant value of ΔR_n at different surface pressures is expected if all molecules of CoPO3 remain at the interface with constant orientation and association. The decrease of ΔR_n indicates a loss of CoPO3 molecules into the bulk aqueous subphase if no reorientation occurs. A possible loss of about 25% during compression is calculated. A stable organisation of the monolayer CoPO3:DMPA = 1:4 under compression, therefore, may require the loss of about 25% molecules into bulk subphase. This tentative conclusion has been confirmed by measuring the π–A and ΔV–A isotherms (shown in Fig. 12) and reflection spectra of the monolayer CoPO3:DMPA = 1:5 (density of CoPO3 reduced by 25% as compared to CoPO3:DMPA = 1:4).

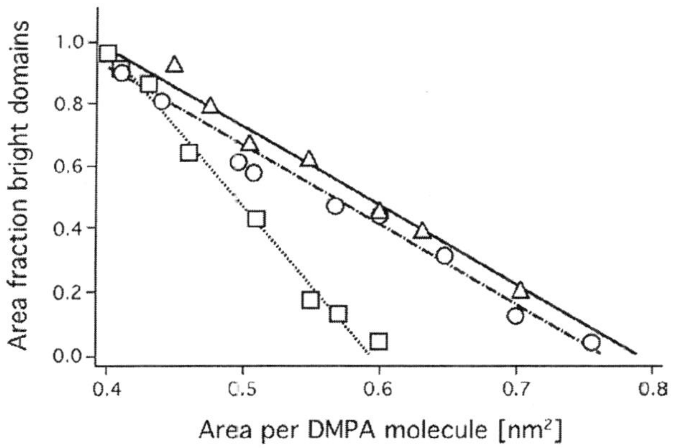

Fig. 15 Area fraction of bright domains vs. area per DMPA molecule as determined from BAM images for monolayers of PO3:DMPA = 1:4 (triangles), and MePO3:DMPA = 1:4, Me = Co^{2+} (squares) and Ni^{2+} (circles).

BAM images of the monolayers MePO3:DMPA = 1:4 (Me = Ni^{2+}, Co^{2+}) indeed show the formation of bright domains PO3 during compression as in the case of the metal-free porphyrin. From the BAM images, the area fraction of the bright domains that are assigned to the monolayer phase with porphyrin dimers has been determined. The results are plotted in Fig. 15 vs. the area per DMPA molecule. For comparison, the results obtained for PO3:DMPA = 1:4 are also shown.

A striking difference is observed between the monolayers with the two metallated porphyrins indicating a different organization of the porphyrin molecules as a function of the central metal ion.

Following a similar analysis used above in the case of PO3, it was concluded that NiPO3 forms dimers very similar to those proposed for PO3 in PO3:DMPA = 1:4 monolayers, where a face-to-face dimer is formed at high surface pressure with the porphyrin rings lying parallel and twisted by 45° with respect to each other underneath the phospholipid matrix, and the second layer of porphyrin still has access to DMPA head groups.

In contrast to NiPO3, CoPO3 molecules associate in a different way. The structure proposed for the CoPO3 associate implies that all Co^{2+} ions in the centre of the porphyrin ring have access to the aqueous subphase allowing the formation of a coordination complex Co(II)–H_2O. The influence of the coordination number of Co(II) ion has a double effect on the structure in the monolayer of CoPO3:DMPA = 1:4. First, a certain fraction of CoPO3 molecules should leave the interface into the subphase. Secondly, the obtained structure should allow access of all Co^{2+} ions to the aqueous subphase. An additional requirement is the access of the porphyrin pyridyl groups to the head groups of the phospholipid. A possible arrangement that meets with all three requirements is shown in Scheme 1.

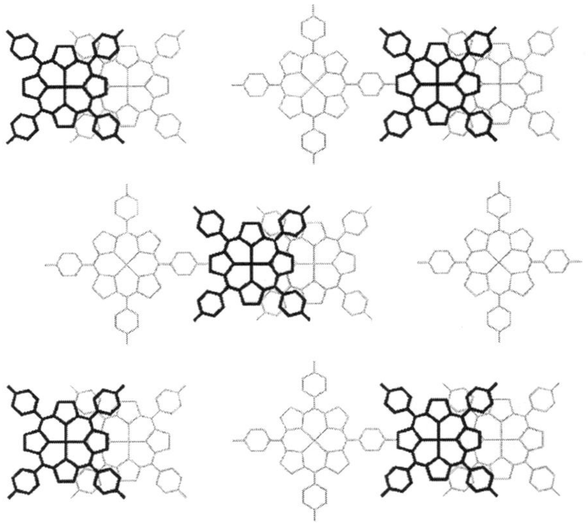

Scheme 1 Organisation of CoPO3 molecules in monolayers of CoPO3:DMPA = 1:4 at the air-water interface at high surface pressure.

As in the case of metal phthalocyanines in Langmuir-Blodgett films [42] the strong influence of the central metal ion on molecular organisation of porphyrins at the air–water interface can be interpreted as a consequence of the coordination number of these central metal ions. Thus, in metal phthalocyanines the interactions between Co(II) or Zn(II), respectively, and water are dominant over π–π interactions favouring the rather horizontal organization in the monolayer, whereas for the Ni(II) compound the π–π interactions between the macrocycles should be more important than in the previous case.

As discussed above, the organization of monolayers formed by co-spreading the lipid anchor and the active, water soluble component is controlled mainly by electrostatic interactions. The electrostatic interactions in monolayers of ionic components may lead to a variety of phenomena, e.g. non-specific effects produced by the presence of the diffuse double layer, effects on the ionisation of components of the monolayer, or specific effects caused by the formation of ion pairs between molecules of the monolayer. Therefore, the influence of different salts in the aqueous subphase (LiCl, $KClO_4$, and tetraethylammonium perchlorate, TEAP, at 10^{-3} M, 10^{-2} M, and 0.1 M concentration) on the monomer-dimer equilibrium of the porphyrin in the mixed monolayer of PO3:DMPA = 1:4 was investigated [43].

In the π–A and ΔV–A isotherms, the effect of the different ions on the behaviour of the monolayer has been expressed (expansion and convergence at low and high surface pressures, respectively; change of the surface pressure values of the LE–LC phase transition; ΔV values at high surface pressures very close to the value for a completely protonated DMPA monolayer, i.e., 530mV). The dependence of the surface potentials of the monolayer of DMPA on the concentration of salt in the aqueous subphase seems to agree with the Gouy–Chapman model, at least for LiCl and $KClO_4$. The value of the slope in a plot ΔV versus log concentration C_{salt} shown in Fig. 16 provides important information about the interactions between the positive ions in the subphase and the negative charges of the phospholipid in the monolayer.

In the case of the DMPA monolayer the three straight lines (Fig. 16a) have nearly the same slope but are shifted to less positive potential in the sequence $K^+ > Li^+ > TEA^+$. This indicates decreasing charge compensation of the lipid matrix in this sequence due to increasing ionic radius. When the mixed monolayer is investigated, the effect of LiCl at low concentrations on the mixed monolayer is similar to that produced by $KClO_4$ (Fig. 16b). However, a substantial decrease of surface potential with 0.1 M LiCl is noted. This may be attributed to a removal of porphyrin due to the presence of Li^+. In the case of TEAP, a decrease of the ΔV values at high surface pressure with increasing salt concentration is observed like in the case of DMPA.

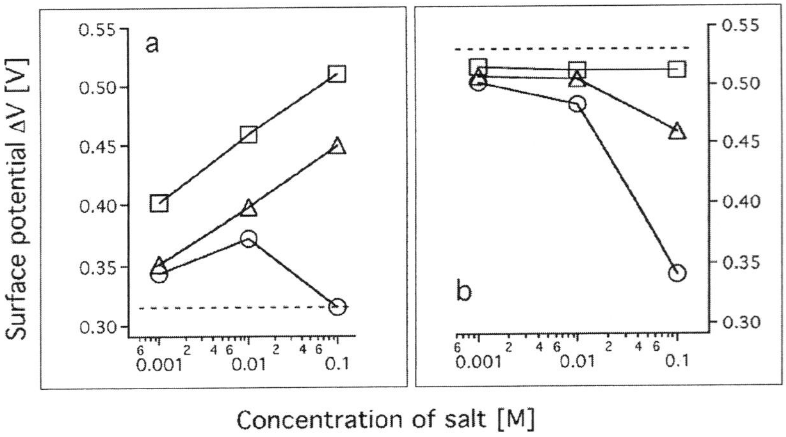

Fig. 16 Surface potential ΔV at $\pi = 35$ mN/m plotted vs. log concentration of salt in the aqueous subphase; the salts are: $KClO_4$ (squares); LiCl (triangles); TEAP (circles).

To infer the molecular organization of molecules incorporated in DMPA monolayer, reflection spectra (ΔR) of complex monolayers of PO3:DMPA = 1:4 have been measured at several surface pressures from 5 to 35 mN/m on different aqueous salt solutions. The reflection spectra of the monolayers are modified depending on the cation. Thus, λ_{max} is shifted to red with increasing the concentration of TEAP suggesting a disappearance of the dimer phase. For the other salts, LiCl and $KClO_4$, respectively, the maximum wavelength is practically constant ($\lambda_{max} = 426 \pm 2$ nm) independent of the salt concentration indicating a constant molar fraction of dimer at the air–aqueous salt solution interface.

The reflection spectra of PO3:DMPA = 1:4 monolayers in the presence of salts have been analysed in detail by the monomer–dimer model described above for this monolayer in the absence of salts, and the total surface density of PO3, Γ(nm^{-2}), i.e. monomer and dimer phases, has been determined. The plot of the surface density of porphyrin, Γ, vs. log C_{salt} in Fig. 17 clearly shows that the surface density of porphyrin decreases with increasing salt concentration independent of the type of salt, although the magnitude of the effect depends on the electrolyte. Thus, porphyrin molecules are removed from the interface by competition with the cation of the subphase as has been indicated by the surface potential values (Fig. 16). In the presence of

TEAP, the area fraction of dimer decreases with increasing salt concentration. This may indicate that the PO3 molecules of the second layer are bound less strongly to the DMPA matrix than those of the first layer. On the other hand, that fraction is nearly independent of the concentration when LiCl or $KClO_4$, respectively, are used. It is amazing that the molar fraction of dimer is kept almost constant and close to 0.5 in the presence of these two salts, moreover since the total surface density of porphyrin decreases strongly with increasing salt concentration. This behaviour may be explained by the formation of the new phase with a fraction of dimer equal to 0.5 that is stabilized by the presence of Li^+ or K^+ ions, including all the porphyrin molecules at the interface. This phase might coexist with a phase where there are no porphyrin molecules and the charges of the headgroups of DMPA are compensated by the cations from the aqueous subphase.

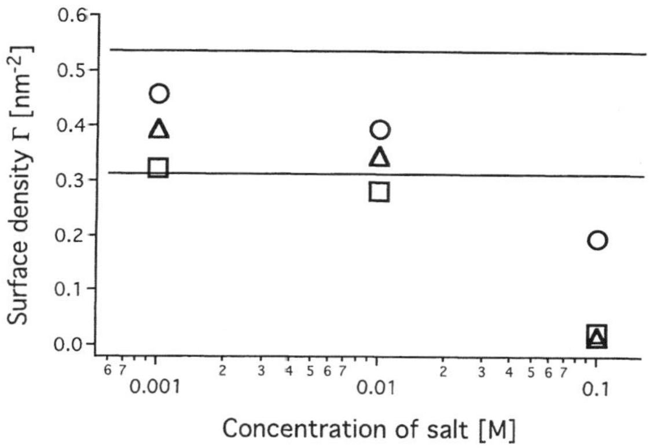

Fig. 17 Total surface density Γ of the porphyrin PO3 at $\pi = 35$ mN/m vs. concentration of salt in the aqueous subphase; TEAP (circles); $KClO_4$ (squares); LiCl (triangles); solid top line: surface density of PO3 in the absence of salt in the subphase ($\Gamma = 0.536\ nm^{-2}$); solid bottom line: surface density of dense-packed flat-lying PO3 monomers ($\Gamma = 0.313\ nm^{-2}$).

2.3 Protonation and Metallation Equilibria

Two pyrrol rings of the porphyrin may be protonated to form a dication. Further, the H-atoms in the centre of the porphyrin ring may be replaced by metal ions, and metal complexes are formed like the Ni^{2+} and Co^{2+} complexes discussed in the preceding section. These protonation and metallation equilibria, respectively, should be modified in organised monolayers at interfaces with respect to the behaviour in bulk solution. Not only the interfacial concentration of metal ions as well as the interfacial pH may differ considerably from those in the bulk, but also the equilibrium constants may be modified due to the different near environment at the interface. In addition, the interfacial dielectric constant of water in the head group region of monolayers with a value around 8 as compared to about 80 in the bulk [44] should strongly influence the intermolecular interactions in particular if charged species are involved.

The incorporation of metal ions in porphyrins at interfaces or bound to enzymes may proceed via two pathways: (1) the displacement mechanism with the porphyrin free base as reactive species or (2) the dissociation mechanism requiring the formation of the dianion as reactive species prior to metal incorporation [45]. Monolayers at the air-water interface have proven to be adequate model systems to elucidate the mechanism [46]. Protonation of the amphiphilic porphyrin PO1 in a matrix of octadecanol (PO1:ODO = 1:20) shifts the Soret band from λ_{max} = 430 nm to λ_{max} = 450 nm as seen in Fig. 18 where a series of reflection spectra measured at different pH of the aqueous subphase is shown. The isobestic point is evidence for an equilibrium between two species. From these spectra, the ratio of the surface densities of the free base and the protonated porphyrin, has been determined. A plot of the logarithm of this ratio vs. subphase pH is shown in Fig. 18, inset, and the straight line represents a linear least squares fit to the experimental data. The slope 2.05 of this linear fit indicates that indeed only the free base and the dication are involved in the protonation equilibrium.

The porphyrin may be metallated by spreading it on an aqueous solution of e.g. $CdCl_2$. The formation of the cadmium complex causes a shift of the Soret band, the maximum being at λ_{max} = 450 nm. Here, the dependence of the ratio complex/base has been determined in analogy to the protonation equilibrium by measuring the reflection spectra of a mixed monolayer of the porphyrin PO1 and eicosanoic acid, PO1:C20 = 1:20, on solutions with increasing concentration of $CdCl_2$ (spectra not shown). In Fig. 19, the logarithm of the ratio complex/base is plotted vs. the logarithm of the concentration of $CdCl_2$ in the subphase (circles). As expected, a linear dependence is observed. In order to elucidate the role of local environment of the reaction site [46], that is characterised in the porphyrin PO1 by the 4 positive charges at the periphery of the ring, a similar amphiphilic porphyrin has been used in which the 4 long hydrocarbon chains are linked to meso-phenyl rings via amide bonds, PO4, structure see Fig. 19.

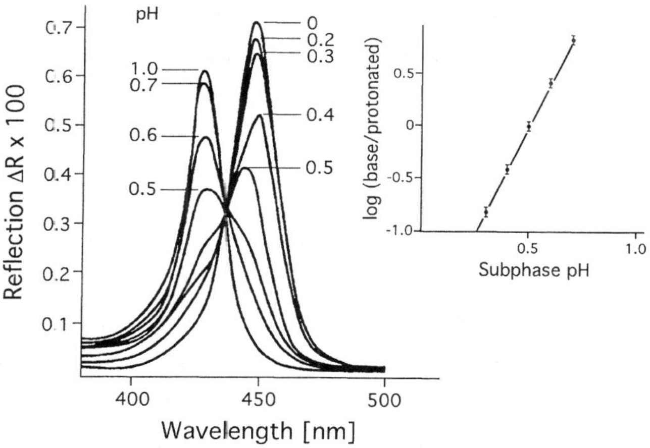

Fig. 18 Protonation equilibrium of the amphiphilic porphyrin PO1 in mixed monolayers PO1:ODO = 1:20 at the air-water interface; reflection spectra measured at the pH of the subphase indicated at the curves; inset: ratio of base/protonated as evaluated from the spectra plotted vs. subphase pH; the straight line is a least squares fit to the experimental data.

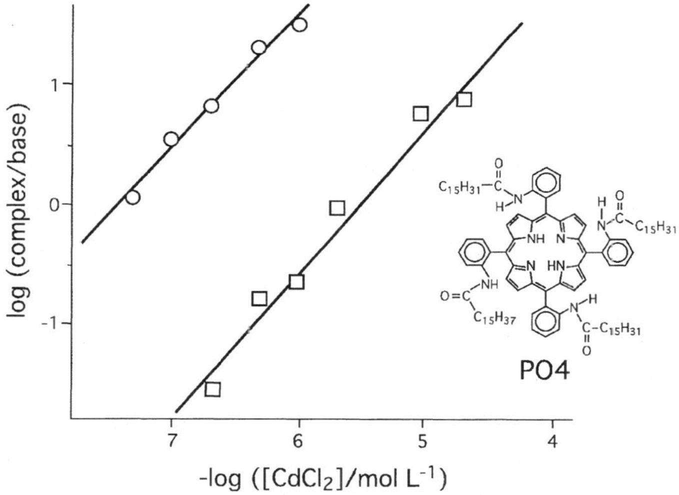

Fig. 19 Metallation of two amphiphilic porphyrins in mixed monolayers at the air-solution interface; logarithmic plot of the ratio complex/base as evaluated from reflection spectra vs. logarithm of concentration of $CdCl_2$ in the subphase; monolayer of PO1:EAM:C20 = 1:4:20 (circles) and of PO4:Opy:C20 = 1:4:20 (squares); inset: structure of PO4

The monolayers to be compared have the compositions PO1: eicosylamide (EAM):C20 = 1:4:20 and PO4:octadecylpyridinium (OPy):C20, respectively, in order to have a closely similar composition of the head group region with EAM mimicking the amide bonds of PO4 and OPy the pyridinium groups of PO1. Nevertheless, the near environment of the porphyrin ring centre, the reactive site, is quite different in both monolayers. From the reflection spectra of the mixed monolayers with PO4 on $CdCl_2$ containing subphases, the ratios of complex/base plotted on a logarithmic scale in Fig. 19, squares, have been evaluated. Again, a linear dependence is observed, however, the data for PO4 are shifted considerably to higher Cd^{2+} concentrations. Obviously, the complex formation is facilitated by the presence of the 4 positive charges near the reactive site, although those charges should cause a depletion of cations near the reactive site. This observation strongly supports the dissociation mechanism that requires the formation of the dianion prior to incorporation of the metal cation.

This conclusion is indirectly supported by a study of the behaviour of the porphyrin PO5 (structure see Fig. 20) at the air-water interface in absence and in the presence of Cd^{2+} ions in the subphase [47]. PO5 has two meso-phenyl rings with a long hydrocarbon chain and a carboxyl group, making the porphyrin amphiphilic and providing it at the same time with two strong hydrophilic groups. According to the π-A isotherms (not shown here), the porphyrin PO5, at a surface pressure of 25 mN/m, has an area per molecule of $A_{PO5} = 0.9$ nm^2 on water and of $A_{PO5} = 1.18$ nm^2 on 10 mM $CdCl_2$. These area values are compatible with an inclined orientation of the porphyrin ring with respect to the water surface. Such inclined arrangement has been found in transferred layers on solid substrates by polarized absorption spectroscopy. Absorption spectra of the monolayers at the air-water interface under different surface pressures measured by the technique of placing a mirror at the bottom of the trough are shown in Fig. 20. On pure water (Fig. 20 left), besides the usual Soret band with maximum at 425 nm, two additional bands with maxima at 458 nm and 665 nm, respectively, are observed. These bands increase with increasing surface pressure and/or decreasing temperature. This behaviour suggests aggregation of the porphyrin, possibly via hydrogen bonds between nitrogens of one porphyrin with the carboxyl groups of a neighbour porphyrin.

Such an interaction should be drastically reduced in the presence of Cd^{2+} ions in the subphase which form the salt with the carboxyl groups and may be incorporated in the centre of the porphyrin ring. Indeed, the band with maximum at 458 nm is seen as shoulder only, and the

Soret band of the porphyrin free base at 425 nm dominates (Fig. 20 right). In contrast to the case of PO1 discussed above, no complex formation with the porphyrin PO5 occurs, supporting the dissociation mechanism. The lack of positive charges at the periphery prevents the formation of the dianion that precedes the incorporation of the metal ion in the porphyrin ring. Of particular interest, however, is the appearance of the band at 458 nm ascribed to an associate of tilted porphyrin rings, since this may be an artificial analogue of the aggregate of bacteriochlorophyll molecules in the light harvesting unit LH2 mentioned above.

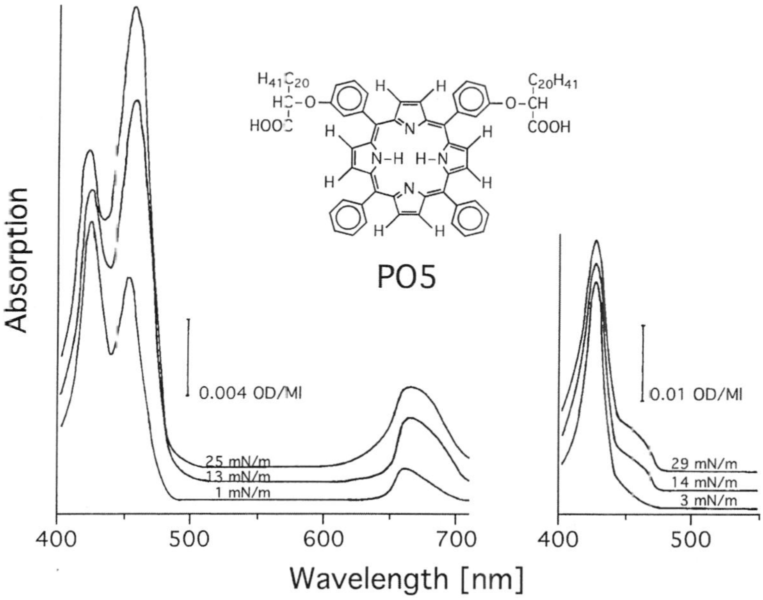

Fig. 20 Absorption spectra of PO5 monolayers at various surface pressures on pure water (left) and on aqueous 0.01 M $CdCl_2$ solution (right).

The complexation equilibria with Cd^{2+} ions of the water soluble porphyrins PO2 and PO3 in the bulk aqueous solution and at the interface bound by electrostatic interactions to appropriate matrix monolayers (EA:AME = 2:3 for PO2 and C20 for PO3) have been investigated. Striking differences of the complex equilibria between the bulk solution and at the interface were observed in both cases [48].

3. MONOLAYER ASSEMBLIES

Ultrathin layered structures may be prepared by various techniques like sequential adsorption of polyanions and polycations or by surface chemical reactions ("self-assembly"). The planned organisation of monolayers, in particular with several components, at the air-water interface and the sequential transfer to solid substrates [1] offers many advantages due to the external control of monolayer organisation, large potential of systematic variation of important parameters, although only a small number of relatively simple and easily accessible compounds is required, and the use of modular subunits in the assembly of complex structures for a large number of purposes. One of the most powerful applications of monolayer assemblies is the construction of model systems to investigate intermolecular interactions and elucidate mechanisms of complex processes. The practical use in devices is another prominent goal.

The basic methods of monolayer transfer from the air-water interface (LB-technique) have been developed long ago [49]. Modern tools like Brewster angle microscopy have been applied in order to understand material flow during transfer and optimise the rheological conditions [50, 51]. The LB technique originally developed for simple amphiphiles like long chain fatty acids has become the way for systematically constructing molecular architectures. Here, systems involving porphyrins will be discussed.

3.1 Porphyrin Organisation in Assemblies

Monolayers that have been organised at the air-water interface may change orientation and/or association upon transfer to a solid substrate since the interactions with the underlying material are different from those with the aqueous subphase. As an example, monolayers of PO3:DMPA = 1:4 at the air-water interface, that consist of a large fraction of dimers at a surface pressure of π = 35 mN/m (see Figs. 9 and 10) have been transferred directly on the hydrophilic surface of a glass plate (by withdrawal through the compressed monolayer at constant surface pressure). The absorption spectra of a single monolayer, given as difference in transmission ΔT between a reference part without the dye monolayer and the sample area with the dye monolayer, measured under different angles of incidence and different polarisation, are shown in Fig. 21. The maximum of the Soret band is at λ_{max} = 432 nm, close to the value for

PO3 measured in methanol solution [38]. Obviously, the porphyrin is transferred as monomer, although the transfer ratio, i.e. the ratio of area change of the monolayer at the air-water interface at constant pressure due to transfer and the monolayer-covered area on the glass plate, is 1. Further, comparing the reflection spectrum under normal incidence at 35 mN/m. (Fig. 9) with the absorption spectrum at 0° in Fig 21, a ratio of $\Delta T/\Delta R \approx 8.5$ is expected, however, the value of 5.73 is obtained. Consequently, the second layer of PO3 molecules forming the dimer is lost upon transfer to the glass surface.

Fig. 21 also shows an example of the determination of the average orientation of transition moments in single monolayers on solid substrates by polarised absorption spectroscopy [22]. This method has been used first for the study of cyanine dyes in monolayers [52] and has been applied in many variations including IR and ESR spectroscopies [53, 54]. Another technique is polarised resonant Raman spectroscopy, that has been used to determine the orientation of the Zn^{2+} complex of PO1 in monolayer assemblies [55]. The angle between the plane of the porphyrin and the substrate surface was smaller than 15°. The ratios $\Delta T_s/\Delta T_p$ and $\Delta T_s/\Delta T_n$ for the angles in Fig. 20 are in agreement with the theoretical values for a flat-lying orientation of the porphyrin rings [38].

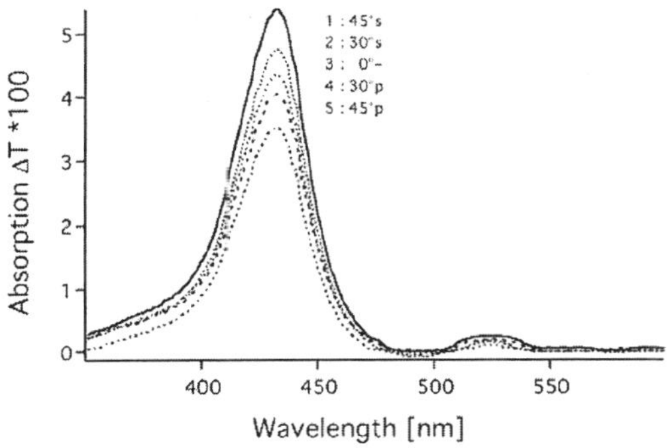

Fig.21 Absorption spectra of a single monolayer of PO3:DMPA = 1:4, transferred from the air-water interface at constant surface pressure of π = 35 mN/m to the hydrophilic surface of a glass plate, measured under different angles of incidence and different polarisations of the light as indicated with the curves.

Monolayers of PO3 bound to hexadecylphosphate as lipid anchor by adsorption from the aqueous subphase have been investigated at the solution-air interface by X-ray reflectivity

[56, 57]. From those measurements it has been concluded, that a single layer of porphyrin is bound to the matrix, however, no direct spectroscopic evidence has been provided. Monolayers transferred to solid substrates also show the porphyrin monomers only, similar to the case of PO3 discussed above.

A fundamental information concerning multicomponent monolayers is the miscibility. Since the monolayers of many simple matrix molecules used for the incorporation of functional molecules like long chain fatty acids undergo phase transitions upon compression, the distribution of dye molecules may change, e.g. the dye may remain in the liquid phase with no dye in the condensed phase. In fact, this has been assumed in the beginning of fluorescence microscopy [58]. However, cyanine dyes may also be present in the condensed phase as dimer, that is non-fluorescent in most cases. The energy transfer test in monolayer assemblies [59] is a totally unambiguous way to discriminate a homogeneous distribution of a dye in the matrix from phase separation. It is based on the energy transfer from excited donor molecules embedded in a monolayer to the dye molecules under investigation in the adjacent monolayer. In most studies of using energy transfer as a tool (e.g. FRET), Förster type transfer of energy is considered fundamentally different from the usual absorption. However, it has been clearly shown, that an absorber in the near field of an emitter, in the classical theory, has exactly the effect of accepting the energy from the excited donor [60] as in the Förster theory. According to energy transfer tests, some cyanine dyes used in fluorescence microscopy are obviously present in the condensed phase, however, often as dimers. The test has been applied to monolayers of the porphyrin PO2 of the composition PO2:DOMA:SME = 1:4:x, (x = 32, 64, 160, 240, 320 and 480) as acceptor layer [61], measuring steady state fluorescence intensity of the donor layer. This fluorescence intensity followed the theoretical dependence expected for a homogeneous distribution within experimental error, clearly showing that no clustering of the porphyrin occurs.

Transfer of monolayers from the air-water interface to a solid substrate by the Langmuir-Blodgett technique, Y-deposition, yields assemblies with interactions between the hydrophilic groups of adjacent monolayers. In the case of the monolayers of PO3:DMPA = 1:4 with porphyrin dimers discussed above this may lead to face-to-face associates with more than 2 porphyrins. When this monolayer is transferred at $\pi = 35$ mN/m directly on the glass surface, only monomers of PO3 are found (see Fig. 21). The monomer can be also obtained, if the glass surface and the porphyrin are separated by spacer layers such as the monolayer sequence EA↑/AA↓/EA↑/AA↓ (↑ monolayer transfer by upstroke, ↓ downstroke).

Porphyrin dimers formed at the air–water interface in monolayers of PO3:DMPA = 1:4 may be assembled by sequential monolayer transfer (PO3/DMPA)↑/[DMPA↓/(PO3/DMPA)↑]$_n$. A linear increase of the absorption ΔT_{max} with increasing the number n of layers transferred to the glass plates is obtained from the absorption spectra (not shown here) of such systems [62]. Also, a blue shift of the Soret band from λ_{max} = 433 nm (for (PO3/DMPA)↑ on glass) to 416 nm for the system (PO3/DMPA)↑/[DMPA↓/(PO3/DMPA)↑]$_5$ with the number of layers transferred is observed in the series of spectra. This is related to the predominant presence of the porphyrin dimer in such assemblies when the number of deposited layers increases. By subtracting the spectrum of the (PO3/DMPA)↑ monolayer in direct contact with the glass surface, corresponding to a monolayer of monomeric porphyrin, from the spectra of the assemblies, the contribution of dimer can be obtained (the small difference between λ_{max} = 416 nm for the dimer absorption in the assemblies and λ_{max} = 420 nm for the dimer at π = 35 mN m^{-1} at the air–water interface is attributed to the change in the environment that is included in the theoretical calculations by a dispersion term of ≈ 0.03eV). Thus, the monomer form of PO3 is only transferred in the first layer of the (PO3/DMPA)↑/[DMPA↓/(PO3/DMPA)↑]$_n$ assemblies, whereas the porphyrin is organized as a dimer in the subsequent monolayers. From the polarised absorption spectra of such assemblies, an inclination of the plane of the porphyrin dimer with respect to the substrate surface of about 10° is obtained.

In the monolayer assemblies above, the PO3/DMPA monolayers were separated by DMPA monolayers preventing interactions between the porphyrin molecules of adjacent monolayers. This is different, when monolayer assemblies with the sequence (PO3/DMPA)↑/[(PO3/DMPA)↓/(PO3/DMPA)↑]$_n$ are built on glass plates. The spectra of such multilayer systems show an increase of the absorption ΔT with increasing number n, see Fig. 22, and an orientation of PO3 molecules approximately parallel to the substrate surface is deduced from measurements with polarised light under inclined incidence. Further, a new absorption band with maximum at 400 nm is observed.

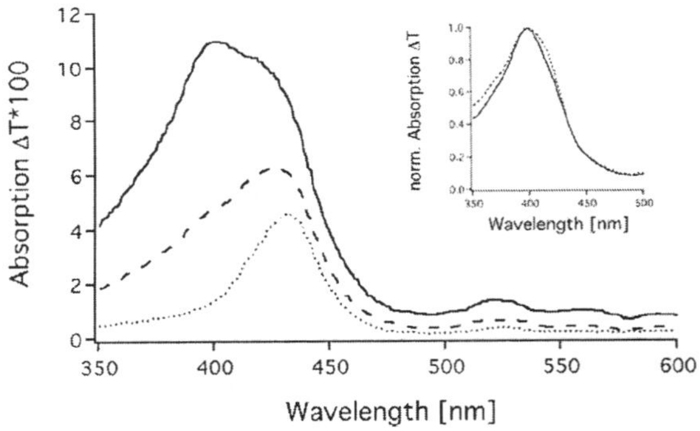

Fig. 22 Absorption spectra of assemblies with monolayers of PO3:DMPA = 1:4, transferred on glass plates at $\pi = 35$ mN/m; monolayer sequence: (PO3/DMPA)↑/[(PO3/DMPA)↓/(PO3/DMPA)↑]$_3$ (solid) and (PO3/DMPA)↑/[(PO3/DMPA)↓/(PO3/DMPA)↑]$_1$ (dashed); the dotted line refers to the spectrum of the initial monolayer on glass; inset: the solid and dashed spectra replotted and normalised to the same ΔT_{max} after subtraction of the dotted spectrum.

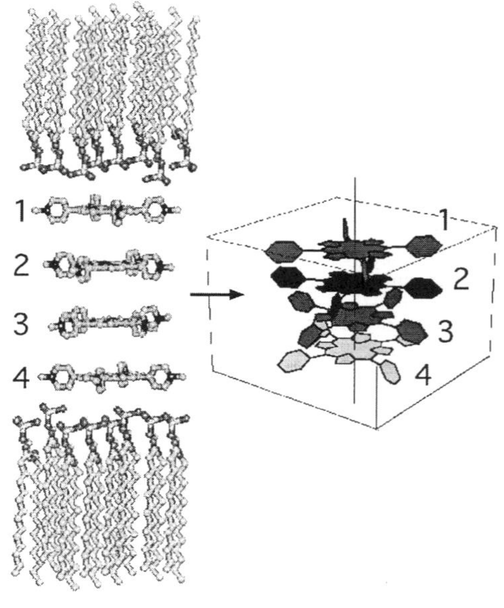

Scheme 2 Tetramers of the porphyrin PO3 between the layers of the amphiphilic anchor DMPA in assemblies of monolayers PO3:DMPA = 1:4; perspective view of the tetramer with the porphyrin rings twisted by 22.5°.

Applying the extended dipole model, the aggregation number can be determined, assuming the formation of a packed tetramer of porphyrin molecules associated face-to-face. In order to have close contact of the porphyrin rings and avoid steric hindrance of the meso-pyridinium rings, as well as to enable access to the negatively charged head groups of DMPA, the central rings of the porphyrin molecules have to be arranged in a helix where the planes between consecutive porphyrin molecules are twisted by an angle of 22.5°, see Scheme 2.

The extended dipole model applied to the arrangement of the porphyrin rings shown in Scheme 2 yields a shift of $\lambda_{max,tetra} - \lambda_{max,mono} = 26$ nm in good agreement with the observed $\Delta\lambda_{max} = 30$ nm [62].

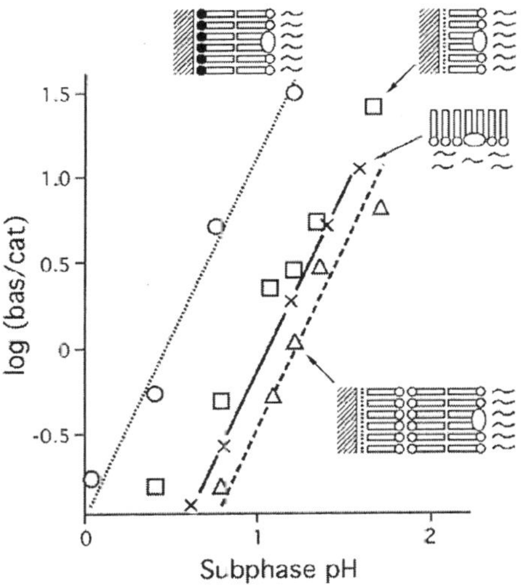

Fig. 23 Protonation equilibrium of the porphyrin PO1 in a monolayer of PO1:C20 = 1:20 at the interface between monolayer assembly and the aqueous phase; log (base/cation) plotted vs. subphase pH for the assembly made by transfer onto glass of eiscosylamine, followed by the monolayer with PO1 (circles and dotted line); transfer of the monolayer of PO1 hydrophobic (squares); transfer of the PO1 monolayer on a bilayer of cadmium eicosanoate that has been transferred on a hydrophilic glass (triangles); assemblies schematically shown; the crosses refer to the equilibrium at the air-water interface.

After transfer to a hydrophobic solid substrate upon immersion (downstroke), the assembly may be left in contact with the aqueous phase. Surprisingly, the monolayer in contact with the aqueous phase may show a behaviour quite different from that when floating on the water under the surface pressure during transfer. This has been demonstrated with an amphiphilic coumarin embedded in a matrix of eicosanoic acid [63]. The apparent pK of the coumarin was shifted by 2 units by variation of the underlying assembly (distance about 2.5 nm). Consequently, the protonation equilibrium of the porphyrin PO1 in a matrix of eiscosanoic acid transferred to a solid substrate and left in contact with the aqueous phase may depend on the underlying structure. This has indeed been observed, see Fig. 23, where the ratio of base/cation, as determined spectroscopically for three different monolayer assemblies is plotted on a logarithmic scale vs. pH of the subphase. For comparison, the result obtained for the monolayer at the air-water interface is also represented.

The equilibrium is clearly shifted due to the influence of the underlying assembly. The physical basis of this phenomenon is not entirely established. The surface potential of the assemblies which depends on the details of the monolayer sequence [64, 65] may modify equilibria at the assembly-water interface.

3.2 Porphyrins as Components of Supermolecules

Supermolecules consist of several functional parts that are chemically linked, in contrast to supermolecular structures where the parts are kept in a particular arrangement by non-covalent bonds. Supermolecules have been synthesized in order to systematically vary parameters like e.g. the distance between an electron donor and an electron acceptor for the investigation of the distance dependence of photoinduced electron transfer (PET) in solution [66, 67]. In monolayer assemblies with porphyrin PO1 as donor and an amphiphilic bipyridinium as acceptor, the formation of the violen radical formed by PET from PO1 to the bipyridinium has been detected by ESR [68]. Also porphyrins have been incorporated in supermolecules for this purpose [69], since they allow to modify the energetic situation by changing the central ion without changing drastically the orbitals involved. An important topic in the studies of photoinduced electron transfer is the influence of the environment and its thermal reorganisation. Therefore, a comparison of the kinetics of PET of the same system of donor and acceptor in solution and in organised monolayer assemblies is of particular interest.

Such an investigation has been done with dyads that consisted of two different Zn porphyrins as donor and two different acceptors chemically linked to form a variety of dyads [70]. The chemical structure of one of the dyads studied is shown in Fig. 24, top. This dyad D was incorporated in monolayers of dioleyl phosphatidylcholin (DOPC) in combination with isonicotinic stearate (INS). INS has been added to DOPC in order to suppress molecular association of the porphyrin by coordination of the pyridin group of INS with the central Zn^{2+} in D. Monolayers were formed on water at 15°C and transferred to quartz plates usually at a surface pressure between 20 and 30 mN/m. Only single monolayers were transferred, since multilayered films were not stable. They have been characterised by measuring absorption and fluorescence spectra as well as fluorescence decay curves and time resolved fluorescence spectra.

The absorption spectrum of monolayers depend on the molar fraction of dyad D and on the surface pressure during transfer, see Fig. 24a. Three different bands can be identified in the range of the Soret band. This becomes more clear after normalisation of the significant spectra to the same height, see Fig. 24b. For comparison, the spectrum of the dyad obtained in solution as well as the excitation spectrum of the fluorescence at 640 nm are also shown.

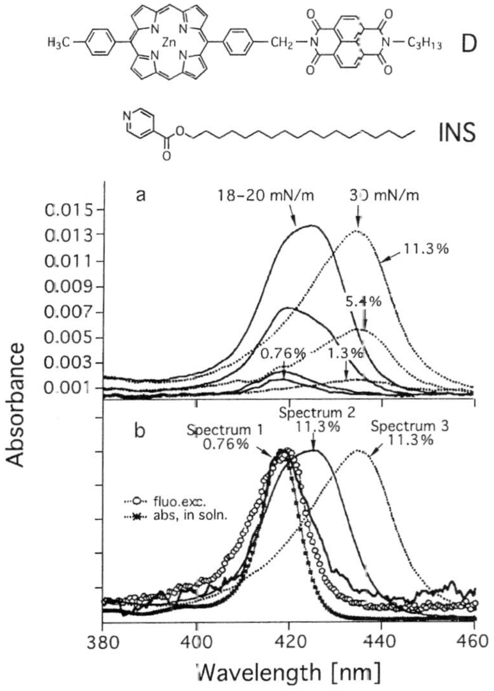

Fig. 24 Absorption spectra (a) of monolayers of the dyad D in various concentrations (mol %) in a matrix of DOPC and INS (40.5 mol%) and surface pressures π during transfer; solid lines: π = 18-20 mN/m; dotted lines: π = 30 mN/m; normalised absorption spectra (b) including the absorption spectrum in solution (stars) and the fluorescence excitation spectrum (circles).

The three bands have the maxima at 419 nm (spectrum 1), 425 nm (spectrum 2), and 434 nm (spectrum 3), respectively. By comparison with the solution spectrum, spectrum 1 is assigned to the monomer of D. According to the concentration and surface pressure dependence spectrum 2 represents dimers and spectrum 3 larger associates. The fluorescence excitation spectrum is independent of surface pressure and concentration of D, indicating that the monomer only is fluorescent.

The shift of the dimer absorption to longer waves with respect to the monomer band shows that the porphyrin moieties in the dimer are not arranged face-to-face but rather slipped. From polarized absorption measurement it is concluded that the distribution of the transition moments is statistical around the surface normal and that the angle between the substrate plane and the long and short axes of the porphyrin part of D in the dimer is 15°.

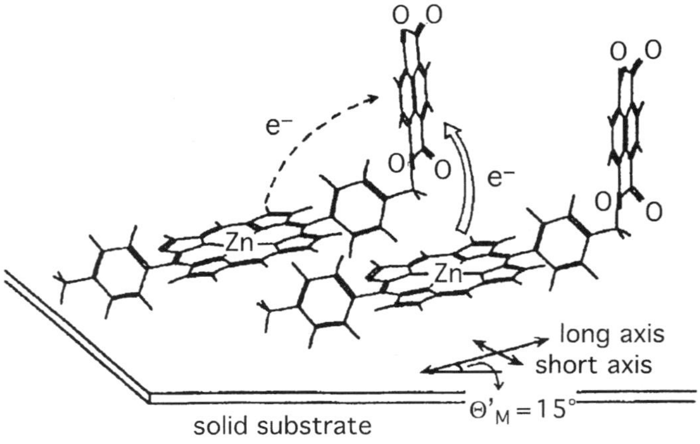

Fig. 25 Spatial organisation of the dimer of the dyad D in INS/DOPC monolayers transferred to glass plates; the tilt of the slipped planes of the porphyrin moieties (long and short axes) with respect to the substrate plane is 15°; photoinduced electron transfer is intermolecular in contrast to intramolecular PET in solution.

From the time-resolved fluorescence measurements several decay times τ_F have been evaluated: 48 ps (main component), 436 ps, and 2500 ps for 0.3 mol % of D in INS/DOPC matrix. Similar results were obtained at 7.4 mol % of D (41 ps, 377 ps and 2000 ps). The fast fluorescence decay in solution, τ_F = 18 ps, is due to intramolecular electron transfer from the excited porphyrin moiety to the naphthalene tetracarboxamide part as acceptor. Considering that the shape of the fluorescence spectrum changes slightly within about 50 ps and that the spectrum at larger time coincides with the steady state spectrum assigned to monomers, it has been concluded, that the fast decay is due to an intermolecular electron transfer occurring at dimer sites. This is in contrast to the situation in solution where the PET is intramolecular. In

Fig. 25, the spatial organisation of the dimer with slightly tilted and slipped porphyrin rings is schematically depicted. PET in this dimer takes place mainly from the excited porphyrin of one molecule D to the acceptor of the adjacent molecule D in the dimer.

This result appears surprising, since "through-bond" electron transfer is considered to be faster than "through-space" electron transfer. However, the overlap of the orbitals of donor and acceptor moieties involved in the PET may be strongly influenced by the particular organisation in the monolayer and thereby modify the efficiency of PET.

3.3 *Potential Applications*

Monolayer assemblies offer a large potential for applications. The chapter entitled "Applications of organised molecular films to electronic and opto-electronic devices" by Petty in this monograph describes in detail some of the most advanced cases. Besides phthalocyanines, that may be chemically more stable, porphyrins have been considered as active components of structures for application e.g. in sensor devices. Examples are sensors for detection of NO by a variation of the electric resistance across a multilayer due to interaction of porphyrin dimers with the gas [33] or specially designed field effect transistors as NO_2 sensors [71,72]. The effect of chlorine gas on the optical properties of porphyrin monolayers on solid support has been used for the design of a sensor [73].

As an example, the optical response of monolayer films of the sulfonamide-porphyrin PO6 (structure see Fig. 26) on the exposure to chlorine gas has been studied for the design of a sensor device [74]. This porphyrin forms stable monolayer on water showing a phase transition at a surface pressure of $\pi = 12$ mN/m between the area per molecule 2.1 nm$^2 > A >$ 1.3 nm^2. At low pressure ($\pi < 12$ mN/m), the porphyrin molecules lie flat at the water surface. At high pressure the molecules are forced into a tilted orientation with an angle of about 30° between the plane of the central porphyrin ring and the surface. Monolayers were deposited directly onto glass plates by withdrawal (upstroke). The absorption spectra of a monolayer of PO6 before and during exposure to chlorine gas at a concentration of 5 ppm are shown in Fig. 26, left. The surface pressure during transfer was 8 mN/m, corresponding to an area per PO6 of about 2.3 nm^2 which is close to the cross-sectional area of the flat porphyrin disk. The Soret band with maximum at 420 nm decreases, and a new band with maximum at 455 nm appears, reaching half of its final intensity after 30 s. Another band at 660 nm also grows in intensity during exposure.

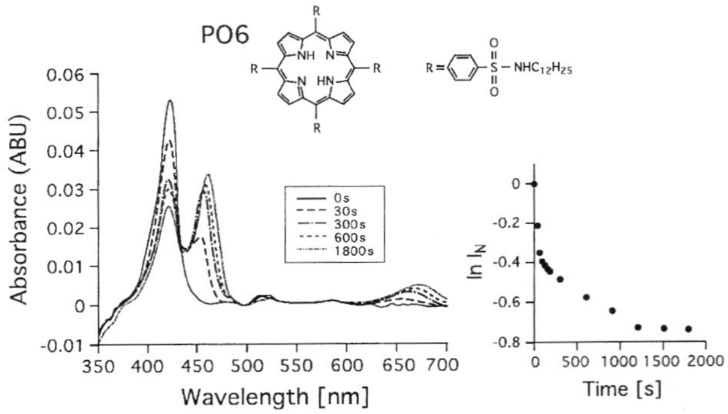

Fig. 26 Absorption spectra after different times (left) and logarithmic plot of the normalised absorbance of the Soret band at 420 nm (right) of a monolayer of PO6 on glass upon exposure to chlorine gas, concentration 5 ppm; transfer of PO6 at $\pi = 8$ mN/m.

The normalised intensity of the Soret band at 420 nm is plotted on a logarithmic scale vs. time in Fig. 26 right. During the first 60 s, a well defined rate constant characterises the response. After this initial period, the kinetics become more complex. Repeated recovery and exposure have shown that reversible and reproducible sensing action occurs if a threshold concentration is not exceeded. The advantage of the optical detection in sensor devices as compared to electrical techniques is the rather simple preparation of samples and avoidance of possible degradation due to electrical currents.

4. ABBREVIATIONS

AA	arachidic acid
AME	methyleicosanoate
BAM	Brewster angle microscopy
C18	octadecanoic acid
C20	eicosanoic acid
DMPA	dimyristoyl-phosphatidic acid
DMPE	dimyristoyl phosphatidyl ethanolamine

DOMA	dimethyl-dioctadecyl-ammonium bromide
DOPC	dioleyl-phosphatidylcholin
EA	eicosylamine
EAM	eicosylamide
INS	isonicotinic stearate
ODO	octadecanol
Opy	octadecylpyridinium tosylate
PET	photoinduced electron transfer
PO1	structure see Fig. 1
PO2	structure see Fig. 3
PO3	structure see Fig. 8
PO4	structure see Fig. 19
PO5	structure see Fig. 20
PO6	structure see Fig. 26
SME	methyloctadecanoate
TEAP	tetraethylammonium perchlorate

5. LIST OF SYMBOLS

A	area per molecule
A_{av}	average area per molecule in multi-component monolayers
A_{POx}	area per porphyrin POx
A_{hc}	area per hydrocarbon chain
Γ	total surface density
π	surface pressure
ΔR	reflectivity difference
σ	surface density
τ_F	fluorescence decay time
ΔT	transmission difference
ΔV	surface potential

6. REFERENCES

1. Kuhn, H. *Pure Appl. Chem.* **1965**, *11*, 345-357.
2. Alexander, A. E. *J. Chem. Soc.* **1937**, *1937*, 1813-1816.
3. Colmano, G. *Biochim. Biophys. Acta* **1961**, *47*, 454-457.
4. Gaines, J., G. L. *Insoluble monolayers at liquid-gas interfaces*; Wiley-Interscience: New York, 1966.
5. Kleuser, H.; Bücher, H. *Z. Naturforsch.* **1969**, *24b*, 1371-1374.
6. Iriyama, K.; Yoshiura, M.; Mizutani, F. *Thin Solid Films* **1980**, *68*, 47-54.
7. Désormeaux, A.; Leblanc, R. M. *Thin Solid Films* **1985**, *132*, 91-99.
8. Chauvet, J.-P.; Agrawal, M. L.; Hug, G. l.; Patterson, L. K. *Thin Solid Films* **1985**, *133*, 227-234.
9. Chauvet, J.-P.; Patterson, L. K. *Thin Solids Films* **1988**, *159*, 149-157.
10. Volkov, A. G.; Gugeshashvili, M.; Zelent, B.; Cote, D.; Munger, G.; Tessier, A.; Blanchet, P. F.; Leblanc, R. M. *Bioelectroche. & Bioenerg.* **1995**, *38*, 333-342.
11. Essounni, A.; Gruszecki, W. I.; Tajmirriahi, H. A.; Zelent, B.; Wang, G. M.; Leblanc, R. M. *J. Colloid Interface Sci.* **1995**, *171*, 134-141.
12. Pockels, A. *Nature (London)* **1891**, *43*, 437-439.
13. Petty, M. C. *Langmuir-Blodgett Films*; Cambridge University Press: Cambridge, 1996.
14. Kuhn, H.; Möbius, D.; Bücher, H. *Spectroscopy of Monolayer Assemblies*; John Wiley & Sons: New York, 1972; Vol. 1 Part 3B.
15. Kuhn, H.; Möbius, D. Monolayer assemblies. In *Investigations of Surfaces and Interfaces*; 2nd ed.; Rossiter, B. W., Baetzold, R. C., Eds.; John Wiley & Sons, Inc.: New York, 1993; Vol. IXB; pp 375-542.
16. Barraud, A.; Leloup, J.; Lesieur, P. *Thin Solid Films* **1985**, *133*, 113-116.
17. Hopf, F. R.; Möbius, D.; Whitten, D. G. *J. Am. Chem. Soc.* **1976**, *98*, 1584-1586.
18. Whitten, D. G.; Eaker, D. W.; Horsey, B. E.; Schmehl, R. H.; Worsham, P. R. *Ber. Bunsenges. Phys. Chem.* **1978**, *82*, 858-867.
19. Grüniger, H.; Möbius, D.; Meyer, H. *J. Chem. Phys.* **1983**, *79*, 3701-3710.
20. Möbius, D.; Grüniger, H. R. *Bioelectrochem. Bioenerg.* **1984**, *12*, 375-392.
21. Ruaudel-Teixier, A.; Barraud, A.; Belbeoch, B.; Roulliay, M. *Thin Solid Films* **1983**, *99*, 33-40.
22. Orrit, M.; Möbius, D.; Lehmann, U.; Meyer, H. *J. Chem. Phys.* **1986**, *85*, 4966-4979.

23. Flörsheimer, M.; Möhwald, H. *Thin Solid Films* **1988**, *159*, 115-123.
24. Bergeron, J. A.; Gaines, J. G. L.; Bellamy, W. D. *J. Colloid Interface Sci.* **1967**, *25*, 97-106.
25. Bardwell, J.; Bolton, J. R. *Photochem. Photobiol.* **1984**, *39*, 735-746.
26. Sartori, E.; Fontana, M. P.; Costa, M.; Dalcanale, E.; Paganuzzi, V. *Thin Solid Films* **1996**, *284-285*, 204-207.
27. Qian, X. P.; Tai, Z. H.; Sun, X. Z.; Xiao, S. J.; Wu, H. M.; Lu, Z. H.; Wei, Y. *Thin Solid Films* **1996**, *284-285*, 432-435.
28. Liang, W.; Nakahara, H. *Chem. Lett.* **1995**, 973.
29. Lehmann, U. *Thin Solid Films* **1988**, *16a*, 257-269.
30. Hada, H.; Hanawa, R.; Haraguchi, A.; Yonezawa, Y. *J. Phys. Chem.* **1985**, *89*, 560-562.
31. Martín, M. T.; Möbius, D. *Thin Solid Films* **1996**, *284-285*, 663-666.
32. Mõbius, D. *Z. phys. Chem., Neue Folge,* **1987**, *154*, 121-144.
33. Arnold, D. P.; Manno, D.; Micocci, G.; Serra, A.; Tepore, A.; Valli, L. *Thin Solid Films* **1998**, *327-329*, 341-344.
34. McDermott, G.; Prince, S. M.; Freer, A. A.; Hawthornthwaite-Lawless, A. M.; Papiz, M. Z.; Cogdell, R. J.; Isaacs, N. W. *Nature* **1995**, *374*, 517-521.
35. Kuhn, H.; Kuhn, C. Chromophore coupling effects. In *J-aggregates*; Kobayashi, T., Ed.; World Scientific: Singapore, 1996; pp 1-40.
36. Porteu, F.; Palacin, S.; Ruaudel-Teixier, A.; Barraud, A. *Thin Solid Films* **1992**, *210/211*, 769-772.
37. Porteu, F.; Palacin, S.; Ruaudel-Teixier, A.; Barraud, A. *J. Phys. Chem.* **1991**, *95*, 7438.
38. Martín, M. T.; Prieto, I.; Camacho, L.; Möbius, D. *Langmuir* **1996**, *12*, 6554-6560.
39. Czikkely, V.; Försterling, H. D.; Kuhn, H. *Chem. Phys. Lett.* **1970**, *6*, 207-210.
40. Lecomte, C.; Baudin, C.; Berleur, F.; Ruaudel-Teixier, A.; Barraud, A.; Momenteau, M. *Thin Solid Films* **1985**, *133*, 103-112.
41. Pedrosa, J. M.; Perez, M.; Prieto, I.; Martín Romero, M. T.; Möbius, D.; Camacho, L. *Phys. Chem.Chem. Phys.* **2002**, *4*, 2329-2336.
42. Fouriaux, S.; Armand, F.; Araspin, O.; Ruaudel-Teixier, A.; Maya, E. M.; Vazquez, P.; Torres, T. *J. Phys. Chem.* **1996**, *100*, 16984-16988.
43. Prieto, I.; Camacho, L.; Martín, M. T.; Möbius, D. *Langmuir* **1998**, *14*, 1853-1860.
44. Vogel, V.; Möbius, D. *J. Colloid Interface Sci.* **1988**, *126*, 408-420.

45. Longo, F. R.; Brown, E. M.; Rau, W. G.; Adler, A. D. In *The Porphyrins*; Dolphin, D., Ed.; Academic Press: New York, 1978; Vol. 5.
46. Loschek, R.; Möbius, D. *Chem. Phys. Lett.* **1988**, *151*, 176-182.
47. Miller, A.; Knoll, W.; Möhwald, H.; Ruaudel-Teixier, A. *Thin Solid Films* **1985**, *133*, 83-91.
48. Loschek, R.; Möbius, D. *J. Chim. Phys. (Paris)* **1988**, *85*, 1041-1045.
49. Blodgett, K. B. *J. Am. Chem. Soc.* **1935**, *57*, 1007-1022.
50. Kurnaz, M. L.; Schwartz, D. K. *Phys. Rev. A* **1997**, *56*, 3378-3384.
51. Knobler, C. M.; Schwartz, D. K. *Curr Op. Colloid & Interface Sci.* **1999**, *4*, 46-51.
52. Bücher, H.; Kuhn, H.; Mann, B.; Möbius, D.; v. Szentpàly, L.; Tillmann, P. *Photogr. Sci. Eng.* **1967**, *11*, 233-241.
53. Messier, J.; Marc, G. *J. Phys. (Paris)* **1971**, *32*, 799-809.
54. Vandevyver, M.; Barraud, A.; Ruaudel-Teixier, A.; Maillard, P.; Gianotti, C. *J. Colloid Interface Sci.* **1982**, *85*, 571-585.
55. Lesieur, P.; Vandervyver, M.; Ruaudel-Teixier, A.; Barraud, A. *Thin Solid Flims* **1988**, *159*, 315-322.
56. Gregory, B. W.; Vaknin, D.; Cotton, T. M.; Struve, W. S. *Thin Solid Films* **1996**, *284-285*, 849-853.
57. Gregory, B. W.; Vaknin, D.; Gray, J. D.; Ocko, B. M.; Stroeve, P.; Cotton, T. M.; Struve, W. S. *J. Phys. Chem. B* **1997**, *101*, 2006-2019.
58. Lösche, M.; Sackmann, E.; Möhwald, H. *Ber. Bunsenges. Phys. Chem.* **1983**, *87*, 848-852.
59. Bücher, H.; v. Elsner, O.; Möbius, D.; Tillmann, P.; Wiegand, J. *Z. Physik. Chem. Neue Folge* **1969**, *65*, 152-169.
60. Kuhn, H. *J. Chem. Phys.* **1970**, *53*, 101-108.
61. Martín, M. T.; Möbius, D. *Supramolec. Sci.* **1997**, *4*, 381-384.
62. Prieto, I.; Pedrosa, J.-M.; Martín-Romero, M. T.; Möbius, D.; Camacho, L. *J. Phys. Chem.* **2000**, *104*, 9966-9972.
63. Möbius, D.; Cordroch, W.; Loschek, R.; Chi, L. F.; Dhathathreyan, A.; Vogel, V. *Thin Solid Films* **1989**, *178*, 53-60.
64. Petrov, J. G.; Möbius, D. *Langmuir* **1993**, *9*, 756-759.
65. Constantino, C. J. L.; Dhanabalan, A.; Oliveira Jr., O. N. *Colloids & Surfaces A* **2002**, *198-200*, 101-105.
66. Gust, D.; Moore, T. A. *Science* **1989**, *244*, 35-41.

67. Gust, D.; Moore, T. A.; Moore, A. L. *Acc. Chem. Res.* **1993**, *26*, 198.
68. Cunningham, J.; Polymeropoulos, E. E.; Möbius, D.; Baer, F. Photoinitiated Electron Transfer Between Donors and Acceptors in Monomolecular Layer Assemblies - EPR and Optical Studies. In *Magnetic Resonance in Colloid and Interface Science*; Fraissard, J. P., Resing, H. A., Eds.; D. Reidel Publ. Comp.: Dordrecht, Netherlands, 1980; pp 603-608.
69. Osuka, A.; Maruyama, K.; Mataga, N.; Asahi, T.; Yamazaki, I.; Tamai, N. *J. Am. Chem. Soc.* **1990**, *112*, 4958.
70. Yamazaki, T.; Yamazaki, I.; Osuka, A. *J. Phys. Chem. B* **1998**, *102*, 7858-7865.
71. Sun, L.; Gu, C.; Wen, K.; Chao, X.; Li, T.; Hu, G.; Sun, J. *Thin Solid Films* **1992**, *210/211*, 486-488.
72. Gu, C.; Sun, L.; Zhang, T.; Li, T. *Thin Solid Films* **1996**, *284-285*, 863-865.
73. Smith, V. C.; Batty, S. V.; Richardson, T.; Foster, K. A.; Johnstone, R. A. W.; Sobral, A.J.F.N.; Rocha Gonsalves, A. M. d. A. *Thin Solid Films* **1996**, *284-285*, 911-914.
74. Richardson, T.; Smith, V. C.; Johnstone, R. A. W.; Sobral, A. J. F. N.; Rocha Gonsalves, A. M. d. A. *Thin Solid Films* **1998**, *327-329*, 315-320.

CHAPTER 5

ENZYMATIC REACTIONS AT INTERFACES

Marie-Hélène Ropers, Gerald Brezesinski and Helmuth Möhwald

Max Planck Institute of Colloids and Interfaces, Am Mühlenberg 1, D-14476 Golm, Germany

1. INTRODUCTION

Lipase is the generic term to denote esterases which can hydrolyse triglycerides. They also include phospholipases which can attack phospholipids, constituting the cell membranes. The interest in such enzymes lies not only in their role in the key metabolic events but also in their flexibility. They can be produced from a lot of systems (animals, plants or microorganisms), act on a wide range of substrates and exhibit a good stability in non aqueous organic solvents. Therefore they were considered in several industrial applications as detergent enzymes [1, 2]. Most of the lipases also exhibit the peculiar behaviour, called interfacial activation, to expose their catalytically active site only in the presence of a lipid phase so that the analysis of lipase catalysis is enhanced in organized systems. A variety of systems were considered such as the liquid/liquid interface with vesicles and emulsions or the air/liquid interface with monolayers. The monolayers spread at the air/water interface and eventually later deposited on solid substrates have proven their applicability and interest in studying enzymatic catalysis. They allow working with a controlled interface (orientation and conformation of the film-forming molecules, molecular and charge densities, surface viscosity). In the present work, our aim is to focus on those model systems and to show the recent developments (the 5 to 7 last years) as well as the main results. In a first part, we describe the methods allowing description of monolayer properties in an enzymatic reaction. Results extracted from these methods are presented in the second part in order to show general trends of enzymatic reactions at interfaces.

2. METHODS

Over the past, several methods have been developed to characterize Langmuir monolayers at the air/liquid interface. We shortly describe the most common ones to analyze enzyme/membrane interactions and we recommend a special review as well as the references therein [3].

a) Surface pressure–area isotherms

Pressure-area isotherms are actually measured using a Wilhelmy film balance with either a Platinum or filter paper plate. Depending on the chain length of the molecules used the monolayer can be fully condensed at low temperatures. Increasing temperature leads to the appearance of a liquid-expanded (LE) phase and to the typical first-order transition from LE to a condensed phase (LC). The transition pressure (π_C) increases with increasing temperature and the area change (ΔA) involved in the transition decreases. The increase of the chain length has the same effect as a decrease in temperature. From the area change ΔA it is possible to derive information about the transition enthalpy ΔH using a modified Clausius-Clapeyron equation [4]. The enthalpy (ΔH) and the pressure (π_C) of the main-transition are linear functions over a wide range of temperature. The linear extrapolation of the π_C values towards $\pi_C = 0$ gives the onset temperature T_0 of the main-transition, the extrapolation towards $\Delta H = 0$ gives the critical temperature T_C. Above this temperature, a condensed phase cannot be achieved by compression of the monolayer.

The change in surface pressure as a function of the area (surface pressure–area isotherms) is the first and basic method to characterize the interaction of an enzyme with the spread monolayer. There are several ways to achieve this. Either the monolayer is spread on the buffer containing the enzyme, or the enzyme is introduced after spreading the monolayer. The first method has the advantage to spread the monolayer on a homogeneous solution. In the latter case, the enzyme may be introduced through the film at several points or into the buffer behind the barrier. The drawback of this method is to break the film and to have uncertainties on the distribution of the enzyme along the monolayer. It enables, however, a better control of the ratio enzyme/spread material. The analysis of the enzyme/membrane interaction is usually performed at a constant area in order to measure the changes of surface pressure.

The necessity to analyze the hydrolytic action of lipases in heterogeneous media led to the development of the zero-order trough which allows to work with a controlled interface and without any added emulsifier. This development, owing due to Verger et al. [5] enables to determine zero-order kinetic constants. The trough is composed of a substrate reservoir and a reaction compartment containing the enzyme solution. Both compartments are covered with the same monomolecular lipid film and connected to each other by a narrow channel in etched glass which prevents the exchange of the two subphases but enables the lipid film to slide on it.

b) Brewster angle BAM and Fluorescence Microscopy

These techniques provide microscopic morphological features of the monolayer, i.e, condensed domain formation as well as transition between different monolayer phases by using the modification of the light reflection conditions due to the monolayer characteristics. In the case of fluorescence microscopy, a fluorescent probe is added to the lipid monolayer (usually 1 mol-%) or the enzyme is labelled in a proper way. Excitation of the fluorescence probe can be achieved using a high pressure mercury lamp or a laser. Discrimination of excitation and emission was regulated by using conventional dichroic mirrors and standard filters. For details see Ref. [6].

In the case of BAM, p-polarized laser light strikes the air/water interface at the exact angle of 53.1° (Brewster angle). If the water surface is bare no light is reflected. The spreading of a monolayer with a refractive index different from that of water or air and a defined thickness d modifies the optical conditions of reflection in a way that the refracted light at the air/monolayer interface and reflected one at the water/monolayer interface interfere in a constructive manner with the incident reflective light, and the coherent superposition gives a more or less strong signal that images the morphology of the monolayer according to its features. Additionally the orientation of molecules gives an anisotropic contribution that modulates the signals and which can be analyzed with an analyzer in front of the CCD camera. BAM images are distorted owing to the observation at the Brewster angle. An image-processing software can be used to correct the digitized images for this distortion. The lateral resolution is in a range of 2 μm. For more detailed information see [7, 8, 9].

c) Infrared Spectroscopy IRRAS and PM-IRRAS

Spectroscopic methods give insight into interactions between the enzyme and the monolayer at the molecular level. Infrared reflection-absorption spectroscopy (IRRAS) provides information about the lipid conformation as well as the protein secondary structure and orientation [10, 11, 12, 13]. IRRAS spectra were recorded using a Bruker IFS66 spectrometer (Bruker, Karlsruhe, Germany) equipped with an MCT detector. The setup consists of an IR-source, a Michelson interferometer and an external reflection unit. The external reflection accessory is enclosed in a container, which maintains a saturated water vapor atmosphere above the sample. The IR beam can be directed onto the water surface at different angles of incidence. The infrared radiation intensity was modulated by the interferometer and polarized with a ZnSe polarizer. The spectral region between 1400 and 1800 cm^{-1} is very important for the investigation of biomolecules. It contains several conformation sensitive bands. However, strong water vapor rotation-vibration bands overlap the IRRAS spectrum in that region. This problem can be solved by using either a shuttle system or the polarization modulation technique. In the first case, the Langmuir trough used consists of two compartments. One compartment contains the surface covered with the monolayer, the other one the bare surface. A motorized shuttle system provides the switch between these two compartments [14, 15]. The second possibility is the use of a photoelastic modulator (PM) [16, 17, 18]. To avoid mirror induced polarization artifacts it is placed between the polarizer and the surface of the trough. The ZnSe photoelastic modulator modulates the IR beam between polarization in the plane of incidence (p) and polarization perpendicular to this plane (s) with a fixed frequency of 73 kHz. From the detected intensity (using electronic filtering and demodulation) the two signals, $(R_p\text{-}R_s)$ and (R_p+R_s), were obtained. Only anisotropic absorptions contribute to the PM-IRRAS signal $S \sim (R_p-R_s)/(R_p+R_s)$. Normalized signals were obtained using the following expression: $\Delta S= (S_d - S_o)/ S_o$, where S_d and S_o are the signals in the presence and absence of a monolayer, respectively. The angle of incidence of the infrared beam with respect to the surface normal was 74°. Spectra were recorded with a spectral resolution of 4 cm^{-1} and collected using 200-400 scans during 5 - 10 min.

For hydrolysis experiments, the monolayer was compressed to a lateral pressure of 40 mN/m and a first control spectrum was measured. At this pressure, the enzyme solution was injected and carefully stirred underneath the monolayer. This surface pressure was used because our

PM-IRRAS experiments have shown that no hydrolysis occurs during 24 hours. Then the film was expanded to a chosen pressure at which the kinetics of the hydrolysis process was investigated by PM-IRRAS. The pressure was kept constant during the reaction by an automatically moving barrier. After finishing the hydrolysis process the monolayer was again compressed to 40 mN/m and a final control spectrum was measured.

d) Grazing Incidence X-ray Diffraction

The organization of molecules in a periodic two-dimensional array in a monolayer at the air/water interface gives rise to X-ray diffraction. Very intense X-ray radiation is needed to study such monomolecular layers. Synchrotron X-ray sources provide very intense, well-collimated beams. The first scattering experiments have been performed at the end of the eighties in the last century. Grazing incidence X-ray diffraction (GIXD) measurements described here were performed using the liquid surface diffractometer (beamline BW1) at HASYLAB, DESY, Hamburg, Germany. The Synchrotron beam was made monochromatic by a Beryllium (002) crystal. The angle of incidence was $0.85*\alpha_C$ (α_C is the critical angle for total external reflection). The Langmuir trough is placed in a Helium-flashed container with windows transparent for the X-rays. The X-ray beam illuminates a monolayer area of approximately 2x50 mm^2. A linear position sensitive detector (PSD) (OED-100-M, Braun, Garching, Germany) with a vertical acceptance $0 < Q_Z < 1.27$ Å^{-1} was used for recording the diffracted intensity as a function of both the vertical (Q_z) and the horizontal (Q_{xy}) scattering vector components. The horizontal resolution of 0.008 Å^{-1} was determined by a Soller collimator mounted in front of the PSD. The Bragg peaks can be indexed by Miller indices h and k. The analysis of the in-plane diffraction data yields lattice spacings according to $d_{hk} = 2\pi / Q_{xy}^{hk}$. From the in-plane and out-of-plane peak positions one can derive information about the tilt angle and the tilt direction. For more information about the principles of GIXD for the study of two-dimensional condensed films at the air-liquid interface, we kindly refer to detailed reviews published recently [19, 20, 21, 22].

e) X-ray and Neutron reflectivity

Specular reflection provides information about the vertical structure of monomolecular films (thickness of the lipid monolayer and of the adsorbed enzyme layer) or interfaces even in lack of positional order in the monolayer. The X-ray reflectivity $R(Q_z)$ can be measured by a NaCl scintillation detector. The normalized reflectivity $R(Q_z)/R_F(Q_z)$ with $R_F(Q_z)$ as the Fresnel reflectivity, calculated for a perfectly sharp interface between air and pure subphase, can usually be inverted into a laterally averaged electron density $\rho(z)$ as a function of the z coordinate. This can be performed either by a model independent method or using a layer model of the interface [22, 23, 24].

In the case of neutron scattering, the interaction is nuclear and the scattering length is dependent on the nature of the nuclei of atoms. Contrast variation techniques can be applied by using isotopic substitution (e.g., hydrogen and deuterium).

f) Ellipsometry

This method does not need highly sophisticated Synchrotron sources and allows to extract information on the monolayer thickness by using appropriate models. An ellipsometric measurement consists in determining the change in polarisation of a polarised light reflected from the interfaces separating the different media air/film/bulk. The time resolution is usually better than 20 ms and the beam area is about 1 mm^2. The phase modulated beam hits the water surface at the Brewster angle air/water of $\Phi_B = 53.1°$. The imaginary part of the measured reflectivity $Im = r_p/r_s$, where r_p and r_s are the amplitudes of p- and s-polarized waves, equals the ellipticity $\bar{\rho}$, if the real part Re of the reflectivity coefficient is zero. The difference between the pure water and the film covered surface is $\Delta\bar{\rho}$ [25, 26]. If the adsorbed films are thicker than about 100 Å, where $Re(r) \neq 0$, the film thickness and the refractive index can be calculated with the complete Fresnel equations [27]. The ellipticity of a multilayer system consisting of a lipid monolayer and a thin adsorption layer can be taken as the sum of the ellipticities of these two single layers.

3. RESULTS AND DISCUSSION

3.1. Polymorphism of phospholipid monolayers

The first step to understand interfacial reactions catalyzed by enzymes is the characterization of the phase state of the substrates. Since we will describe mainly results from monolayer experiments, the first part deals with structures of phospholipid monolayers at the air/liquid interface. The results were obtained using grazing incidence X-ray diffractions.

Phospholipids as major building blocks of biological membranes have been continuously studied over the last decades using a variety of approaches. From a physico-chemical point of view, their ability to form various structured mesophases such as multilayers, non-lamellar phases or monolayers at interfaces is of particular interest.

Monolayers at the air/water interface exhibit a rich polymorphism that can be easily examined by numerous techniques. Pressure-area isotherm measurements provide a straightforward physicochemical characterization since molecular area and lateral pressure are directly accessible. At certain temperatures typical lipids reveal a gas-analogous phase at very large molecular areas and with compression liquid-expanded (LE) and condensed (LC) phases can be observed. The first order phase transition between LE and LC is connected with a plateau region in the pressure-area isotherms, while kinks indicate higher order transitions between condensed phases. Using GIXD the structures of condensed phases can be characterized with Å-resolution. The monolayers are powders within the horizontal plane. The diffraction pattern is averaged over all domain orientations in the monolayer plane. Therefore, the in-plane components Q_x and Q_y of the scattering vector cannot be determined separately. Using phospholipid monolayers, the first-order peaks are the only observed ones since the peak intensities decay rapidly with increasing momentum transfer due to lattice fluctuations. In the simplest model, a domain of the monolayer is treated as a two-dimensional crystal consisting of uniformly oriented rigid rod-like molecules. The different condensed phases can be distinguished by the tilt, the tilt direction and the lattice distortion [28]. Up till now seven condensed phases have been observed. At high pressure and low temperature the two phases CS (2D crystal) and S (periodic in one direction) exhibiting a centred rectangular packing of non-tilted molecules can be observed. Increasing temperature leads to the LS phase that exhibits only one diffraction peak indicating hexagonal packing of non-tilted molecules. At lower pressures several tilted phases can be observed: L_2'', L_2', L_2 and Ov.

As expected, it was shown that these structural parameters depend systematically on the chemical structure of the amphiphilic molecules [29, 30, 31, 32, 33, 34]. Therefore, chemically modified lipids have been prepared with variations in the head group and chain regions. Branched-chain lipids are of particular interest in such investigations not only because they are found in nature, but also because they allow systematic studies of general features determining the structure in model membranes. Phospholipids with two hydrophobic chains (saturated or unsaturated, carbon number ranging usually between 12 and 20) attached to the glycerol backbone and a hydrophilic head group (zwitterionic or ionic) form stable monolayers at the air/water interface in a wide range of temperature.

Comparing the characteristic temperatures found in monolayers (T_0, T_C) and bilayers (T_m) shows that phospholipids differing only in the chain length can be directly compared by shifting the temperature. Using this approach allows the construction of generic phase diagrams for different phospholipids [35]. Fig. 1 displays such a generic phase diagram for triple-chain glycero-3-phosphocholines (rac. 1-*O*-hexadecyl-2-(2-tetradecylhexadecanoyl)-glycero-3-phosphocholine (1H-2(2C_{14}-16:0)-PC) and rac. 1-*O*-hexadecyl-2-(2-hexadecyloctadecanoyl)-glycero-3-phosphocholine (1H-2(2C_{16}-18:0)-PC)).

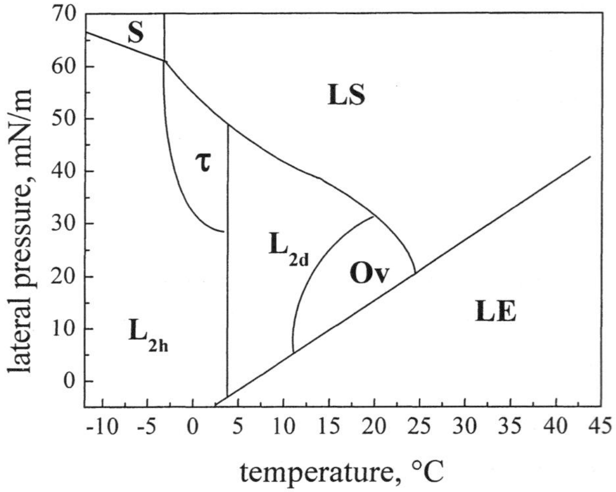

Fig. 1. Generic phase diagram of glycerol-3-phosphocholines.

In a small range of temperature and pressure, the NNN-tilted Ov-phase has been observed. This phase is characterized by an intensity distribution with two diffraction peaks at non-zero Q_z-values with $Q_z^n = 2 \cdot Q_z^d$. The point of intersection between the tilting transition and the main phase transition was taken from the $\Delta H = f(T)$ plots where a change of slope was found. This slope change can now be interpreted as a change in the phase sequence (LE to L_{2d} or LE to LS) leading to changes in the transition enthalpy derived from the isotherm measurements. At lower lateral pressures and lower temperatures, the NN tilted rectangular phase L_2 is observed. For symmetry reasons, this phase was subdivided into two phases L_{2d} and L_{2h}, possessing disordered and herringbone-ordered backbone planes, respectively. Plotting the signed unit cell distortion $d = ((l_1^2 - l_2^2)/(l_1^2 + l_2^2)\cdot\cos 2(\beta - \omega)$, where l_1 and l_2 are the major and minor axes of the ellipse passing through all six nearest neighbours of a hydrocarbon chain, and β and ω are the tilt and distortion azimuths, respectively, as a function of $\sin^2(t)$, a linear relation is expected according to the Landau theory prediction. The extrapolation to zero tilt yields a d_0 value, which gives information about the different contributions to the distortion (e.g. chain tilt, backbone ordering). The plot of d_0 versus effective temperature yields the transition temperature between L_{2h} and L_{2d}. Starting from the L_{2h} phase (NN tilt and NNN distortion) at low temperature, increasing pressure should lead to the S phase. This transition occurs at high pressure and was therefore not observed experimentally. However, since the condensed isotherms exhibit linear relations between pressure and molecular area: $A_{xy} = K_1 - K_2\pi$, where $A_{xy} = A_0/\cos(t)$ with A_0 as cross-sectional area of the chains and t as chain tilt angle, plotting of $1/\cos(t)$ as a function of π allows the determination of the tilting transition pressure π_t by extrapolation towards $1/\cos(t) = 1$. At low pressures and higher temperatures, the L_{2h} phase exhibits NN distortion. In the case of single-chain amphiphiles, the pressure increase usually leads to a change in distortion from NN to NNN passing through the undistorted state at one defined pressure. In contrast, an unusual behaviour was found for 1H-2(2C_{16}-18:0)-PC: increasing pressure leads to an undistorted in-plane lattice that remains undistorted on further pressure increase despite of a change in tilt angle. Such a phase was named τ.

In the case of methyl-branched phosphatidylcholines, no distinct Bragg peaks are observed but the diffracted intensity is distributed along a curve [36]. It is not possible to describe such a scattering profile as a superposition of only three peaks resulting from an oblique lattice. Therefore, such diffraction patterns are assigned to be a superposition of diffraction peaks from several lattices within the monolayer. It was shown that the distributed intensity results from a variation of the tilt azimuth [37]. The tilt azimuth is the angle between the projection of the tilted chain on the surface plane and the lattice axis. A possible variation of the tilt angle could be excluded. Two rectangular lattices (NN and NNN tilt) and several oblique lattices with the same tilt angle but different tilt azimuths contribute to the observed scattering profile. The coherence length derived from the FWHM of the Bragg peaks shows that the methyl branches disturb the packing and increase the number of defects.

3.2. Interfacial Reaction

3.2.1. The binding step

Due to the location of the active site of an enzyme in its three-dimensional structure, the presentation of the substrate to the enzyme constitutes itself an important step in the hydrolysis process. Consequently, the first step of the reaction analysis is to understand the interaction of the enzyme with the monolayer and its effects on its properties. This is usually combined with the understanding of the interfacial properties of the enzyme itself.

It is generally accepted that the transport of proteins to the interface is entirely governed by diffusion [38]. Monitoring the process of diffusion of proteins to the air/water interface as a function of time reveals its surface activity and/or its eventually synergetic interaction with the membrane. Such experiments can be performed at several pressures by compressing the monolayer before enzyme injection. This procedure allows to determine the limiting pressure of enzyme incorporation.

Several factors influence the adsorption of an enzyme at the interface and they are described for esterases and oxidases as representative examples.

3.2.1.1. Adsorption of esterases

Esterases are enzymes hydrolyzing ester bonds of amphiphilic molecules. They are termed triacylglycerol hydrolases or most commonly lipases when their substrates are triacylglycerides and are termed phospholipases when they attack phospholipids [1]. All investigated lipases exhibit structural and functional similarities. The crystallographic data reveal the presence of an amphiphilic peptide loop covering the active site of the enzyme in the inactive state, like a lid [39, 40]. This lid could undergo a conformational rearrangement, rendering the active site accessible to the substrate. Even if some contradictory works exist on the reality of this mechanism since some lipases, carrying a lid, were discovered to catalyze efficiently distant from an interface, it received confirmation by X-ray crystallography with the determination of structures of lipases covalently complexed to hydrophobic inhibitors [41, 42]. After the lid has opened, a large hydrophobic surface is created to which the hydrophobic part of the substrate binds.

The enzyme can itself adsorb at the air/water interface and/or undergo attractive interactions with the monolayer. The main factors influencing the adsorption are the concentration of the enzyme and the presence of a monolayer, as the recent example of butylcholinesterase shows [43]. This enzyme adsorbs faster as its concentration increases in solution and if the air/water interface is already covered by a monolayer of octadecylamine, even if the latter has the same charge as the enzyme [43]. The composition of the buffer plays also a role. Salt (KCl) enhances the spreading of the organophosphorous hydrolase at the air/water interface [44]. Usually the electrostatic interactions are the major forces involved in the interactions as observed by studying oppositely charged membranes on an enzyme solution and recording the lateral pressure [43, 45]. They usually correspond to a positively/negatively charged pair [44, 45] but some examples offer the apparent contradiction of positively/positively charged interactions as the positively charged didocecyldimethylammonium bromide and the amphoteric surfactant sodium N-(2-hydroxydodecyl) sarcosinate, which has a net positive charge at pH 3 as well as the enzyme *Rhizomucor miehei* lipase [46]. PM-IRRAS experiments coupled to surface potential and surface dipole moment measurements prove that acetylcholinesterase adsorbs at the air/water interface and reorients as the monolayer is compressed [47, 48, 49]. This adsorption has the particularity to be perfectly reversible [47, 49]. In presence of DPPA or DPPC monolayers, PM-IRRAS spectra demonstrate that acetylcholinesterase influences more strongly DPPA than DPPC but no changes occur in the acetylcholinesterase secondary structure, which adopts the same orientation as without the monolayer. However, in presence of its substrate, dissolved in the subphase, the acetylcholinesterase changes its conformation to adapt to the substrate for intramolecular interactions [47]. The determination of the number of enzyme molecules bound to the monolayer is not easily reachable. It must be pointed out that the quantification of the enzyme binding was possible trough a new method involving surface-immobilized phospholipid vesicles and surface plasmon resonance [50]. The binding constant of PLA_2 to partially biotinylated DOPC/DOPE vesicles could be precisely determined with this method (1 PLA_2 per 74 ±16 outer leaflet phospholipids).

Upon compression the adsorbed enzyme layer passes through several states where the mixed film undergoes reorganization of the surfactant chains and enzyme molecules. A further increase in the lateral pressure squeezes the enzyme out. The squeezed out enzyme can also drag some surfactant molecules into the solution (Fig. 2). As a consequence, the isotherms asymptotically approach that of the pure phospholipid as the film is compressed. This description is valid in many cases and for example with the butyrylcholinesterase adsorbed on octadecylamine [43].

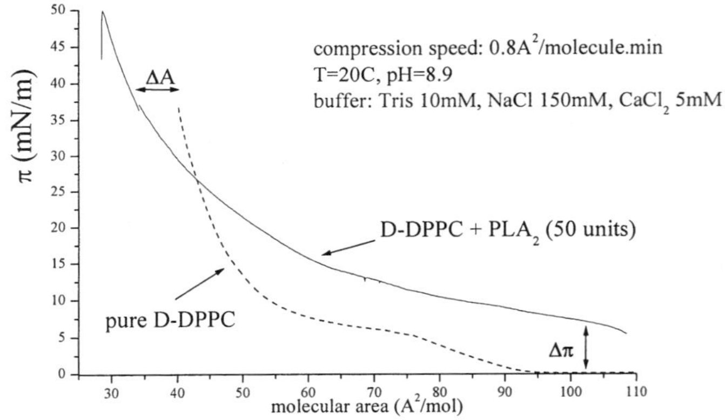

Fig. 2. DPPC isotherms in the presence and absence of PLA_2 in the buffer at 20 °C.

As soon as the enzyme is adsorbed, the layers can be characterized. To investigate the adsorption of the *Humicola lanuginosa* lipase to sodium dodecylsulfate SDS or tetradecyltrimethylammonium bromide TTAB at the air/water interface or on a solid wafer, neutron reflectivity and ellipsometry measurements were carried out. Below the CMC, the addition of TTAB increases the lipase fraction in the adsorbed layer, i.e., its packing, due to a good neutralization of charges between the cationic surfactant and the negatively charged lipase. The adsorbed layer may become very thick with a low packing density compared to the situation at a concentration close to the cmc where the adsorption layer disappears. The adsorption is limited for the anionic SDS and the adsorption layer is more diluted and extended [45, 51].

In the particular case where the enzyme is surface active and not its substrate (for example the system organophosphorous hydrolase OPH/organophosphorous compound), it is the substrate itself that modifies the interfacial properties of the enzyme at the air/water interface by strongly decreasing the collapse pressure as well as the area occupied per molecule [44]. If the enzyme is alone at the air/water interface, the change in slope of the π-A isotherms coupled to the surface potential measurements indicates three states of enzyme ordering. The coupling substrate/enzyme reduces the changes in the π-A isotherms and the substrate is likely to introduce a disorder in the protein ordering.

When the substrate is partially soluble, Tanaka et al. introduced a method to analyze solely the interfacial (air/water) activation of a lipase and not the activation in water [52]. A monolayer of a lipid, not substrate of the enzyme, is formed at the air/water interface on a trough containing two separate subphases. The lipase is introduced into the subphase under this monolayer and can adsorb to the monolayer. After complete adsorption, the monolayer is gently shifted to the other compartment in order to dispose the enzyme only at the air/water interface. Then the substrate is introduced into the subphase of this second compartment and a circulating flow allows to analyze the hydrolysis products through a fluorimetric cell by fluorescence spectrometry. Hydrolysis products (umbelliferone and stearate) are not surface active and the head group of the substrate (umbelliferone) is fluorescent only after cleavage from the stearate group.

When the monolayer is a substrate of the enzyme, only in special cases it is possible to analyze the suspected adsorption of the enzyme to the monolayer. Phospholipase A_2 is a small calcium-dependent enzyme which hydrolyzes stereo-selectively the *sn*-2 ester linkage of phospholipids leading to a fatty acid and a lysophospholipid. PLA_2 is an interfacially active enzyme that has a 10^4-fold higher activity for substrate aggregates, such as micelles, vesicles, or membranes, as compared with substrate in a monomeric dispersion [53, 54]. PLA_2 plays an important role in the arachidonyl cascade, since arachidonyl is the main group found in the 2-position at the glycerol of phospholipids in mammal cells. The first step of the reaction is the adsorption of enzyme at the aggregate interface, followed by the catalytic action. Both steps are dependent on the "quality" of the aggregate, defined by important parameters such as composition and phase structure. Since PLA_2 exhibits a stereospecific behaviour, D-enantiomers of phospholipids are known to be recognized but not cleaved [55]. The monolayer behaviour of L- and D-enantiomers is exactly the same. In the first-order phase transition region between LE and LC, seen as a plateau region in the isotherm, condensed phase domains of a characteristic kidney shape can be observed. The PLA_2 adsorption to such a monolayer leads to a pressure increase at large molecular areas indicating the penetration of the enzyme (Fig. 2).

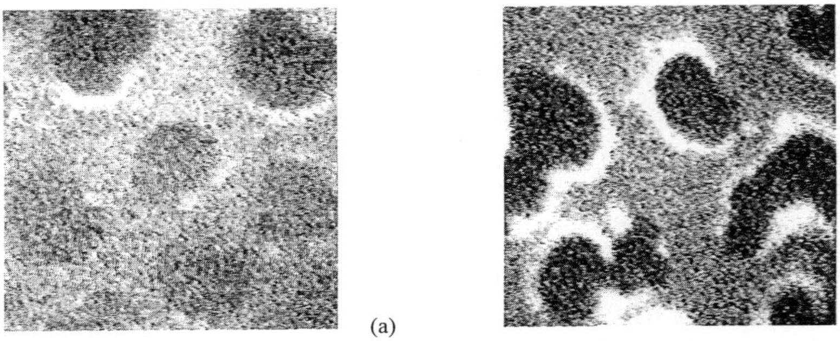

Fig. 3. Fluorescence microscopy images of the distribution of labelled-PLA_2 enzyme injected beneath a D-DPPC monolayer compressed to the LE/LC transition. Images were taken 10 min (a) and 30 min (b) after enzyme injection [56].

On compression, smaller molecular areas in the high pressure region indicate the squeezing-out of the protein. If one uses no fluorescence label in the phospholipid monolayer but a fluorescein- labelled enzyme injected into the subphase at a lateral pressure corresponding to the LE/LC phase transition, after a certain adsorption time the same type of domains becomes visible (Fig. 3) [56]. Fluorescence intensity gives a qualitative measure of the local enzyme concentrations. The experiments have shown that the enzyme adsorbs preferentially at the phase boundaries and this is also the location where the reaction is first detectable [56, 57]. On the buffer used for high PLA_2 activity, D-DPPC exhibits three distinct diffraction peaks indicating an oblique lattice structure with only translational symmetry.

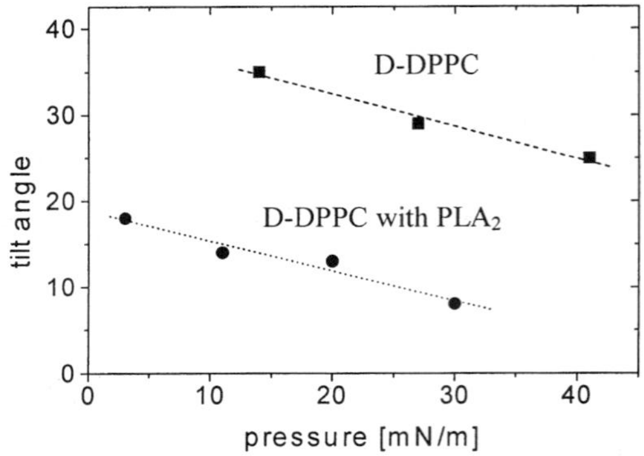

Fig. 4. Modification of the DPPC chain tilt angle in monolayers due to the adsorption of PLA_2 at the air/water interface [56].

Enzyme adsorption leads to an orthorhombic unit cell (two diffraction peaks) with a smaller tilt angle of the aliphatic chains (Fig. 4). The specific interaction of the enzyme with the phospholipid changes obviously the head group structure (conformation, hydration) in such a way that allows a tighter lipid packing.

3.2.1.2. Adsorption of oxidases and other enzymes

Lipases have received a great attention for their key roles in the metabolism of triglycerides, but also other enzymes have attracted attention, e.g. oxidases. The behaviour of xanthine oxidase XO adsorbed on shingomyelin or sphingolipid is one example [58]. The expansion of the compressed film does not reach the zero value, suggesting an incomplete repulsion of the enzyme from the air/water interface [58]. For the sphingomyelin monolayer, the pressure reached after 3 hours of adsorption is less than that for the monolayer-free XO film. Two explanations are possible: either the monolayer viscosity increases with the formation of intra- and intermolecular disulfide bridges in the adsorbed layer or the lipid/protein complex facilitates the solubilization of the lipid in the subphase. The larger increase in surface pressure for sphingomyelin compared with sphingolipid and the higher pressure at which mixed systems and pure sphingomyelin isotherms coincide suggest a stronger interaction of XO with sphingomyelin [58].

Interestingly the enzyme/membrane interaction of the β-galactosidase has been observed in two kinds of model membranes: vesicles and monolayers at the air/water interface. The monolayer technique proves the higher tendency of the enzyme to be localized at the lipid/water interface than at the free air/water interface with a cut-off pressure of 30mN/m [59]. The detection of the adsorbed glucose oxidase GOx (from *Aspergillus niger*) at the air/water interface was achieved using the resonant enhancement of the π- π^* transition of the flavin adenine dinucleotide (FAD) chromophores, naturally present in the structure of GOx. The change in depolarization of the second harmonic signal indicates either a reorientation or an internal rearrangement of the protein at the interface [60]. At pH 7, GOx is negatively charged and exhibits an enhanced adsorption to the interface in presence of a positively charged surfactant such as CTAB and is repelled from it with a negatively charged surfactant (arachidic acid) [60]. Dai et al. have analyzed the folding behavior of GOx at the air/water interface with Circular Dichroism (CD) measurements. They were conducted for samples deposited on quartz plates at different surface pressures. The content in α-helix is higher in Langmuir-Blodgett films than in solution and increases with the surface pressure, suggesting a higher enzyme activity [61]. The study of Gidalevitz et al. indicates that GOx as well as urease and alcohol dehydrogenase denature during adsorption to the air/water interface according to X-ray reflectivity measurements [62]. The spreading of a lipid (dibehenoyl-PC) does not prevent proteins from unfolding while a covalent cross-linking with glutaraldehyde is quite efficient. GOx penetrates into the lipid monolayer film by creating holes accompanied with an increase of surface pressure [62]. Previously, Rosilio et al. had shown that GOx goes to the air/water interface by penetrating the lipid bilayer, even using concentrations at which no changes in the surface pressure were recorded at the monolayer-free surface [63]. The change in pressure, i.e. the penetration into the lipid bilayer, had been interpreted to be the result of hydrophobic interactions between the hydrophobic chains of the enzyme and the lipid [64]. The adsorption of GOx at the interface can have a stabilizing effect on a behenic acid film and the stability increases with the concentration of GOx in the subphase. 28mN/m is the limiting pressure of GOx incorporation. Below 28mN/m, GOx is present both at the air/water interface and underneath the behenic acid film [65]. The kind of interaction is unclear since at pH 5.6 behenic acid is half negatively charged and GOx molecules are globally negatively charged. However, GOx is a polyampholyte and positive charges exist in the molecule too [65]. It is interesting to note that the kinetic profile of adsorption exhibits three distinct steps attributed to a diffusion-penetration-rearrangement process and that the first step disappears at high enzyme solution concentrations [63]. Even after compression of the mixed film (GOx/behenic acid), the adsorption of the GOx to the behenic acid monolayer continues as shown by lateral force microscopy and spectrophotometric measurements [66]. The penetration of GOx into monolayers of a series of glycolipids with different head groups (glucoside or maltoside) and different hydrocarbon chain lengths is enhanced in the case of glycolipids with $n = 15$ and $n = 17$ compared to glycolipids with shorter chains ($n = 11$ or 13) [64]. The idea that a compact packing prevents enzyme penetration has received here a contradictory example. Indeed, dialkyl-glycerylether-β-D glucosides or maltosides with $n = 15$ and 17 exhibit fully condensed isotherms whereas for those with $n = 11$ and 13 a liquid-expanded to liquid-condensed transition was observed. The facility of GOx to penetrate in 1 hour into the monolayer follows the order $C_{17}>C_{15}>C_{13}>C_{11}$ irrespective of the head group and stopped above an initial pressure of 10mN/m [64].

Another example of interfacial activation is the system hydroxynitrile lyase/mandelonitrile with the enzyme at the oil/water interface and the substrate dissolved in the oil phase [57]. Both the substrate and the product are soluble in the organic (diisopropylether) phase. Although the enzyme is interface active, it does not strongly affect the interfacial tension at which the activation reaction was performed. Nevertheless, the higher the enzyme concentration the stronger the influence on the surface pressure at the oil/water interface, as measurements with the pendent-drop technique showed [67].

In some cases, the enzyme does not exist in the native state but is anchored to a lipid (glycosylphosphatidylinositol anchored alkalinephosphatase AP-GPI) [68] or to a protein as the recombinant phosphatidylinositol-3-kinase fused to the glutathion transferase, GST-PI3K [69]. The interactions with the monolayer are thus depending on the properties of both the enzyme and the anchoring part. In the case of AP-GPI, the enzyme obviously adsorbs at the air/water interface due to the surface active glycosylphosphatidylinositol and keeps its active sites turned outside the membrane. The adsorption of this enzyme to the monolayer is stronger in the case of DPPS than DPPC, and PM-IRRAS indicates a strong enzyme-induced disorder in the DPPS chain organization in contrast to DPPC where only the polar groups are disturbed [68]. In vesicles, GST-PI3K binds not only to its substrate (phosphoinositols) but also to other anionic phospholipids such as PS and PA, whereas the binding to neutral phospholipids is very weak. This could be performed through its N-terminal binding site located away from the kinase domain. The anchorage to the membrane seems to be a pre-requisite for the phosphorylation process. Besides the nature of the head group, the fatty acid composition of phospholipids influences also the extent of binding to PI3K. The binding to anionic phospholipids containing unsaturated fatty acid is significantly higher than to phospholipids composed of saturated fatty acids, suggesting a penetration into the phospholipid layer. Therefore, the interactions of this enzyme with an unsaturated phosphatidic acid, β-arachidonoyl γ-stearoyl PA, have been analyzed at the air/water interface [70]. First results indicate that the enzyme adsorbs to the air/water interface only when it is covered by a monolayer. The interaction with the enzyme (binding with charged head groups and penetration into the monolayer) is a long-term process lasting several hours to reach a plateau at 12mN/m. At 30mN/m, reflectivity curves suggest a squeezing-out of the enzyme into the subphase or at least into a sublayer underneath the lipid monolayer.

3.2.2. The catalyzed reactions

The mechanism of enzymatic lipolysis depends strongly on the mode of organization of the lipid substrate in interfacial structures such as monolayers, micelles, liposomal dispersions or oil-in-water emulsions. However, some common kinetic features exist reviewed by Panaiotov et al. [71]. Recently the kinetics of hydrolysis reaction catalyzed by *Bacillus cereus* sphingomyelinase SMase in monolayers of pure SM at the air/water interface was investigated [72]. It is consistent with a mechanism involving three steps: i) an initial partition of the enzyme into the interface, ii) a subsequent step that can be operationally simulated as a bimolecular surface process depending on the interfacial concentration of enzyme and iii) a subsequent step of activation [72]. Since the interface may play such an important role in these enzymatic reactions, results obtained using monolayers and dispersed systems will be described in the following sections.

3.2.2.1. Enzymatic reactions in monolayer systems

The denaturation of enzymes at the air/water interface has already been mentioned [61, 62] and is the sign of a more or less large lose of activity. This has been quantified with the Horseradish peroxidase enzyme adsorbed at the air/water interface. The substrate, ABTS (2,2'–azinobis (3-ethylbenzthiazolinesulphonate)), leads to spectrophotometrically recordable signals. After 15 minutes of enzyme adsorption to the air/water interface, the relative activity has decreased by 10% [73]. The development of the zero-order trough was dedicated to follow and to measure the kinetics of enzymatic lipolytic activity at the air/water interface. The hydrolysis rate is followed according to the decrease of the area corresponding to the solubilization of hydrolysis products in the subphase. This method has recently received attention [74, 75, 76, 77, 78, 79]. Following the activity with the monolayer technique is rendered difficult when products have also a surface activity at the air/water interface. This is the case with the previous system DPPC/PLA_2. For sphingomyelin and dihydrosphingomyelin, the sphingomyelinase leads to ceramides occupying a significantly lower molecular area at the air/water interface so that the reaction can be followed at a constant pressure by analyzing the decrease in area [80]. The hydrolysis reaction is stereospecific, i.e. the hydrolysis rate is higher for the D-erythro compound than for the racemic mixture [81]. Except the molecular analysis with IRRAS which allows to differentiate signals from substrates and products, quantitative measurements are difficult. This could however be solved by complexing the hydrolysis products (palmitic acid and L-lysophosphatidylcholine or oleic acid and monolein) with β-cyclodextrin, which moreover does not act on the substrate [76, 82]. Furthermore, the efficiency of β-cyclodextrin is higher in alkaline conditions [82]. This problem is also easily overpassed when products and/or substrates are spectrophotometrically active in a different range such as organophosphorous compounds. The monitoring of the hydrolysis reaction of Paraoxon organophosphates by organophosphorous hydrolase OPH was performed by an original method based on UV-visible spectroscopy. The reaction is followed *in situ* by monitoring the absorbance through the subphase and a quartz window inserted in the bottom of the trough. Measurements seem to indicate a faster reaction when the monolayer is compressed [44]. The kinetics of the umbelliferonstearate/DLPC/lipase system was investigated in presence of cholesterol, known for its role in the stiffness of cell membranes. The hydrolysis rate monotonically decreases with the increasing cholesterol content in the mixed lipase/DLPC monolayer. Since the lateral diffusion of a lipid probe decreases with increasing cholesterol fraction, the catalytic reactions in monolayers are diffusion-controlled [83]. This could correlate with results obtained with DPPC/phospholipase systems. The hydrolysis is stopped at high lateral pressures where lipid molecules are tighter packed [84]. On the other hand, the enzymatic reaction of the pancreatic porcine lipase is reported not to take place when the local concentration of the substrate at the air/water interface is too low, corresponding to a lateral pressure below 15 mN/m [79]. The hydrolysis rate of 1,2-dicaprin films spread at the air/water interface on a subphase containing the porcine pancreatic lipase is observed to be independent of the lipid substrate concentration (zero-order kinetic) [79]. Most of the studies analyze the enzymatic reaction with only the substrate and/or the enzyme at the interface. Arimoto et al. used a more real membrane with triglycerides as the substrate and phosphatidylcholine PC or shingomyelin SM [85]. The rate of lipolysis of the short chain triglyceride tricaprylin TC catalyzed by lipoprotein lipase LPL activation is surprisingly enhanced in presence of either PC or SM in fractions larger than 0.5 by a factor 2. The effect of PC and SM are however opposite for trimyristin TM. SM inhibits the hydrolysis of long chain triglycerides TM and tripalmitin TP whereas PC increases it at a constant rate whatever the composition. Lipid-lipid interactions, evaluated by means of the collapse pressure measurements indicate for TM a higher collapse pressure with SM than PC. It suggests that the orientation of the acyl chains of TC molecules in a monolayer is crucial for the LPL activity and the strong interaction between SM and TC retards the substrate transfer from the mixed monolayer to the catalytic pocket of LPL [85]. Similarly, Fanini et al. have shown that SMase- and PLA_2-mediated hydrolytic pathways are mutually modulated by the presence of their respective substrates and products at a lateral pressure of 15 mN/m [86]. On one hand, the activity of SMase is unaltered by the presence of DPPC or equimolar mixtures of its hydrolysis products (fatty acid and lysoPC) but is inhibited by the presence of only one of them or of the liquid-expanded DLPC. On the other hand, PLA_2 is inhibited by SM and by palmitic acid, one of the products of its own hydrolysis reaction [86]. The presence of these substrate/product mixed films increases the cut off point of both SMase and PLA_2 [87]. In the case of cholesterol oxidase ChOx, the hydrolysis rate measured with the zero-order trough indicates that monolayers containing mono-unsaturated phospholipids (POPC and N-oleoyl-D-sphingomyelin O-SPM) have a higher oxidation rate as compared with saturated phospholipids (DPPC and N-palmitoyl-D-sphingomyelin P-SPM). Furthermore, as the concentration in SM decreases, the average oxidation rate increases [88] inversely as mentioned before with the hydrolysis of triglycerides [85]. Similarly to the ChOx case, saturated sphingomyelins are more slowly converted than the monounsaturated SM. However, it must be also pointed out that saturated SMs undergo a phase transition at room temperature that influences the lag-time whereas the monounsaturated ones remain in a fluid state [89]. The difference of reactivity due to the phase behaviour was previously encountered with cholesterol oxidase [90]. The oxidation of cholesterol by cholesterol oxidase has been also visualized using monolayer fluorescence microscopy since the lipid fluorophore, tetramethylrhodamine-labelled phosphatidylethanolamine is highly soluble in a cholestenone containing monolayer as compared with a cholesterol monolayer. The cholesterol oxidase penetrates into the monolayer at random sites and initiates the reaction. At 22 °C, the mixed cholesterol/DMPC monolayers compressed to 5 mN/m exhibit cholesterol-rich domains immersed in a more expanded phospholipid phase. Then the enzyme-catalyzed reaction starts from the expanded domains and leads eventually to the dissipation of the phase boundary lines between expanded and condensed phases. This is confirmed with the labelled enzyme [90]. According to the work of Grandbois et al. [91], the PLA_2 activity depends on the phospholipid organization (increased activity in presence of defect regions) and orientation of the bilayer. The hydrolysis reaction creates small channels with a width comparable to the diameter of a single PLA_2 suggesting a nanolithographic control [91]. The evaluation of the activity can also be electrochemically performed. A monolayer on an electrode gives rise to measurable signals which disappear with the hydrolysis of the monolayer. The substrate L-α-DOPC is adsorbed on a mercury electrode and immersed in a porcine pancreatic PLA_2 solution. The rate-limiting step is the chemical step itself rather than the binding step of the enzyme from the bulk to the interface. As L-α-DOPC is converted into water-soluble hydrolysis products, PLA_2 replaces the substrate at the electrode surface without any effect of the Ca^{2+} concentration [92]. Another method developed to determine quantitatively the catalytic activity of an enzyme is performed at the solid/air interface. The β-galactosidase is adsorbed to a methylated silica surface and catalyses the conversion of a nonfluorescent fluorogenic substrate into a soluble highly fluorescent product, fluorescein [93].

The work of Cajal et al. raises the question of the validity of some previous interpretations in the literature and the use of the monolayer technique to measure kinetics [94]. In the kinetic measurements, Verger et al. have observed a lag time in the reaction progress at the air/water interface which was attributed to the penetrating power of PLA_2 [95]. Since this delay to the steady-state has not been seen with micellar and anionic bilayer interfaces [94 and references therein], authors claim it is questionable. The lag time seen during the reaction on PC bilayers is eliminated by the presence of the hydrolysis products or other anionic amphiphiles. The same measurements performed with a mixed monolayer of didecanoyl-PC and its hydrolysis products indicate that in the PLA_2-catalyzed reaction the hydrolysis products have a control on the E to E^* step. They claim that the so-called penetration power is not true and the enzyme binding to the interface depends on the magnitude of the interfacial anionic charge [94]. They also question the assumption that all hydrolysis products leave the monolayer. The partioning of the hydrolysis products of didecanoyl-PC between the bulk phase and the monolayer was ascertained by isotherms and surface potential measurements [94]. Another example concerns the caproic acid which is a water soluble compound. The complete solubilization of the short chain caproic acid, hydrolysis product of 1-palmitoyl-2-caproyl-PC, is not effective since PLA_2 was demonstrated to form domains at the interface that bind to caproic acid [96]. The decreased area as a measure of the reaction progress can be misleading and care must be taken by using the zero-order trough [94]. The use of β-cyclodextrin is one example to overpass these problems [76, 82], especially for long chain lipids [97]. Another drawback of the measured decreasing area is that the hydrolysis product can be miscible with the substrate so that it may modify the arrangement of the substrate (as in the case of the PLD catalyzed hydrolysis, see below) and enhances virtually the area decrease due to tighter packing. This can be especially important for SMase which catalyzes the hydrolytic cleavage of SM into ceramides which stay at the air/water interface with the substrate but occupy a smaller area than the substrate [89].

Since the lateral pressure in biological membranes is unknown or assumed to be close to 30 mN/m, the pressure effect was analyzed in many enzymatic reactions [98, 99, 100, 101, 102]. Yield and velocity of the DPPC hydrolysis decrease drastically at higher lateral pressures. This effect can be explained as follows: at high lateral pressures, the phospholipid molecules are tightly packed and the penetration of the active centre of the enzyme is not possible, whereas at lower pressures the enzyme penetrates more easily into the monolayer. It is widely assumed that phospholipases cannot hydrolyze the substrate unless at least partial penetration of the enzyme into the lipid layer occurs. Such threshold pressure was observed for different phospholipases [98, 99, 100, 101]. The threshold pressure depends on the structure of the enzyme, the nature of its active centre and the electrostatic conditions and physical-chemical state of the substrate. For example, it appears that the Bee venom phospholipase A_2 has a Gaussian distribution in activity centred at 20 mN/m while the maximum activity of phospholipase C is found at 32 mN/m. The phospholipase A_2 activity of recombinant human lecithin-cholesterol acyltransferase has a similar behaviour, its activity is pressure-sensitive and maximal at 29 mN/m, just below its exclusion pressure. On the other hand, its activity is reduced below 26 mN/m, suggesting a critical phospholipid packing density for optimum access of the enzyme to the sn-2- fatty acid bond [103]. The enzyme activity seems to be depending on the enzyme sources. Using 1-palmitoyl-2-caproyl-PC or 1-caproyl-2-palmitoyl-PC, Cobra and bee venom PLA_2 are active at any pressure up to the monolayer collapse pressure at 37 mN/m whereas the bovine pancreatic activity is blocked at about 18 mN/m [96]. Similarly, the porcine pancreatic lipase is efficient at low pressures around 12 mN/m whereas the acetylcholinesterase is efficient at high pressures close to 35 mN/m. The hydrolysis of phospholipids catalyzed by *Bacillus cereus* phospholipase C in presence of the anesthetics halothane or isoflurane has the same effect as the compression in the sense that the change of the monolayer surface pressure induced by halothane, isoflurane or physical compression leads to the same slow hydrolysis rate below 20 mN/m, a high hydrolysis rate around 27-32 mN/m and a virtually stopped hydrolysis above 32 mN/m [98]. In this sense, the lateral pressure can be viewed as a parameter of inhibition or activation.

The characterization of hydrolysis reactions at a molecular level can be performed with PM-IRRAS, which is a versatile tool to analyze interfacial enzymatic processes [104]. The possibility to measure kinetics irrespective of the solubility behavior of hydrolysis products is of special interest. The hydrolysis of DPPC catalyzed by *Bacillus cereus* PLC at 8 mN/m was followed by PM-IRRAS. The most important change is in area of the phosphate band at 1228 cm^{-1} due to the cleavage of the PC head group. It enables to monitor quantitatively the hydrolysis of DPPC. The shift in the ν(C=O) ester vibration from 1732 to 1739 cm^{-1} is characteristic for both DPPC and DPG and the increased intensity of the $\delta(CH_2)$ bending mode reveals a change in the packing of acyl chains [105]. The same study was performed with PLA_2 on DPPC [106]. In this case, the decrease in the ν(C=O) ester band of DPPC and the appearance of two new bands associated to the COO^- vibration of the palmitic acid enables to follow the reaction (Fig. 5).

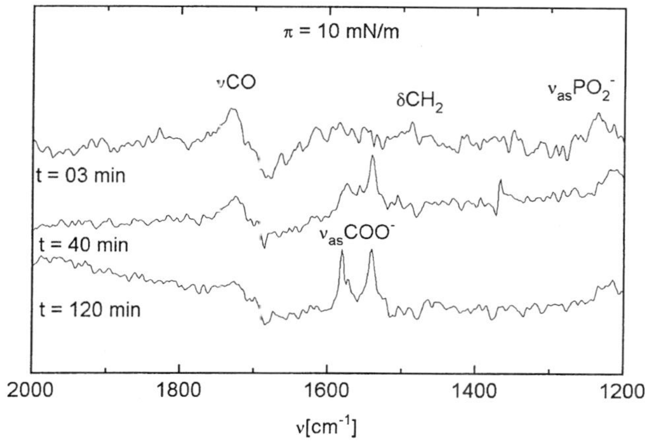

Fig. 5. PM-IRRAS spectra of a L-DPPC monolayer versus time of hydrolysis catalysed by PLA_2 at 10 mN/m.

Investigating the decrease in intensity of the carbonyl band at different lateral pressures shows that PLA_2 exhibits maximal activity in the coexistence region between condensed and liquid-expanded phases (Fig. 6) [84]. Taking into account that the enzyme adsorbs preferentially at the phase boundaries and leads to a tighter lipid packing, two important features for PLA_2 activity were deduced from these findings: The pre-orientation of the substrate in the ordered phase facilitates the enzyme reaction, and the reaction can only take place at defects within the membrane (Fig. 7). In biological membranes such defects could be created by local differences in lipid composition. Additionally to the dependence on the phase state, the hydrolysis efficiency depends also on the chemical structure of the phospholipid [36, 107]. Measurements using chemically modified phospholipids show that phospholipids with un-branched acyl chains are better hydrolyzed than lipids with side-branches located at different position along the main chains. The long side-chains are more effective in reducing the hydrolysis efficiency than the shorter ones. The strongest effect was observed if the *sn*-2 chain is branched. An ether connection of the *sn*-1 chain also reduces the hydrolysis efficiency, obviously due to changes of the lipid orientation at the interface.

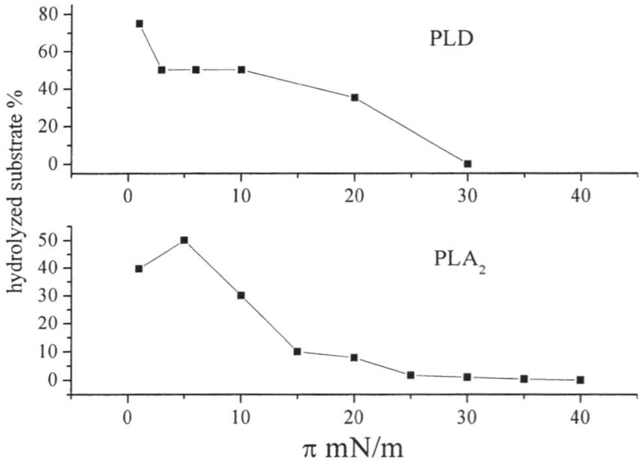

Fig. 6. Percentage of hydrolysed substrate after 100 min (PLD) or 60 min (PLA_2) of hydrolysis reactions in DPPC monolayers at the air/water interface versus lateral film pressure.

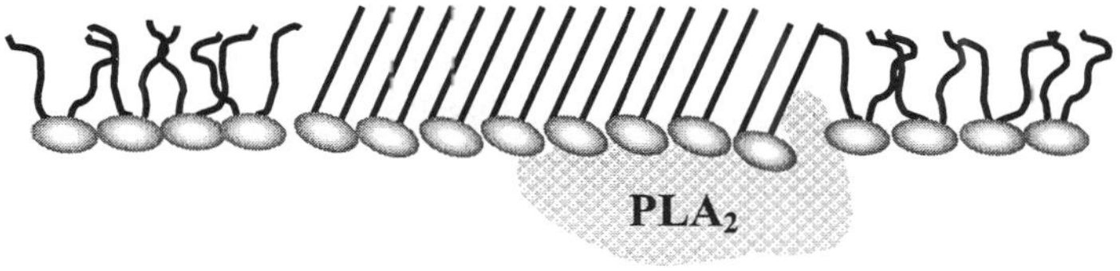

Fig. 7. Schematic diagram of the structural changes induced upon PLA_2 adsorption into the DPPC film.

As mentioned above, the value of the optimal hydrolysis pressure seems to be a characteristic parameter of both the enzyme and the lipid [84, 108, 109, 110, 111]. Comparing different phospholipases, interacting at different positions in the hydrophilic and hydrophobic regions, allows getting a more detailed insight in the mechanisms of hydrolysis. In the case of PLD, the most convenient band to follow the hydrolysis reaction would be the vibration of νCN^+C because it is absent in the spectrum of the reaction product DPPA. However, this band has not been used because the response of the detector drastically decreases at wavenumbers smaller than 970 cm^{-1}. Therefore spectra of pure DPPC as the substrate and of pure DPPA as the hydrolysis product have been recorded at 40 mN/m. DPPC and DPPA can be distinguished by the spectroscopic parameters of carbonyl and phosphate vibrational bands: νCO, $\nu_s PO_2^-$ and $\nu_{as} PO_2^-$ [112, 113]. The νCO band has a pronounced asymmetry on the low frequency side. It consists of two overlapping components, one at 1740 cm^{-1} and another at approximately 1726 cm^{-1}. The high frequency component of this band is assigned to the non-hydrogen-bonded (free) carbonyl group and the lower frequency component to the hydrogen-bonded carbonyl group. An increase of the intensity of the free carbonyl band can be seen for DPPA compared to DPPC. Therefore a high frequency shift of the overall CO band was observed. This can be explained with a tighter packing of DPPA at 40 mN/m and hence with a reduction of the hydration of the carbonyl group. An even more pronounced difference between the spectra of DPPA and DPPC is observed for the integral intensities of bands in the region of the phosphate stretching vibrations. Also high-frequency shifted antisymmetric stretching phosphate vibrations were found for DPPA. This can be explained by a partial dehydration of these groups due to the interaction with Ca^{2+} ions.

The phosphate vibrations were chosen for the quantitative estimation of the amount of hydrolysis product. The integral intensities of the ($\nu_s PO_2^-$ + $\nu CO(P)$) and $\nu_{as} PO_2^-$ vibrations linearly depend on the mole fraction of DPPC. Therefore, it was concluded that the head group conformation of DPPA remains the same in all mixtures and the intensity of these vibrations depends only on the quantity of DPPA. Such linearity is expected for ideal mixtures or phase separated systems. The fraction of substrate hydrolyzed during 100 min was studied at different surface pressures: a) below the plateau region in the liquid-expanded phase LE, b) in the coexistence-region of LE and a liquid-condensed (LC) phase, c) in the LC phase. The results show that the highest hydrolysis efficiency can be found in the LE phase (Fig. 6) [113]. In contrast to PLA_2, PLD acts in the hydrophilic region of the phospholipid. The observation that the maximum activity of PLD is found in a more disordered phase indicates that fluidity and defects in the monolayer structure are more important than a pre-orientation of the substrate induced by enzyme adsorption.

The analysis of enzymatic reactions also concerns other systems than lipids. The hydrolysis of poly(lactic acid)–poly(glycolic acid) monolayers catalyzed by *Humicola lanuginosa* lipase leads to the appearance of charged fragments at the interface produced according to a random scission mode of fragmentation. Moreover the catalytic activity of the lipase is dependent on the lactic acid and glycolic acid composition and the molecular weight [114]. The cleavage of the hydrophilic part of the β-casein by the specific proteolytic enzyme endoproteinase was followed by neutron reflectivity at the solid/air interface as a function of salt composition in the buffer [115]. The reaction is stronger when the β-casein is adsorbed from a solution containing $CaCl_2$. Another example of interfacial activation is given by the hydroxynitrile lyase that adsorbs at the oil/water interface. The enzyme activity (mandelonitrile cleavage) is measured by following with UV-visible measurements benzaldehyde product formation, which is soluble in the oil phase (diisopropylether). As expected for an interfacial reaction, the reaction rate increases with increasing interfacial area and is independent of the aqueous volume [67]. The enzymatic hydrolysis of poly (D,L- lactide) spread monolayers by cutinase was investigated at acidic and basic pH. At acidic pH, the hydrolysis rate reaches a maximum value due to the competition with the well-known reversibility of the reaction at low pH [116].

3.2.2.2. Enzymatic reactions in dispersed systems

Enzymatic reactions were also investigated in dispersed systems (micelles, vesicles) or at the oil/water interface. The enzymatic activity of β-galactosidase has been observed in two kinds of model membranes: vesicles and monolayers at the air/water interface. The interaction with the model membranes occurs in both systems but the activity of the enzyme seems to be modulated by the surface curvature and/or the change in interface [59]. In such dispersed systems, the ratio of interfacial area/volume is very large and the contribution of the phenomena taking place at the interface is amplified [71]. The phosphorylation of phosphoinositols PI on the third position of the inositol ring by PI3K is shown to be strongly dependent on the curvature of liposomes containing PC and PI whereas the phosphorylation at the fourth position catalyzed by PI4K enzyme shows a weak dependence on vesicle size. These results suggest different locations of these enzymes and especially PI3K that could induce deformation of the membrane after binding [117]. The phosphatidylinositol-(4,5)-biphosphate hydrolysis catalyzed by PI-phospholipase C catalysis is virtually complete in monolayers whereas the hydrolysis rate is limited to only 10% in micelles [118]. The activation energies of the hydrolysis of glycerol trioctanoate and glycerol dodecanoate have been determined in emulsions and monolayer systems [119]. They decrease with increasing chain length and are lower in the case of emulsions. They however decrease with increasing surface pressure, and at values close to the monolayer collapse they are similar in both emulsions and monolayer systems. The difference can be explained in terms of molecular densities and substrate orientations [119]. By using a mixture of polymerized 10,12-tricosandiynoic acid and DMPC, the hydrolysis of DMPC produces changes in the π-conjugated polymer backbone and thus a color change from blue to red [120]. In the case of PLD from *Streptomyces chromofucus*, hydrophobic beads coated with a monolayer of phospholipid or unilamellar vesicles have been investigated to analyze the enzyme/phospholipid binding. PLD exhibits a higher affinity to PA than PC surfaces especially at pH above 7 and the cleavage of PC can be enhanced by the addition of the product phosphatidic acid PA whereas other anionic lipids such as phosphatidylinositol, oleic acid or phosphatidylmethanol have no effect [121]. The Ca^{2+} modulation of PA/PLD binding, as a factor weakening the PA/PLD interaction, could involve protein aggregation. This could be correlated with the bimolecular activation deduced from kinetic measurements in monolayers [72]. The oil-drop tensiometer was also considered as a potential device to analyze the interfacial activity at the oil/water interface. The catalytic rate constants of hydrolysis of soybean oil by human pancreatic lipase in presence of a co-lipase measured at the oil drop surface were found to be of the same order of magnitude as those obtained with monomolecular films spread at the air-water interface [122]. Similarly, the activity of *Fusarium solani pisi* cutinase, adsorbed at the triolein/water interface is similar to that obtained in triglyceride emulsions [123]. When the maximal amount of enzyme is bound to the oil droplet, hydrolysis products are generated at a constant rate. Generation of hydrolysis products, oleic acid, at the interface causes a further decrease of the surface tension as observed by the oil drop technique. Since they rapidly dissolve into the oil phase, the interfacial tension gets to a plateau value, which is a measure for the steady state turnover of the enzyme [123]. The interest in multicomponent-systems concerns also the dispersed systems. The activity of PLA_2 from *Agkistrodon piscivorus piscivorus* on large unilamellar vesicles composed of DPPC:SMPC and DMPC:DSPC:SMPC was recently investigated [124]. PLA_2 preferentially hydrolyses the short chain DMPC instead of DSPC. This discriminating effect is, however, less pronounced in presence of the asymmetric lipid SMPC, which on the other hand prolongs the lag time of DPPC and DMPC:DSPC vesicle hydrolysis. The activity of PLA_2 is moreover enhanced in the temperature range of the coexistence region where the lipid mixtures exhibit lateral gel-fluid phase separation [124], similarly to the results obtained with monolayers. Moreover, this correlates with the efficiency of PLA_2 on DPPC substrate forming gel and liquid-crystalline phases (Fig. 8).

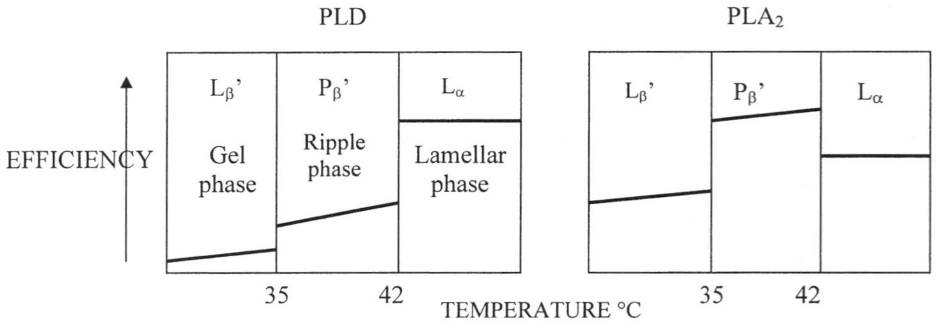

Fig. 8. Schematic diagram of the PLA_2 and PLD efficiency in bulk systems of DPPC with 80 wt.-% buffer as a function of temperature. L_{β}', P_{β}' and L_{α} refer to the gel, rippled and liquid crystalline lamellar phases.

3.2.3. Inhibition – Activation

Many parameters influence an enzymatic reaction. The effect of the lipid composition has been already described above [36, 83, 85, 86, 87, 121, 107]. A simple increase in temperature may induce a denaturation of the enzyme by unfolding at the air/water interface, as seen by PM-

IRRAS with the change in the α-helix and the β-sheet signals [47]. The chemical composition of the buffer plays also a role. For example, Ca^{2+} is usually present in the buffer containing lipases since it acts as a cofactor of the lipolytic reaction. PLA_2 prepared in a buffer containing Zn^{2+} instead of Ca^{2+} ions does not induce the hydrolysis of DMPC in vesicles [120]. It was also recently demonstrated that the hydrolytic rate of porcine pancreatic lipase PLA_2 exhibits a dependence on the Ca^{2+} concentration with a maximum at about 6 mM. Above 6mM, Ca^{2+} appears to have an inhibitory effect on the hydrolysis [92]. Similarly to the PLA_2/Ca^{2+} system, the SMase activity is markedly enhanced by Mg^{2+} with a maximal activation at 5 mM [89]. Several studies were developed to understand the role of molecular inhibitors. For example, the 1-hexadecyl-3-trifluoroethylglycerol-2-phosphomethanol inhibits the hydrolysis of DMPC catalyzed by PLA_2 and the vesicles remain blue instead of turning red during the reaction [120]. Tanaka et al [125 and references therein] have pointed out that cholesterol or a synthetic polymer having surface activity have a retardation effect on the lipase kinetics as measured at the air/water interface on a phospholipid substrate. The polymer is not likely to exert the same sort of retardation effect on monolayer dynamics by specific interactions that result in condensed complexes of phospholipid and cholesterol. Cholesterol has however an opposite effect on the hydrolysis of SM catalyzed by SMase. Its inclusion even in a small proportion markedly increases the SM degradation and shortens the lag-time of the reaction [89]. In the case of the system polyethylene glycol PEG and poly(lactide-glycolide) PLAGA, PEG has almost no effect on the hydrolysis rate of PLAGA whatever the investigated process: at alkaline pH or under the enzymatic action of cutinase [75]. In other cases, it seems to reduce the rate of deactivation of a lipase at the air/water interface [126]. The inhibiting effect of synthetic 2-oxo-amide-triacylglycerol analogues was investigated with the zero-order trough. At a lateral pressure of 15 mN/m, the hydrolysis rate of 1,2-dicaprin monolayer spread at the air/water interface is independent of the chain length of the triacylglycerol analogue and its chirality. In contrast to the porcine pancreatic lipase, the human gastric lipase is stronger inhibited by the (R) configuration of the 2-oxo-amide-triacylglycerol than by the (S) configuration [127]. The inhibitor β-lactoglobulin is miscible in all proportions with 1,2-dicaprin and the resulting monolayer is more stable than the two compounds alone. The inhibition effect of β-lactoglobulin is not directed on the enzyme but on the lipid substrate by modifying the substrate properties [79]. The same inhibition effect was observed in the didecanoylglycerol DDG/porcine pancreatic lipase system. The β-lactoglobulin modifies the substrate DDG on which the enzyme should act [128]. With PM-IRRAS, it was possible to detect the origin of the acetylcholinesterase inhibition. The presence of the inhibitor Paraoxon in the subphase destroys the secondary structure of the enzyme by forming a covalent complex with the inhibitor at the active site of the enzyme. It leads to its unfolding at the air/water interface. This process is, however, reversible with the addition of a nucleophilic agent TMB-4. After 30 minutes, PM-IRRAS shows the recovery of the secondary structure at the air/water interface [47]. Some other potentially efficient synthetic inhibitors were considered and tested [129]. The effect of complex glycosphingolipids (gangliosides) on the activity of phopsholipase C from *Bacillus cereus* was investigated using lipid monolayers, mixed micelles and small unilamellar vesicles containing phosphatidylcholine as substrate. In all artificial systems assayed, gangliosides exhibit qualitatively similar inhibitory properties. Gangliosides decrease the enzyme activity irrespective of the aggregation structure in which the substrate is offered to phospholipase C, and they do not affect the adsorption process of the enzyme. The modulating effect of gangliosides occurs at the level of the interface, affecting both the maximum rate of catalysis of the enzyme already adsorbed and the availability of the substrate in a suitable organization for the enzymatic catalysis [130]. Peptides have also their role in the enzymatic activity. Melittin is an amphiphilic peptide that acts as an activator of the DPPC hydrolysis catalyzed by PLA_2 by creating defects [131]. This sensitivity to defects has already been encountered for PLA_2 by Dahmen-Levison et. al [84] and for ChOx by Slotte [90]. It appears also that the PLA_2 is moreover enhanced by peptide mimics bearing positive charges [131]. Maggio contributes also in this field by analyzing the modulation of the PLA_2 activity by electrostatic fields and dipole potential of glycosphingolipids in monolayers [132]. The changes of enzymatic activity in mixed monolayers of DLPA or DLPC with glycosphingolipids by applying external constant electrostatic fields are correlated to the depolarization or hyper-polarization of the film caused by specific dipolar properties of glycosphingolipids. In the pure DLPA film, hyperpolarizing fields enhance the enzymatic activity while depolarizing fields induce a decrease of activity. The effect of glycosphingolipids depends on their proportion in the monolayer. The interface of mixed films containing neutral glycosphingolipids or gangliosides is already partially depolarized. The magnitude of activation induced by an external hyperpolarizing field is decreased compared to the pure substrate and depolarizing fields cause an increased inhibition of activity. This shows that the activity of PLA_2 is modulated by the polarity and magnitude of the lipid polar head group dipole moments [132]. This work confirms previous findings on the role of charged lipids. It was observed that changing the electrostatic conditions by adding, for example, charged phospholipids as phosphatidic acid into the film facilitates the penetration of the active centre and hence enhances the enzymatic activity [100, 101]. Depending on the nature of the enzyme and the added protein, the latter is thought to inhibit the lipase activity by either i) its concentration in the bulk phase or ii) its ability to penetrate into the monolayer (probably due to an increase of the lateral surface pressure instead of a chemical inhibition) or iii) the long-range interactions even at low surface densities [111].

In most experiments, a lag phase or slow hydrolysis was observed after the addition of the enzyme. The so-called lag-burst behaviour is a general phenomenon described in detail for hydrolysis reactions catalyzed by phospholipase A_2 [5, 133, 134, 135]. In monolayer experiments, the duration of the lag period depends to a large extent on the monolayer preparation and on the enzyme distribution in the subphase. Lipid packing defects, which are related with structural micro-heterogeneity and composition heterogeneity, have been correlated with the duration of the lag period [136, 137]. The increase of the amount of the defects decreases the duration of this period. During the lag phase the hydrolysis product accumulates and finally initiates the process of fast hydrolysis. However, the percentage of DPPC hydrolyzed during the lag phase must be below 5% and cannot be detected with high enough accuracy in the time-resolved scans at low surface pressure. In bilayer experiments it was demonstrated that the hydrolysis products segregate in the bilayer plane towards the end of the lag period [133, 135]. At surface pressures below the plateau in DPPC/DPPA mixtures with small DPPA mole fraction, small domains can be already seen using Brewster angle microscopy [138]. This indicates a phase separation in the monolayer, where DPPA rich domains are surrounded by fluid DPPC.

There is a certain DPPA concentration at which the PLD catalyzed hydrolysis stops. This inhibiting DPPA concentration is a function of the monolayer pressure. This inhibitory effect of DPPA at higher concentrations is believed to be partially caused by a surface pH shift. The same quantity of DPPA, which inhibits PLD at higher pressure, is not capable to stop the hydrolysis at low pressure. This finding cannot alone be explained by a pH effect. It may also be connected with the miscibility behaviour of substrate and hydrolysis product. The hydrolysis product DPPA is only slightly miscible with DPPC [138]. However, DPPA incorporated into DPPC domains reduces the tilt angle of the DPPC chains. If this tilt angle decreases to approximately 16° the hydrolysis reaction stops. This means that the reaction product DPPA changes the membrane structure and affects drastically the enzyme activity (Fig. 9). The parameter tilt angle was used because it can be precisely determined, however, it may not be directly relevant. It is related to the lateral lipid density, probably also to head-group conformation and hydration, and this may be the more direct structural parameters which shall be elucidated in dedicated experiments. The observed high-frequency shifts of the symmetric and antisymmetric PO_2^- vibrations as well as of the carbonyl vibrations during the formation of DPPA indicate also a reduced hydration or ion binding due to tighter head group packing. This can reduce the accessibility of the POC bond for an attack by PLD.

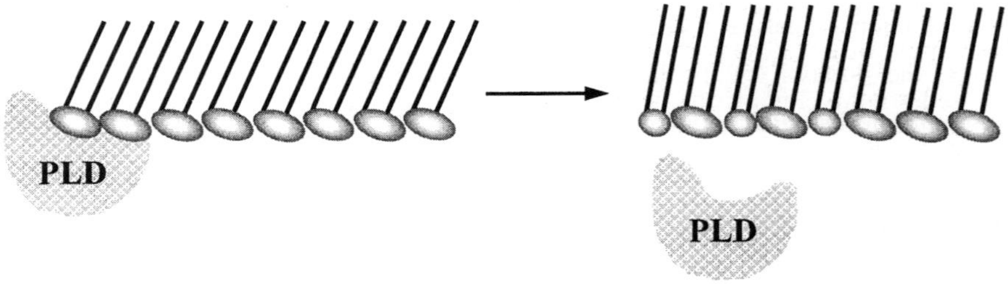

Fig. 9. Schematic diagram of the PLD inhibition induced by hydrolysis products into the DPPC film.

4. CONCLUDING REMARKS

In conclusion, the results of the last 7 years show or confirm that esterases and oxidases exhibit some common features:

- The adsorption at the air/water interface of enzymes is effective and may result in its denaturation by unfolding of its structure and consequently in a lose of activity.
- The chemical structure of the polar head groups and of the hydrophobic chains of amphiphiles is important for the kinetics and efficiency of the adsorption process. Usually the electrostatic interactions are the major forces involved in these interactions as observed by studying oppositely-to-enzyme charged membranes.

- Esterases and oxidases are sensitive to the ordering of the monolayer. They are squeezed-out from the interface at high pressures when the monolayer exhibits a well-ordered state. Moreover, they seem to adsorb preferentially at defect regions.
- Additional to the packing effect, their activity is modulated by the presence of natural or synthetic inhibitors and by the presence of other lipid components or products of the enzymatic reactions.
- The mode of organization of the lipid substrate (monolayer or dispersed systems) may influence the enzymatic catalysis.

In summary, the knowledge about the effect of the environment on the activity of enzymes studied in model membrane systems has broadened. Actually, numerous works are directed towards the characterization of enzymes in layer-by-layer assemblies [139], which avoid the denaturation of the enzyme at the air/water interface and improve its catalytic activity [61, 140, 141]. We expect that the knowledge just gain with well-defined models will be extremely useful to understand biology but also to design systems of technological relevance like biosensors and biocatalysts.

5. ABBREVIATIONS

Methods

PM-IRRAS	polarization-modulated infrared reflection absorption spectroscopy
BAM	Brewster angle microscopy

Phase behaviour

LE	liquid-expanded phase
LC	liquid-condensed phase

Lipids

PA	phosphatidic acid
PC	phosphatidylcholine
PS	phosphatidylperine
DLPA	1,2-dilauroylphosphatidic acid
DLPC	1,2-dilauroylphosphatidylcholine
DMPC	1,2-dimyristoylphosphatidylcholine

DOPC	1,2-dioleylphosphatidylcholine
DOPE	1,2-dioleylphosphatidylehtanolamine
DPG	1,2-dipalmitoylglycerol
DPPA	1,2-dipalmitoylphosphatidic acid
DPPC	1,2-dipalmitoylphosphatidylcholine
DPPS	1,2-dipalmitoylphosphatidylserine
DSPC	1,2-distearoylphosphatidylcholine
POPC	1-palmitoyl 2-oleyl phosphatidylcholine
SM	sphingomyelin
SMPC	1-stearoyl 2-myristoyl-phosphatidylcholine

Enzymes

ChOx	cholesterol oxidase
GOx	glucose oxidase
PLA_2	phospholipase A_2
PLD	phospholipase D
PLC	phospholipase C
SMase	sphingomyelinase
PI3K	phosphatidylinositol-3-kinase

6. REFERENCES

1. R. D. Schmid, R. Verger, *Angew. Chem. Int. Ed.* 37 (1998) 1608.
2. C. Nicolini, *Ann. NY Acad. Sci.* 864 (1998) 435.
3. P. Dynarowicz-Łątka, A. Dhanabalan, O. N. Oliveira Jr., *Adv. Colloid Interface Sci.* 91 (2001) 221.
4. O. Albrecht, H. Gruhler and E. Sackmann, *J. Phys.* 39 (1978) 301.
5. R. Verger and, M.C.E. Mieras, G.H. DeHaas, *J. Biol. Chem.* 248 (1973) 4023.
6. M. Lösche and H. Möhwald, *Rev. Sci. Instrum.*, 5 (1984) 1968.
7. D. Hönig, G.A. Möbius and D. Möbius, *J. Phys. Chem.*, 95 (1991) 4590.
8. S. Henon and J. Meunier, *Rev. Sci. Instrum.*, 62 (1991) 936.
9. D. Vollhardt and V. Melzer, *J. Phys. Chem. B*, 101 (1997) 3370.
10. R.A. Dluhy, *J. Phys. Chem.*, 90 (1986) 1373.

11. R.A. Dluhy, M.L. Mitchell, T. Pettenski and J. Beers, *Appl. Spectrosc.*, 42 (1988) 1289.
12. R. Mendelsohn, J.W. Brauner and A. Gericke, *Ann. Rev. Phys. Chem.*, 46 (1995) 305.
13. S. Widayati, S.M. Stephens and R.A. Dluhy, *Mikrochimica Acta*, Suppl 14 (1997) 679.
14. C.R. Flach, J.W. Brauner, J.W. Taylor, R.C. Baldwin and R. Mendelsohn, *Biophys. J.*, 67 (1994) 402.
15. B. Patranarios, S. Taneva, K.M.W. Keough, A.J. Mautone and R. Mendelsohn, *Biophys. J.*, 69 (1995) 2631.
16. T. Buffeteau, B. Desbat and J.M. Turlet, *Appl. Spectrosc.*, 45 (1991) 380. Turlet, *Appl. Spectrosc.*, 47 (1993) 869.
17. D. Blaudez, T. Buffeteau, J.C. Cornut, B. Desbat, N. Escarfre, M. Pezolet and J.M. Turlet, *Appl. Spectrosc.*, 47 (1993) 869.
18. D. Blaudez, J.M. Turlet, J.C. Cornut, B. Desbat, N. Escarfre, M. Pezolet and J.M. Turlet, *Thin Solid Films*, 242 (1994) 146.
19. D. Jacquemain, S.G. Wolf, F. Leveiller, M. Deutsch, K. Kjaer, J. Als-Nielsen, M. Lahav and L. Leiserowitz, *Angew. Chem. Int. Ed. Engl.*, 31 (1992) 130.
20. J. Als-Nielsen, D. Jaquemain, K. Kjaer, M. Lahav, F. Levellier and L. Leiserowitz, *Phys. Rep.*, 246 (1994) 251.
21. V. M. Kaganer, H. Möhwald and P. Dutta, *Rev. Mod. Phys.*, 71 (1999) 779.
22. T.R. Jensen and K. Kjaer, In: *Novel Methods to study interfacial layers*, (Eds.) D. Möbius and R. Miller, Elsevier, Amsterdam, vol. 11, 2001, pp.205.
23. K. Kjaer, *Physica B*, 198 (1994) 100.
24. J.S. Petersen, *J. Appl. Crystallogr.*, 25 (1992) 129.
25. D. Beaglehole, *Physica B*, 100 (1980) 163.
26. C. D. Bain, *Curr. Opin. Colloid Interface Sci.* 3 (1998) 287.
27. J.-A. Lekner, *Theory of Reflection*, Martinus Nijhoff Publishers, Dordrecht, 1987.
28. V. M. Kaganer, H. Möhwald, P. Dutta, *Rev. Mod. Phys. 71* (1999) 779.
29. F. Bringezu, G. Brezesinski, H. Möhwald, *Chem. Phys. Lipids* 94 (1998) 251.
30. F. Bringezu, G. Brezesinski, P. Nuhn, H. Möhwald, *Thin Solid Films* 327-329 (1998) 28.
31. U. Dahmen-Levison, G. Brezesinski, H. Möhwald, *Colloid Surface A* 171 (2000) 97.
32. F. Bringezu, G. Rapp, B. Dobner, P. Nuhn, G. Brezesinski, *Phys. Chem. Chem. Phys.* 2 (2000) 4509.

33. F. Bringezu, G. Brezesinski, *Colloids Surfaces A* 183-185 (2001) 391.
34. F. Bringezu, G. Rapp, B. Dobner, P. Nuhn, G. Brezesinski, *J. Phys. Chem. B* 105 (2001) 1901.
35. F. Bringezu, B. Dobner, G. Brezesinski, *Chem. Eur. J.* (2002) in press.
36. U. Dahmen-Levison and G. Brezesinski, *Phys. Chem. Chem. Phys.* 2 (2000) 4605.
37. G. Weidemann, G. Brezesinski, D. Vollhardt, H. Möhwald, *Langmuir* 14 (1998) 6485.
38. G.E. Graham, M. C. Philips, *J. Colloid Interface Sci.* 70 (1979) 403.
39. F.K. Winkler, A. D'Arcy, W. Hunziker, *Nature* 343 (1990) 771.
40. L. Brady, A. M. Brzozowski, Z.S. Derewenda, E. Dodson, G. Dodson, S. Tolley, J.P. Turkenburg, L. Christiansen, B. Huge.Jensen, L. Norskov, L. Thim, U. Menge, *Nature* 343 (1990) 767.
41. U. Derewenda, Z.S. Derewenda, *Biochem. Cell. Biol.* 69 (1991) 842.
42. Z.S. Derewenda, A. M. Sharp, *TIBS* 18 (1993) 20.
43. J.M. Chovelon, K. Wan, N. Jaffrezic-Renault, *Langmuir* 16 (2000) 6223.
44. S.V. Mello, C. Coutures, R.M. Leblanc, T.C. Cheng, V.K. Rastogi, J.J. DeFrank, *Talanta* 55 (2001) 881.
45. L.-T. Lee, B.K. Jha, M. Maltsen, K. Holmberg, *J. Phys. Chem. B* 103 (1999) 7489.
46. B. Folmer, K. Holmberg, M. Svensson, *Langmuir* 13 (1997) 5864.
47. L. Dziri, B. Desbat, R.M. Leblanc, *J. Am. Chem. Soc.* 121 (1999) 9618.
48. L. Dziri, S. Boussaad, N. Tao, R.M. Leblanc, *Langmuir* 14 (1998) 4853.
49. L. Dziri, K. Puppala, R.M. Leblanc, *J. Colloid Interface Sci.* 194 (1997) 37.
50. L.S. Jung, J.S. Shumaker-Parry, C.T. Campbell, S.S. Yee, M.H. Gelb, *J. Am. Chem. Soc.* 122 (2000) 4177.
51. K. Holmberg, M. Nyden, L.T. Lee, M. Malmsten, B. K. Jha, *Adv. Colloid Interface Sci.* 88 (2000) 223.
52. K. Tanaka, P.A. Manning, H. Yu, *Langmuir* 16 (2000) 2665.
53. W. A. Pieterson, J. C. Vidal, J. J. Volwerk, G. H. de Haas, *Biochemistry* 13 (1974) 1455.
54. M. Menashe, D. Lichtenberg, C. Guiterrezmerino, R. L. Biltonen, *J. Biol. Chem.* 256 (1981) 4541.
55. P.P. Bonsen, G. H. de Haas, W. A. Peterson, L. L. M. van Deenen, *Biochim. Biophys. Acta* 270 (1972) 364.

56. U. Dahmen-Levison, G. Brezesinski, H. Möhwald, *Thin Solid Films* 327-329 (1998) 616.

57. J. Li, Z. Chen, X. Wang, G. Brezesinski, H. Möhwald, *Angew. Chem. Int. Ed.* 39 (2000) 3059.

58. D. Kristensen, T. Nylander, J.T. Rasmussen, M. Paulsson, A. Carlsson, *Langmuir* 12 (1996) 5856.

59. J.M. Sanchez, M.A. Perillo, *Colloids Surfaces B* 24 (2002) 21.

60. J. Rinuy, P.F. Brevet, H.H. Girault, *Biophys. J.* 77 (1999) 3350.

61. G. Dai, J. Li, L. Jiang, *Coll. Surf. B* 13 (1999) 105.

62. D. Gidalevitz, Z.Q. Huang, S.A. Rice, *Proc. Natl. Acad. Sci. USA,* 96 (1999) 2608.

63. V. Rosilio, M.M. Boissonnade, J.Y. Zhang, L. Jiang , A. Baskin, *Langmuir* 13 (1997) 4669.

64. Y.K. Du, J.Y. An, J. Tang, Y. Li, L. Jiang, *Colloids Surfaces B* 7 (1996) 129

65. N. Dubreuil, S. Alexandre, C. Fiol, J.M. Valleton, *J. Colloid Interface Sci.* 181 (1996) 393.

66. S. Alexandre, N. Dubreuil, C. Fiol, D. Lair, F. Sommer, T.-M. Duc, J.M. Valleton, *Thin Solid Films* 293 (1997) 295.

67. A. Hickel, C.J. Radke, H.W. Blanch, *Biotechnol. Bioeng.* 65 (1999) 425.

68. F. Ronzon, B. Desbat, J.P. Chauvet, B. Roux, *Biochem. Biophys. Acta* 1560 (2002) 1.

69. C. Kirsch, R. Wetzker, R. Klinger, *Biochem. Biophys. Res. Comm.* 282 (2001) 691.

70. M.H. Ropers, C. Kirsch, R. Klinger, G. Brezesinski, 2002

71. I. Panaiotov, M. Ivanova, R. Verger, *Curr. Opin. Colloid Interface Sci.* 2 (1997) 517.

72. M.L. Fanani, B. Maggio, *J. Lipid. Res.* 41 (2000) 1832.

73. A. Tronin, T. Dubrovsky, S. Dubrovskaya, G. Radicchi, Nicolini C, *Langmuir* 12 (1996) 3272.

74. A.E. Ivanov, T.A. Volchenkova, B. Aha, S.Y. Zaitsev, *Colloids Surfaces B* 23 (2002) 349.

75. T. Ivanova, P. Saulnier, A. Malzert, F. Boury, J.E. Proust , I. Paianotov, *Colloids Surfaces B* 23 (2002) 7.

76. F. Gambinossi, M. Puggelli, G. Gabrielli, *Colloids Surfaces B* 23 (2002) 273.

77. A. Tiss, S. Ransac, H. Lengsfeld, P. Hadvàry, A. Cagna, R. Verger, *Chem. Phys. Lipids* 111 (2001) 73.

78. A. Ben Salah, A. Sayari, R. Verger, Y. Gargouri, *Biochimie* 83 (2001) 463.

79. F. Nannelli, M. Puggelli, G. Gabrielli, *Mat. Sci. Eng. C-Bio S* 8-9 (1999) 445.

80. M. Kuikka, B. Ramstedt, H. Ohvo-Rekila, J. Tuuf, J.P. Slotte, *Biophys. J* 80 (2001) 2327.

81. B. Ramsted, J.P. Slotte, *Biophys. J.* 77 (1999) 1498.
82. M. Ivanova, R. Verger, I. Panaiotov, *Colloids. Surfaces. B* 10 (1997) 1.
83. K. Tanaka, S.P. Mecca, H. Yu, *Langmuir* 16 (2000) 2672.
84. U. Dahmen-Levison, G. Brezesinski, H. Möhwald, *Progr. Colloid Polym. Sci.* 110 (1998) 269.
85. I. Arimoto, M. Fujita, H. Saito, T. Handa, K. Miyajima, *Colloid Poylm. Sci.* 275 (1997) 60.
86. M.L. Fanani, B. Maggio, *Mol. Memb. Biol.* 14 (1997) 25.
87. M.L. Fanani, B. Maggio, *Lipids* 33 (1998) 1079.
88. P. Mattjus, J.P. Slotte, *Chem. Phys. Lipids* 81 (1996) 69.
89. M. Jungner, H. Ohvo, J.P. Slotte, *Biochem. Biophys Acta* 1344 (1997) 230.
90. J.P. Slotte, *Biochem. Biophys. Acta* 1259 (1995) 180-186
91. M. Grandbois, H. Clausen- Schaumann, H. Gaub, *Biophys. J.* 74 (1998) 2398.
92. S. Chen, H.D. Abruña, *Langmuir* 13 (1997) 5969.
93. E.F. Tsung, R.D. Tilton, *J. Colloid. Interface Sci.* 213 (1999) 208.
94. Y. Cajal, M.A. Alsina, O.G. Berg, M.K. Jain, *Langmuir* 16 (2000) 252.
95. S. Ransac, M. Ivanova, R. Verger, I. Paianotov, *Methods Enzymol.* 286 (1997) 263-292.
96. K.M. Maloney, M. Grandbois, D.W. Grainger, C. Salesse, K.A. Lewis, M.F. Roberts, *Biochem. Biophys Acta* 1235 (1995) 395.
97. M.G. Ivanova, T. Ivanova, R. Verger, I. Paianotov, *Colloids Surfaces B* 6 (1996) 9.
98. R. H. Qarles, R. M. C. Dawson, *Biochem. J.* 113 (1969) 697.
99. R. A. Demel, W. S. M. Geurte van Kessl, R. F. A. Zwaal, B. Roelofson, L. L. M. Van Deenen, *Biochim. Biophys. Acta* 406 (1975) 97.
100. K. Hirasawa, R. F. Irvine, R. M. C. Dawson, *Biochem. J.* 193 (1981) 607.
101. J. Chen, G. Barton, *Canadian J. Biochem.* 49 (1971) 1362.
102. D.M. Goodman, E.M. Nemoto, R.W. Evans, P.M. Winter, *Chem. Phys. Lipids* 84 (1996) 57.
103. R.B. Weinber, J.B. Jones, P.H. Pritchard, A.G. Lacko, *Biochem. Biophys. Res. Comm.* 211 (1995) 840.
104. A. Gericke, H. Huhnerfuss, *Chem. Phys. Lipids* 74 (1994) 205.
105. M.Grandbois, Desbat B., D. Blaudez, Salesse C., *Langmuir* 15 (1999) 6594.
106. M.Grandbois, Desbat B., Salesse C., *Biophys. Chem.* 88 (2000) 127.

107. U. Dahmen-Levison, G. Brezesinski, H. Möhwald, J. Jakob, P. Nuhn, *Angew. Chem. Int. Ed.* 39 (2000) 2775.

108. A. J. Slotboom, R. Verger, H. M. Verheij, P. H. Baartmans, L. L. M. van Deenen, G. H. de Haas, *Chem. Phys. Lipids* 17 (1976) 128.

109. R. Verger, J. Rietsch, M. C. Mieras, G. H. de Haas, *J. Biol. Chem.* 251 (1976) 3128.

110. S. Ransac, H. Moreau, C. Riviere, R. Verger, *Methods Enzymol.* 197 (1991) 49.

111. G. Piéroni, Y. Gargouri, L. Sarda, R. Verger, *Adv. Colloid Interface Sci.* 32 (1990) 341

112. I. Estrela-Lopis, G. Brezesinski, H. Möhwald, *Phys. Chem. Chem. Phys.* 2 (2000) 4600.

113. I. Estrela-Lopis, G. Brezesinski, H. Möhwald, *Biophys. J.* 80 (2001) 749.

114. T. Ivanova, A. Svendsen, R. Verger, I. Panaiotov, *Colloid Polym. Sci.* 278 (2000) 719.

115. T. Nylander, F. Tiberg, T.-J. Su, J.R. Lu, R.K. Thomas, *Biomacromolecules* 2 (2001) 278.

116. T. Ivanova, I. Panaiotov, F. Boury, J.E. Proust., Benoît J.P., R. Verger, *Colloids Surfaces B* 8 (1997) 217.

117. S. Hübner, A.D. Couvillon, JA Käas, V.A. Bankaitis, R. Vergners, C.L. Carpenter, P.A. Janmey, *Eur. J. Biochem.* 258 (1998) 846.

118. S.R. James, S. Smith, A. Paterson, T.K. Harden, C.P. Downes, *Biochem. J.* 214 (1996) 917.

119. E. Cernia, L. Battinelli, S. Soro, *Thin Solid Films* 284-285 (1996) 727.

120. S.Y. Okada, R. Jelinek, D. Charych, *Angew. Chem. Int. Ed.* 38 (1999) 655.

121. K. Stieglitz, B. Seaton, M.F., Roberts, *J. Biol. Chem.* 274 (1999) 35367.

122. S. Labourdenne, O. Brass, M. Ivanova, A. Cagna, R. Verger, *Biochemistry* 36 (1997) 3423.

123. (a) J.A.C. Flipsen, R.A Kramer, J.P.M. van Duijnhoven, H.T.W.M. van der Hijden, M.R. Egmond, H.M. Verheij, *Chem. Phys. Lipids* 95 (1998) 169 (b) J.A.C. Flipsen, H.T.W.M. van der Hijden, M.R. Egmond, H.M. Verheij, *Chem. Phys. Lipids* 84 (1996) 105.

124. P. Høyrup, O.G. Moritsen, K. Jørgenson, *Biochem. Biophys Acta* 1515 (2001) 133.

125. K. Tanaka, H. Yu, *Langmuir* 18 (2002) 797.

126. M. Mohanty, R.S. Ghadge, N.S. Patil, S.B. Sawant, J.B. Joshi, A.V. Deshpande, *Chem. Eng. Sci.* 56 (2001) 3401.

127. G. Kokotos, R. Verger, A. Chiou, *Chem.-Eur. J* 6 (2000) 4211.

128. G. Gabrielli, G. Caminati, M. Suggelli, *Adv. Colloid Interface Sci.* 87 (2000) 75.

129. A. Chiou, R. Verger, G. Kokotos, *Lipids* 36 (2001) 535.

130. J.J. Daniele, B. Maggio, I.D. Bianco, F.M Goni, A. Alonso, G.D. Fidelio, *Eur. J. Biochem.* 239 (1996) 105.

131. M. Grandbois, J Dufourcq, C. Salesse, *Thin Solid Films* 284-285 (1996) 743.

132. B. Maggio, *J. Lipid Res.* 40 (1999) 930.

133. W. R. Burack, Q. Yuan, R. L. Biltonen, *Biochemistry* 32 (1993) 583.

134. W. R. Burack, R. L. Biltonen, *Chem. Phys. Lipids* 73 (1994) 209.

135. L. K. Nielsen, J. Risbo, T. H. Callisen, T. Bjørnholm, *Biochim. Biophys. Acta* 1420 (1999) 266.

136. T. Hønger, K. Jørgensen, R. L. Biltonen, O. G. Mouritsen, *Biochemistry* 35 (1996) 9003.

137. T. H. Callisen, Y. Talmon, *Biochemistry* 37 (1998) 10987.

138. I. Estrela-Lopis, G. Brezesinski, H. Möhwald, *Chem. Phys. Lipids* (2002) submitted.

139. M.K. Ram, P. Bertoncello, H. Ding, S. Paddeu, C. Nicolini, *Biosens. Bioelectron.* 16 (2001) 849

140. J.P. Santos, E.R. Welsh, B. P. Gaber, A. Singh, *Langmuir* 17 (2001) 5361

141. L. Pastorino, T.S. Berzina, V.I. Troitsky, M.P. Fontana, E. Bernasconi, C. Nicolini, *Colloids Surfaces B* 23 (2002) 357

CHAPTER 6
ELECTRON SPIN RESONANCE SPECTROSCOPY OF LANGMUIR-BLODGETT FILMS CONTAINING FUNCTIONAL MOLECULES

Shin-ichi Kuroda

Department of Applied Physics, Nagoya University, Nagoya 464-8603, Japan

1. INTRODUCTION

The Langmuir-Blodgett (LB) technique has been attracting much attention as a tool for arranging various kinds of molecules into the form of monolayer assemblies [1]. In the course of the development of new LB films, their characterization is indispensable. ESR spectroscopy is a powerful tool in elucidating microscopic properties of LB films, when the system contains unpaired electrons [2]. Unpaired electrons in LB films come from isolated radical molecules produced by charge transfer, paramagnetic metallic ions such as Cu^{2+}, Mn^{2+} etc., strongly interacting spins in the mixed valence states in charge-transfer salts and so on. These spins often manifest the functions of materials. They can also act as microscopic probes in the ESR analysis devoted for the elucidation of structure and electronic states of molecular aggregates formed in LB films. This is the subject discussed in this chapter with specific examples in LB films containing functional molecules.

The first application of ESR spectroscopy to LB films was reported in 1971 by Messier and Marc who studied molecular orientational order in Cu-stearate LB films [3]. A similar study was carried out on other Cu-containing films [4]. In the case of Mn- stearate LB films, ESR served to clarify the phase transition and the magnetic dimensionality of the magnetic monolayers formed by Mn ions [5]. Other early examples are the studies of the photoinduced charge-transfer processes in the heterostructured LB films containing donor and acceptor monolayers [6, 7]. In these systems, photoinduced radical species were detected. In the 1980s and afterwards, LB films composed of various kinds of functional molecules were developed. ESR spectroscopy has made contribution in revealing the nature of structural and electronic properties of newly developed films.

In structural studies, ESR is of particular importance in the analysis of molecular orientation in LB films [2, 8]. ESR can unambiguously determine the orientation of molecules through g-value anisotropy: different g value, different resonance field. That is, the distribution of molecular orientation is readily visualized as the distribution of resonance magnetic field. Two types of new control methods of molecular orientation in LB films originated from the ESR analysis. The first one is the study of in-plane orientation of stable radicals in mixed LB films of a merocyanine dye and arachidic acid which led to the discovery of flow-orientation effect during dipping process of LB films to control the in-plane orientation [2, 9-12]. The other is the observation of drastic change of orientation of Cu-porphyrin derivatives in LB films using the trigger molecule, n-hexatriacontane [13-15]. Added component, the trigger molecule can control the orientation of host molecules, porphyrins, through molecular interactions even with considerably smaller molar ratio than unity to the host molecule. These examples and others indicate the high orientational order can be achieved in LB films.

In the studies of electronic properties, hyperfine interactions between electron and nuclear spins provide information about molecular orbitals and local structures. Stable isotope has been successfully applied to the stable radicals in merocyanine LB films to identify hyperfine couplings [16, 17]. LB films composed of charge-transfer complexes provide another important class of materials whose electronic as well as structural properties were studied by ESR [2, 18-20].

In the following, the next section briefly describes the parameters obtained from ESR analysis in LB films. In section 3, the case of mixed LB films of merocyanine dyes and arachidic acid and that of the lightly-substituted porphyrins with triggers, respectively, are discussed in some detail as typical examples. Other examples are also briefly touched upon. Concluding remarks are given in section 4.

2. PARAMETERS DETERMINED BY ESR SPECTROSCOPY

In the case of diluted spins, with negligible interactions with other spins, the ESR spectra may be reproduced by the following spin Hamiltonian [21].

$$\mathrm{H} = \mu_B \mathrm{S} \cdot \mathrm{g} \cdot \mathrm{H} + \sum_i \mathrm{S} \cdot \mathrm{A}_i \cdot \mathrm{I}_i \qquad (1)$$

The first and second terms in the right-hand side represent the Zeeman energy and the hyperfine coupling, with $\boldsymbol{g}$ and $\boldsymbol{A}_i$ being the g tensor and hyperfine tensor with i^{th} nuclear spin, respectively. Other parameters, μ_B, $\boldsymbol{S}$, $\boldsymbol{H}$ and $\boldsymbol{I}_i$ denote Bohr magneton, electron spin, the external magnetic field, and i^{th} nuclear spin, respectively. Contrasted with the single-crystal cases, most LB films are more or less associated with a distribution of the molecular orientation. When the molecules in LB films have g-value anisotropies, ESR provides high accuracy in determining the orientational order because the orientation distribution is readily visualized as the distribution of the resonance field. The resonance field is given as $h\nu/g\mu_B$, where g shows the g value of the given direction of the paramagnetic species and $h\nu$ shows the microwave photon energy. Hyperfine anisotropy also contributes to the determination of molecular orientation. The distribution function of the molecular orientation is usually determined by comparing the observed spectra with the curves simulated using Eq.(1) together with an appropriate form of the angular-distribution function. Complete determination of the orientation is possible by taking the three-dimensional degree of freedom in the orientation into account using Eulerian angles. Actual examples will be described in the next section.

In the case of concentrated spins, the interaction between spins smears out the hyperfine and fine structures. Instead, the spin susceptibility, and the spectral line shape become important sources of microscopic information. In the case of well-localized magnetic moments, both quantities are known to be crucially dependent on the dimensionality of the system. They can be deduced from the spin Hamiltonian [2],

$$\mathrm{H} = \sum_i \mu_B S_i \cdot g \cdot H + \sum_{i<j} J_{ij} \cdot S_i \cdot S_j + \sum_{i<j} S_i \cdot D_{ij} \cdot S_j \qquad (2)$$

The second and third terms of the right-hand side of Eq. (2) represent the exchange and dipolar interactions between spins. The values of J_{ij}, the exchange integral between i^{th} and j^{th} spins, can be obtained from the temperature-dependent spin susceptibility, $\chi(T)$. D_{ij} represents the magnetic dipolar interaction between i^{th} and j^{th} spins. As for the spectral line shape, one-dimensional magnets show a characteristic angular variation resulting from an exchange-narrowed secular dipolar width, where the minimum of the width occurs at the 'magic angle'. Thus the analysis of the linewidth possibly gives the parameters J_{ij}, when D_{ij} are calculated either known or assumed crystal structures. In the spectra of LB films, however, a complication arises from the fact that the linewidth contains not only the above intrinsic width but also the width due to the g-value distribution caused by the orientation distribution. The separation of the two components can be achieved by measurements employing two different microwave frequencies. It should be noted that the linewidth minimum at the magic angle can be resolved in the angular-dependent ESR spectral turning point. The analysis of the spectra then determines not only the parameters J_{ij}, and D_{ij} but also the orientation of the molecules.

Sample preparation is also briefly discussed in this section. ESR samples of LB films can be prepared either by standard vertical dipping method or horizontal lifting method on appropriate substrates. The substrates for ESR measurements require the following conditions. Firstly, they should not lower the Q-value of ESR cavity. Secondly, no significant background ESR signals should be observed from substrates. For these purposes, the classical substrates employed are quartz-glass substrates. Normally the substrate forms are rectangular in shape with dimensions to fit into the space available in the sample cavity.

An alternative powerful substrate has been found to be mylar (polyethylene terephthalate) substrates [22]. Usually they are precoated with 5 layers of cadmium arachidate films to give hydrophobic surface. The advantage of this substrate is that it can be cut into any desired shape after sample deposition. This enabled us to carry out the first detailed measurements of in-plane molecular orientation of LB films, which could be made possible by setting the appropriately cut pieces of mylar film inside the sample holder. Another advantage of this substrate is that cut pieces can be stacked inside the sample tube to enhance the signal intensity. As for the sensitivity of ESR, standard ESR spectrometer can detect 10^{11} spins per linewidth of 1 gauss and hence the signal of monolayers with concentrated spins can be detected when the linewidth is sufficiently narrow.

3. EXAMPLES OF ESR ANALYSIS OF LB FILMS

3.1. Characterization of in-plane molecular orientation in merocyanine dye LB films

3.1.1. Anisotropic ESR spectra of stable radicals - Characterization of in-plane and out-of-plane molecular orientation

Mixed LB films of merocyanine dyes, DS in Fig. 1, with arachidic acid exhibit a characteristic red-shifted band, the so-called J-band in the optical absorption spectra (peak wavelength, $\lambda_{peak} = 593$ nm for DS), indicating the formation of J-aggregates of dye molecules [23]. Definition of the molecular axes is also shown in Fig. 1. In the case of DS films, the absorption intensity of the J-band shows a dichroism for the light polarization in the film plane, suggesting that the transition dipole moment (nearly parallel to the y-axis of the molecule) of dyes is preferentially oriented along the dipping direction of the substrate. Figure 2 shows an example of the anisotropy of optical absorption in a DS film. The solid and dashed lines correspond to the electric vector parallel and perpendicular to the dipping direction. No explanation, however, was given to the origin of such preferential orientation of transition dipole moment.

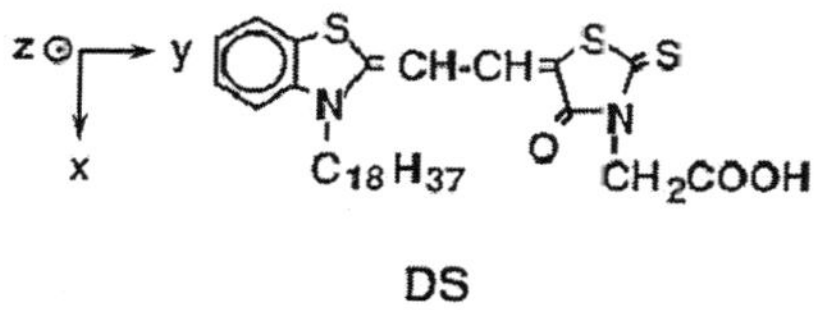

Fig. 1. Chemical structure of surface-active merocyanine dye. Definition of molecular axes is also shown.

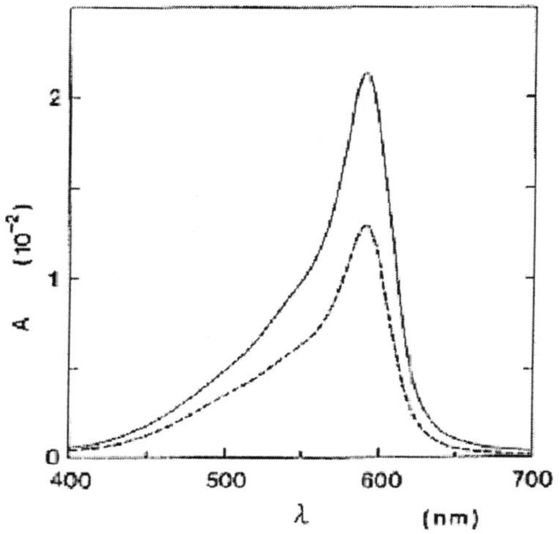

Fig. 2. In-plane optical dichroism of a DS film. The solid and dashed lines correspond to the electric vector parallel and perpendicular to the dipping direction, respectively.

On the other hand, DS films contain high-concentration stable π-electron radicals in the dark condition (1 spin/ several hundred dye molecules), which possibly arises from an intermolecular charge transfer in J-aggregates [22]. Detailed analysis of the molecular orientation of the stable radicals has revealed a characteristic in-plane orientation of the radicals, which led to the discovery of the flow orientation effect as the origin of the in-plane preferential orientation [2, 9-12]. The solid curves in Fig. 3 show the orientation dependence of the first-derivative ESR spectra at the X-band (9.47 GHz) in a DS film with a molar ratio of DS:C_{20}=1:2. The direction of the external magnetic field is normal to the film plane in Fig. 3(a) whereas it lies in the film plane and makes an angle of 0° and 90° with the dipping direction of the substrate in Figs. 3(b) and (c), respectively. Both the in-plane and out-of-plane anisotropy of the film is clearly observed. As seen in the figure, there are resolved structures for the in-plane spectra.

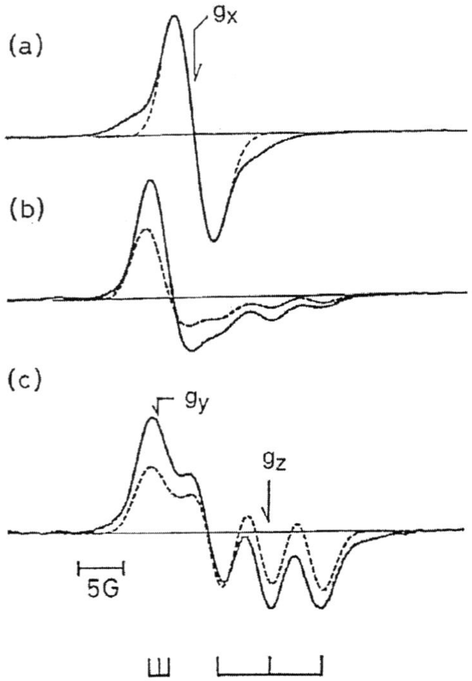

Fig. 3. Orientation dependence of the first-derivative ESR spectra of a DS film at the X band at room temperature (solid line). The external field is normal to the film plane in (a), whereas it lies in the film plane and parallel and perpendicular to the dipping direction of the substrate for (b) and (c), respectively. The g_i's mark the resonance fields due to the molecular axis in Fig. 1. The dashed line shows the spectra calculated for the primary radical species. The nitrogen triplet hyperfine structures along the y- and z-axes are shown at the bottom of the figure

By considering the preferential orientation of the dye molecule (area per molecule of 60 $Å^2$), together with the nearly uniaxial anisotropy of ^{14}N hyperfine coupling of a π-electron, the observed triplet structures in Fig. 3(c) have been predicted and identified to be the nitrogen hyperfine structure due to one of the two nitrogens in the molecule [9, 24]. That is, the direction of the pπ orbital axis (the z-axis which is perpendicular to the molecular plane in Fig. 1(a)) along which the nitrogen splitting becomes the largest, is expected along the direction perpendicular to the dipping direction of the substrate. This is because, firstly, the hydrophobic chain axis of the molecule (x-axis) is expected to be perpendicular to the substrate plane from the lamellar structure of LB films. This is consistent with the area per molecule, that is both y and z axes are confined in the substrate plane. Secondly, the chromophore axis (y-axis) in the substrate plane is indicated to be parallel to the dipping direction of the substrate, as already mentioned from the optical dichroic behaviour of the J-band. Thus the z-axis lies in the plane and perpendicular to dipping direction. Then the spectrum in Fig. 3(c), where the external field is parallel to this direction, certainly shows the largest resolution of the triplet structures compared with other curves. This is consistent with the anisotropy of the nitrogen coupling with nuclear spin 1 with the largest splitting along the z-axis. As to be discussed below, later studies identified the observed nitrogen site to be the one in the benzothiazole ring (the left-hand side of the molecule in Fig 1) by using the isotope substitution technique. This nitrogen atom belongs to the chromophore part of the dye, N-C=C-C=C-C=O, responsible for the functional property of the dye molecule [9].

The observed curves contain information of the molecular orientational order. As mentioned above, in Fig. 3(c), the external magnetic field is parallel to the film plane and perpendicular to the dipping direction. In the figure, g_i shows the g-tensor components for the respective molecular axis. As discussed above, the pπ orbital component (g_z component) is most prominently observed for this direction where ^{14}N nitrogen hyperfine structure shows the largest splitting, as indicated by triplet stick lines centred at g_z. On the other hand, the smaller hyperfine splitting at g_y is not resolved due to the linewidth. Simultaneous observation of g_y and g_z components in a spectrum directly reflects the distribution of molecular orientation, thus giving rise to the prominent lines at respective g-tensor components [9]. Of course, their relative intensities are different between the curves (b) and (c), showing the orientation distribution is not isotropic in the plane. Thus the observed curves provide the basis for the analysis of the molecular orientation by using a spectrum simulation method.

Dotted curves show the simulated curves, which reproduce the observed features considerably well. The remaining discrepancy between the observed and calculated curves mainly comes from the overlapping broad and structureless signals due to the second radical species. In the actual calculation, the three dimensional degree of freedom of the molecular orientation is taken into account by using Eulerian angles. The procedure of the simulation is as follows.

Figure 4 shows the Eulerian angles (θ, ϕ ,ψ), in the usual notation to define the molecular orientation. *X*, *Y*, and *Z* represent the coordinates fixed to the substrate, where the *X* axis and the *Z* axis are parallel to the dipping direction and the plane normal, respectively. *x*, *y*, and *z* are the coordinates fixed to the molecule defined in Fig. 1. θ and ϕ define the orientation of the *z* axis (the pπ orbital axis) for the out-of-plane and the in-plane orientation of the substrate. As pointed above, the ^{14}N hyperfine splitting shows a clear splitting due to nuclear spin 1 along the *z* axis, and a much weaker splitting along the direction perpendicular to the z axis. The lack of a triplet structure in the spectrum with the external field along the plane normal in Fig. 3(a) shows that the z axis is nearly confined in the film plane, that is, $\theta \approx 90°$. In this case, ψ, which shows the rotation of the y axis (the long axis of the chromophore) around the z axis represents the out-of-plane deviation of the y axis.

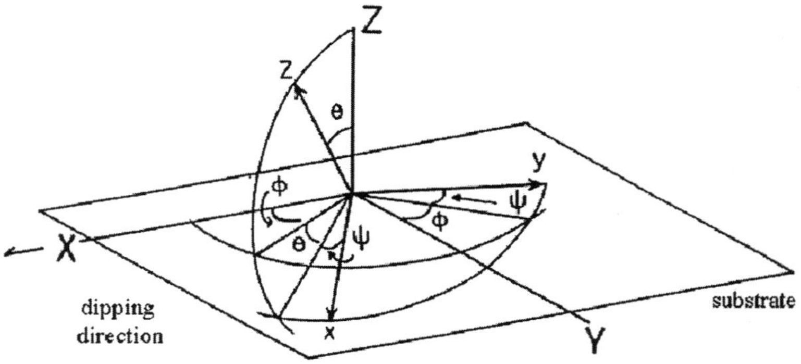

Fig. 4. Eulerian angles to define the molecular orientation of DS molecules in the substrate. The X, Y, and Z axes are the coordinates fixed to the substrate, where the X axis is parallel to the dipping direction. The x, y, z axes show the coordinate fixed to the molecule, as defined in Fig. 1.

The direction cosines (l_i, m_i, n_i) with $i = x, y$, or z of those molecular axes with respect to the substrate coordinate can be expressed using the Eulerian angles according to the standard procedure. When the direction cosines of the external field to the substrate are given by (L, M, N), the direction cosines (l, m, n) of the external field with respect to the molecular coordinate are expressed as

$$l = \hat{H} \cdot \hat{x} = Ll_x + Mm_x + Nn_x \quad (3)$$

and similar expressions for m and n. Then the molecule shows the g and hyperfine coupling values given as follows

$$.g^2(\theta,\phi,\psi) = g_x^2 l^2 + g_y^2 m^2 + g_z^2 n^2 \quad (4)$$

$$A^2(\theta,\phi,\psi) = A_x^2 l^2 + A_y^2 m^2 + A_z^2 n^2 \quad (5)$$

Here g_i and A_i with $i = x, y$, or z show the principal tensor components. The resonance field for the molecule pointing (θ, ϕ ,ψ) direction is given as

$$H_{res}(\theta,\phi,\psi,m_I) = [h\nu - m_I A(\theta,\phi,\psi)] / g(\theta,\phi,\psi)\mu_B \tag{6}$$

Here $h\nu$ shows the energy of the microwave photon. m_I is the magnetic quantum number of the nuclear spin. The ESR spectrum $I(H)$ for the given orientation of the external magnetic field is obtained by summing the contributions from all orientations of molecules considering the orientation distribution as

$$I(H) = \sum_{m_I} \int_0^{\pi} \int_0^{\pi} \int_0^{\pi} F[H - H_{res}(\theta,\phi,\psi,m_I)] P(\theta,\phi,\psi) \sin\theta d\theta d\phi d\psi \tag{7}$$

Here the function F is the line-shape function, which is assumed to be Gaussian with isotropic linewidth, for simplicity. The function P is the distribution function for the molecular orientation. As pointed above, the z and y axes are nearly confined in the plane. Considering this point we assume that the function P can be decoupled into the product of the distribution functions of each angle, the functional form being assumed as follows. The in-plane distribution function for ϕ, for example, becomes

$$\begin{aligned} P(\phi) \propto & \exp[-\sin^2(\phi+\phi_0)/2\sin^2\delta] \\ & + \exp[-\sin^2(\phi-\phi_0)/2\sin^2\delta] \end{aligned} \tag{8}$$

Here $\pm\phi_0$ shows the most probable orientation. δ shows the width of distribution. In the case of $P(\phi)$, the physical basis has been given by the flow orientation theory as the Boltzmann function, to be discussed later.

Employing the above procedure, the spectra shown in Fig. 3 could be analyzed. For g values and hyperfine splittings; g_x=2.0068, g_y=2.0093, g_z=2.0023, and A_z=5.4 G. A_z=A_y=1.1.G and the peak-to-peak width of 3.1 G of the Gaussian line-shape function in Eq. (7) was estimated from the width of the spectrum along the plane normal in Fig. 3(a) and the widths of structures of the in-plane spectra in Figs. 3(b) and 3(c). Using those values, good agreement between the calculated spectra of primary radical species, shown by broken lines in Fig. 3, and the observed spectra was obtained for the distribution functions of the form in Eq. (8) with the parameter values as listed below. The difference between the observed and calculated curves is due to the overlapping signal from the second radical species as mentioned above.

In each figure, g_i ($i = x,y,z$) shows the g value of the corresponding molecular axis. Stick lines at the bottom of Fig. 3(c) show triplet hyperfine structures along the y axis (left-hand side) and the z axis (right-hand side). From the simulation, the angular distribution parameters have been obtained as follows. For the in-plane orientation, $\phi_0 = 60°\text{-}65°$ with $\delta = 25°\text{-}30°$, where ϕ_0 denotes the angle between the z-axis and the dipping direction of the substrate. This characteristic in-plane orientation of the pπ orbital axis of $\phi_0 \approx 60°$ with respect to the dipping direction is schematically shown in Fig. 5.

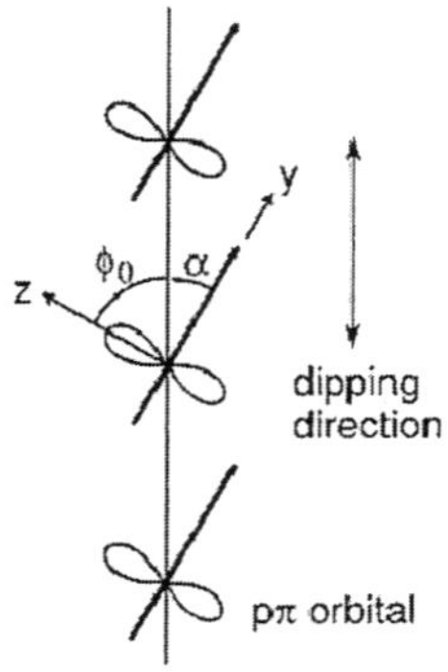

Fig. 5. Orientation of the stable radical molecules in the film plane. ϕ_0 denotes the average angle between the pπ orbital axis (z axis in Fig. 1) and the dipping direction of the substrate. The long axis of the dye chromophore (y axis) is expressed by a thick line.

This result shows an interesting correlation with the low-dimensional structure of J aggregates proposed by Kuhn and co-workers using extended dipole model [25]. In their brick stone model of the two-dimensional structure of the dye aggregates, one of the possible structures is such that the angle between the direction connecting the centre of molecules and the long axis of the dye molecule, defined as α becomes about 30°. If the direction connecting the centres of adjacent molecules of the J-aggregates, or the long axis of J-aggregates in other words is oriented along the dipping direction by a flow orientation as in the figure, the angle between the dipping direction and the z axis of the molecule is given and $\phi_0=90°-\alpha=60°$, which is in remarkable agreement with the angle obtained by the present analysis. This result strongly suggests that the origin of the in-plane orientation is the flow-orientation effect.

The result also suggests that the radicals originate from the molecules which form J-aggregates. In fact this has been shown to be the case from the detailed correlation between the observed dichroic behaviours of optical absorption and the ESR spectra, which has been well explained by the later developed theory of flow orientation. In the theory, the monolayer at the air-water interface is treated by an ideal fluid approximation [26]. The physical basis has been given for the functional form of the distribution function in Eq. (3), resulting from the Brownian motion of the dye aggregates elongated in shape. The distribution width δ can be related to the local stress ζG by the following equation.

$$1/2\sin^2\delta = \varsigma G/k_BT \tag{9}$$

Where ζ is the effective rotatory coefficient and G corresponds to the eigenvalue of the velocity gradient tensor. It is seen that a higher degree of orientation (smaller δ) is obtained for larger G. If we introduce the two parameters dipping velocity v_d and substrate width a, it can be shown that G increases for larger values of v_d / a. That is, higher degree of orientation can be obtained for a higher dipping velocity and narrower substrate width. It can be also shown that G depends on the position in the substrate and a higher degree of orientation is obtained toward the edge of the substrate. All these effects have been confirmed quantitatively for the behaviour of both the optical in-plane dichroism and ESR in-plane anisotropy. Thus the control of the in-plane orientation by flow orientation is well established by these experiments and the theory provides effective guide line to control the parameters governing the flow orientation, not only in Dye films but also other systems like polymer LB films.

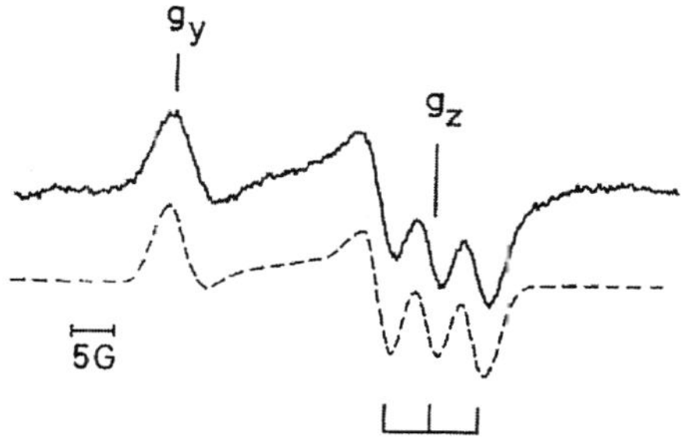

Fig. 6. K-band ESR spectrum of a DS film at room temperature (solid line) with the external field parallel to the film plane and perpendicular to the dipping direction. The microwave frequency is 23.4 GHz. Dashed line shows the calculated spectrum.

As for the out-of-plane distribution parameters, they are θ_0 -90° ≈ 7° with δ = 6°-8° for the z-axis and ψ_0 ≈ 5° with δ ≈ 10° for the y-axis. That is, both the z-axis and y-axis are considerably well confined in the plane, showing that higher orientational order exists along the plane normal. The values of δ are comparable with those reported for the out-of-plane parameters in fatty-acid salts with high orientational order along the plane normal [3], resulting from well-developed lamellar structure.

In the preceding discussions, nitrogen hyperfine structure has played a crucial role in the ESR analysis. Thus it would be of interest to present further confirmation of triplet structure using the microwave frequency dependence of the spectra and isotope substitution study. Figure 6 shows the K-band ESR spectrum of a DS film with the external magnetic filed being parallel to the films surface and perpendicular to the dipping direction, that corresponds to the case of Fig. 3(c). If we compare Figs. 3(c) and 6, we can discriminate the origin of the structures in the observed spectra whether they are associated with the different g-values or the hyperfine structure. That is, the hyperfine splitting does not depend on the frequency of measurements, and thus the triplet hyperfine splitting is the same for both the X- and K-band spectra. On the other hand, the separation between the lower field peak at g_y and the centre of the triplet at g_z is seen to be proportional to the microwave frequency, that is, the separation at the K-band is about 2.5(=23/9.3) times larger than that at the X-band. This ensures that the lower filed peak is associated with the different g value, g_y. Thus, in the LB system where the distribution of molecular orientations exists, the experiments with multiple microwave frequencies is useful in uniquely identifying the origin of the structures of the ESR spectra. The dashed line shows the calculated spectrum using the same parameter sets to reproduce the X-band spectra except the value of the microwave photon energy $h\nu$. Good agreement between the observed and calculated spectra assures the validity of the present simulation analysis.

3.1.2. ESR spectroscopy using isotope-substituted dyes

Since there are two nitrogen atoms present in a DS molecule, it would be of quite importance which nitrogen is responsible for the observed triplet splitting. The identification is possible through the isotope substitution technique [12, 16, 17]. Figure 7 shows the comparison of the films using unsubstituted and ^{15}N-substituted dyes. The direction of the external field is the same as that in Fig. 3(c).

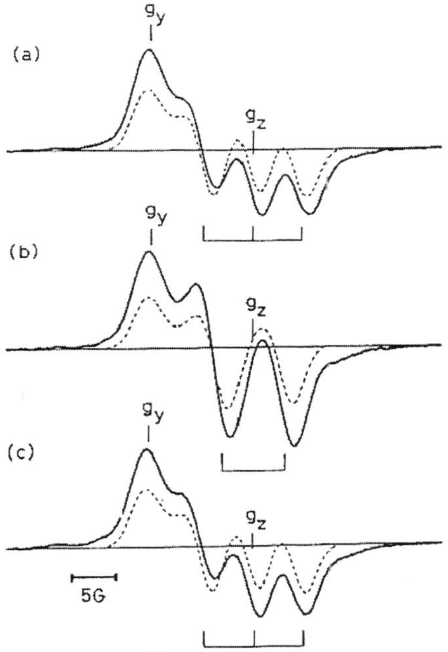

Fig. 7. Effect of ^{15}N-substitution in DS films. All the spectra were taken at room temperature with the external field parallel to the film plane and perpendicular to the dipping direction. Solid line corresponds to (a) no substitution; (b) ^{15}N-substitution in benzothiazole (left-hand side of DS in Fig. 1); (c) ^{15}N-substitution in rhodanine acetic acid (right-hand side of DS), respectively. Triplet or doublet stick lines show ^{14}N or ^{15}N hyperfine structure along the z axis. Dashed line shows the calculated spectra.

Solid curves in Figs. 7(a)-(c) correspond to the cases of (a) no substitution, (b) ^{15}N-substitution in benzothiazole (left-hand side of DS in Fig. 1) and (c) ^{15}N-substitution in rhodanine acetic acid (right-hand side of DS), respectively. While the resolved triplet splitting is preserved in Fig. 7(c), it is completely replaced by a doublet in Fig. 7(b) with a splitting 40% larger than that of the triplet, as compared with Fig. 4(a). This change in Fig. 4(b) is consistent with the difference of nuclear spin I and gyromagnetic ratio γ between $^{14}N(I=1)$ and $^{15}N(I=1/2)$ with the ratio of $\gamma(^{15}N)/\gamma(^{14}N)=1.403$. Therefore it is concluded that the ^{14}N hyperfine structure of the stable radical in the present system is associated with the nitrogen in benzothiazole. That is, the radical wave function resides on the chromophore part, N-C=C-C=C-C=O, which is an important part in determining the electronic states of the dye molecule. Thus the present confirmation provides the first case which directly identified the radical of dye molecules forming the J-aggregates in LB films. The observed hyperfine splitting A=5.4 G of the ^{14}N nucleus gives a spin density ρ_N=0.17 on the pπ orbital of the nitrogen atom, using the relation $A=A_N\rho_N$ with A_N=32G for ^{14}N.

More recently, the left-hand side carbon of the central methine chain of the chromophore has been substituted by ^{13}C having nuclear spin I=1/2, which resulted in the additional doublet hyperfine splitting [17]. Observed splitting yielded the spin density of 5-10%. This indicates the radical wave function has significant density on the chromophore part. The spectrum simulation study has successfully reproduced the observed X- and K-band spectra of ^{13}C-substituted films. From the analysis, the principal axes of ^{14}N and ^{13}C tensors were found to be parallel, showing the planar structure of the radical molecule and thus excluding the possibility of non-planar spyropiran form as the origin of the radical.

3.1.3. Light-induced ESR of merocyanine dye LB films

As described in section 3.1.1, the analysis of the ESR spectra of the stable radicals in DS films has revealed characteristic orientation of radicals, which explains well the orientation of the majority of dye molecules forming the J-aggregates. The results suggest that the radicals are generated in J-aggregates by intermolecular charge transfer. In this respect, light-induced ESR (LESR) attracts much attention because it can directly detect the radicals generated by photoinduced charge transfer, which is one of the most fundamental processes in molecular aggregates. Early examples of LESR studies of LB films are reported on hetero-structured LB films containing donor and acceptor monolayers [6, 7]. Donor species were chlorophyll *a* (ref. 6) or derivatives of cyanine and porhyrin (ref. 7) with acceptors of chloranil and bromanil in the former and derivatives of paraquat and quinone in the latter cases. In DS films, the enhancement of ESR signal intensity under illumination of visible light has already been observed, however, no detailed LESR studies have been carried out so far. LESR spectra successfully detected in pristine DS films are discussed in this section [27]. The LESR spectrometer with variable excitation energy, which can observe the action spectra, is described [28, 29]. The spectrometer has been successfully applied to the recent study of optically-generated polarons in conjugated electroluminescent polymers, poly(paraphenylene vinylene) and its derivatives. The wave function of the polaron, extending over about 4 phenyl rings has been revealed from the study of hyperfine coupling.

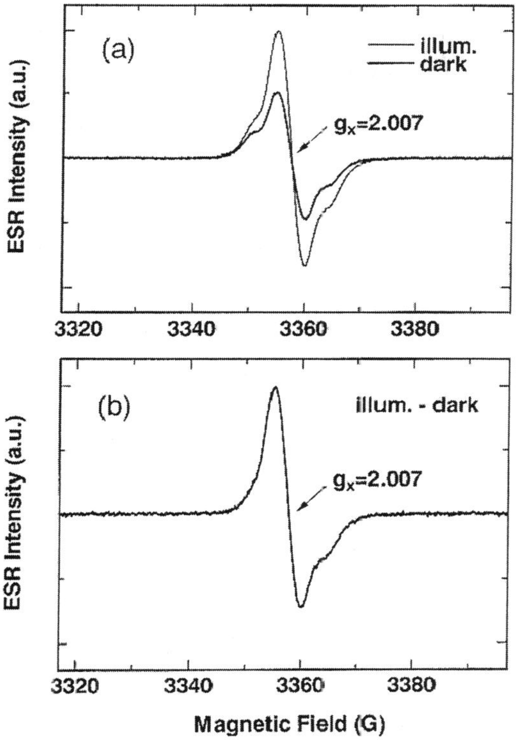

Fig. 8. (a) ESR signal of a DS film at 20K with and without illumination of 400-nm wavelength light. The external field is perpendicular to the film plane. (b) Light-induced ESR (LESR) signal, obtained by the subtraction between the dark and the illuminated signals of (a).

Figures 8 shows the light-induced ESR signal in a DS film at 20K with the external field perpendicular to the film plane. Figure 8(a) shows the ESR signals in the dark (thick line) and illuminated conditions at wavelength of 400 nm (thin line) and Fig. 8(b) shows LESR signal obtained by subtracting the dark curve from that under illumination. Spectral line shapes of the dark and LESR spectra resemble considerably each other and the same component as the dark one may be recognized in the LESR signal. LESR spectrum, however, exhibits more asymmetrical line shape compared with the dark one. This point will be discussed later in relation to possible discrimination of positive and negative ion radicals photogenerated.

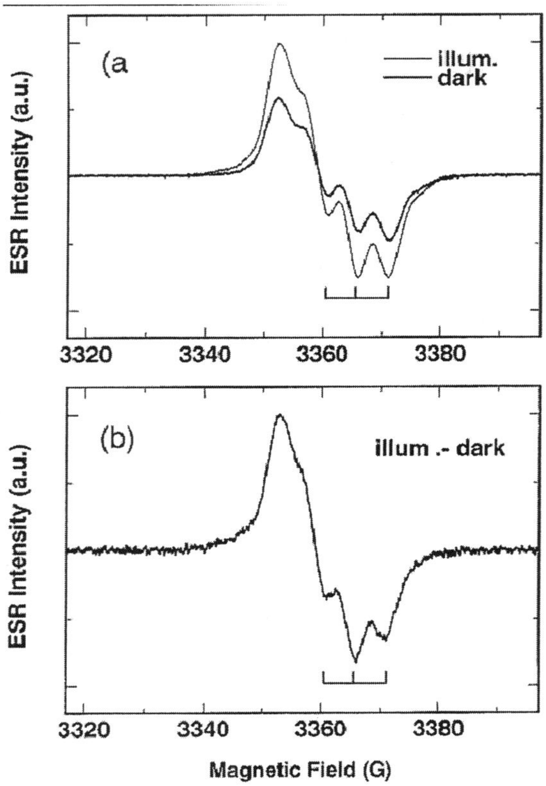

Fig. 9. (a) ESR signal of a DS film at 20K with and without illumination of 400-nm wavelength light. The external field is parallel to the film plane and perpendicular to the dipping direction of the substrate. Triplet lines show the nitrogen hyperfine structure. (b) LESR signal, obtained by the subtraction between the dark and the illuminated signals of (a).

Figure 9 shows the similar curves as those in Fig. 8 except that the external field is parallel to the film plane and perpendicular to the dipping direction. It is remarkable that the signal enhancement under illumination closely follows the spectral span of the dark signal and LESR signal in Fig. 9(b) in fact shows well resolved triplet nitrogen hyperfine structure similar as that in the dark signal. Anisotropic LESR signals with a resolved hyperfine structure provide the first example which directly identified the dye radicals photogenerated in the J-aggregates in LB films. The line shape of LESR signal in this direction is also slightly different from corresponding dark signal, which suggests that the rather broad structure-less component is overlapped because of the lower resolution of the triplet structure in the LESR signal.

Study of temperature dependence of LESR signals showed the LESR can be detected at temperatures below 200K. Time response study showed that LESR signal is completely transient in nature, indicating that the LESR signal appears by illumination due to charge separation and subsequent trapping and decays by recombination after cutting off the light. More details will be described elsewhere. Observed LESR action spectra also showed interesting behaviour. Figure 10 shows a preliminary result of LESR action spectrum of a DS film at 40K (closed circles) under the condition of the constant photon flux. Dotted line shows an optical absorption spectrum associated with a J-band peak at 593 nm. It can be seen that the LESR signal increases above the optical absorption peak energy despite of the lower optical intensity. Such an increase above the optical absorption peak is also observed in recent LESR study of conjugated polymers [28, 29] and it may suggest that the efficiency of charge separation (photoinduced charge transfer) increases above a certain gap energy correlated with the HOMO-LUMO band gap of a dye molecule in the aggregates.

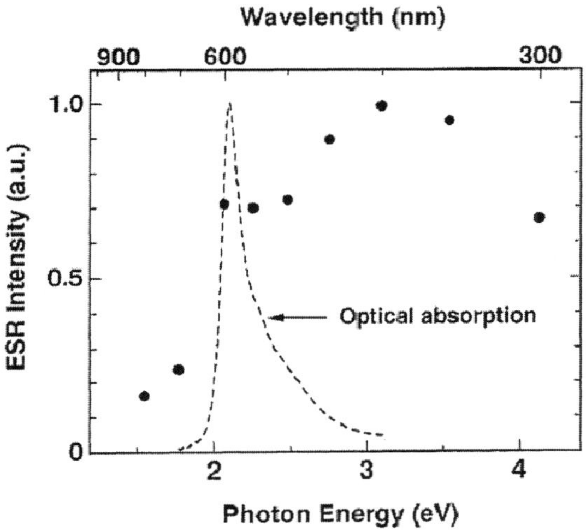

Fig. 10. Excitation spectrum of the LESR signal at 20K for a DS film. Dotted curve shows the optical absorption spectrum of a DS film.

Finally, the wave function of photogenerated radicals is discussed in relation to possible discrimination of positive and negative ion radical species. It should be pointed out that the radical species with triplet nitrogen splitting has been observed both for dark and LESR signals. On the other hand, the difference between the dark and LESR signals pointed above suggests that there exists another radical species in the case of light-induced signals, which could be possibly related to the existence of positive and negative ion radicals expected for photoinduced charge transfer. As for the stable radicals, its origin has not been clarified so far. One possibility from the above observations would be that the dark radical may be a cation radical which generated by intermolecular charge transfer between the DS molecule and molecular oxygen, as often observed in organic molecular aggregates. Then the other radical species observed in LESR would be anion radicals. In this case, since the wave function of positive and negative ions should be considerably different, then, it is reasonably expected that spectra show the different g-value and linewidth due to hyperfine structures (in fact unresolved), as observed. The resemblance of the spectra between the dark and photogenerated species with the resolved nitrogen structure indicates that the latter species are also highly oriented in the J-aggregates. High orientational order of they dye LB films is also consistent with the well-defined height profile of dye films using AFM observation [27, 30]

3.2. ESR studies of Cu-porphyrin films mixed with "trigger molecules"

3.2.1. Monomer porphyrin with hexatriacontane, the trigger

Since the pioneering work by Messier and Marc, LB films containing Cu^{2+} ions continue to provide an important class of materials for ESR investigation. By analyzing the out-of-plane anisotropy of the ESR spectra of Cu-stearate films, Messier and Marc have shown the high orientational order existing in these films [3]. In this case, the system contains high-concentration ions, where the exchange couplings between the ions wash out the hyperfine structures due to Cu nuclei having spin 3/2. On the other hand, in the diluted spin cases, the anisotropic hyperfine coupling also becomes useful probe for analyzing the molecular orientations. The case of lightly-substituted Cu-porphyrin LB films is discussed here as a typical example. An introduction of long-chain alkanes as a trigger molecule has been found to change drastically the orientation of Cu-porphyrin molecules in LB films [13-15]. This phenomenon can be considered to provide a new type of controlling method of molecular orientation in supramolecular assemblies through the specific microscopic interactions.

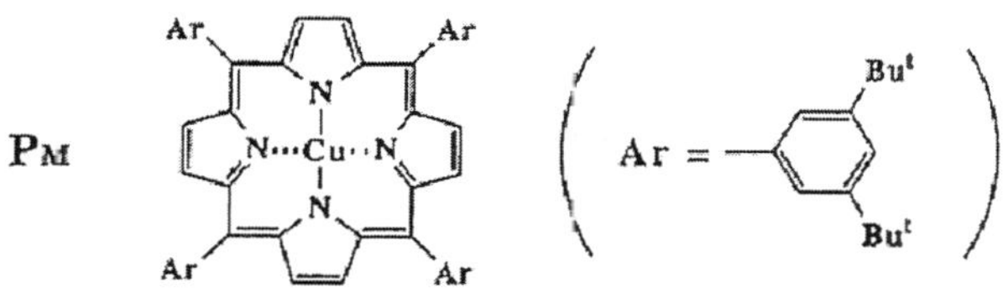

Fig. 11. Chemical structure of lightly-substituted porphyrin, PM.

Figure 11 shows the chemical structure of lightly-substituted porphyrin (hereafter abbreviated as PM). In ESR studies of mixed LB films of this molecule and arachidic acid together with an long-chain alkane, n-hexatriacontane($C_{36}H_{74}$, abbreviated as HA), it has been found that the orientation of PM changes drastically upon mixing of HA. The mixing ratio was PM/C_{20}/HA=1.5/10/x with $0 \leq x \leq 5$. When mixed with no HA (x=0), porphyrin molecular plane was found to tilt with respect to the normal direction of the substrate. If we define the angle between the normal direction of the molecular plane (defined as the z-axis) and that of the substrate to be θ_0, $\theta_0 = 58°$ was obtained. On the other hand, when mixed with HA, the orientation of the porphyrin plane drastically changed to be nearly perpendicular to the substrate, that is, took nearly edge-on configuration ($\theta_0 = 80°$).

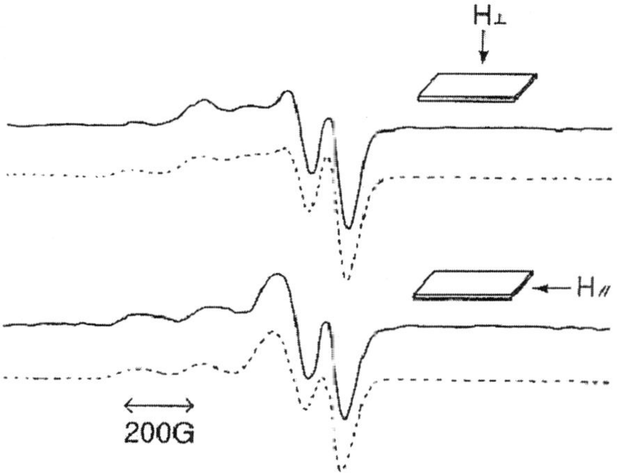

Fig. 12. Orientation dependence of the X band ESR spectra of a PM films without hexatriacontane at room temperature (solid line). The external magnetic field is perpendicular and parallel to the film surface for the upper and lower curves, respectively. Dashed line shows the calculated spectra.

Figure 12 shows the orientation dependence of the ESR spectra in the case of x=0. The external magnetic field is perpendicular and parallel to the film plane in (a) and (b), respectively. In this system, no in-plane anisotropy is observed. Dotted lines show the calculated spectra using Eq. (1), which reproduce the observed curves fairly well. The specific values of g-value and A tensor components are given in [13]. The hyperfine tensor of Cu nuclei show uniaxial anisotropy and the principal value takes the maximum value perpendicular to the molecular plane and gives a quartet hyperfine splitting due to nuclear spin 3/2. On the other hand, the splitting in the molecular plane is nearly an order of magnitude smaller and washed out by the linewidth. In the figure, the structures of the lower-field side (the left-hand side) are due to the hyperfine structures. The fact that these structures are observed both in the spectra of (a) and (b), directly reflects that the molecular plane is tilted with respect to the normal direction of the substrate plane. In the calculation, the orientation distribution function is adopted as the form in Eq. (3) and the value of distribution width δ is taken as 14°. In addition, about 20% of the molecules were found to be in the configuration with their molecular plane nearly parallel (θ_0=20°) to the film plane, indicating the existence of misoriented fraction of molecules.

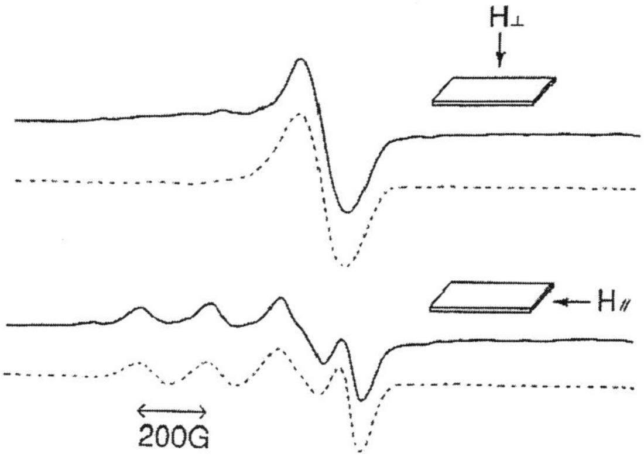

Fig. 13. Orientation dependence of the X band ESR spectra of a PM film with n-hexatriacontane (HA) with a molar ratio, PM:HA=3:1 (solid line) The external magnetic field is perpendicular and parallel to the film surface for the upper and lower curves, respectively. Dashed line shows the calculated spectra.

Figure 13 shows the orientation dependence of the ESR spectra in the case of x=0.5. This concentration of HA corresponds to 1 HA molecule for every three porphyrin molecules. The direction of the external magnetic field in (a) and (b) is the same as those in Fig. 12. The drastic change of the spectra compared with Fig.12 is clearly seen. Hyperfine splitting is only observed for the in-plane spectrum shown in (b). The separations of hyperfine lines are larger than those in Fig. 12. This situation reflects the edge-on configuration of porphyrin molecules. Dotted lines show the calculated spectra with $\delta=12°$. In the simulated curves is also included the contribution from misoriented falt-on fraction of about less than 10% of the total molecules. Thus the disordered fraction is also smaller compared with the case of x=0 or in other words, the introduction of HA results in the better orientational order of the system.

The above shown drastic change of the orientation of porphyrins due to the introduction of HA observed by ESR is also complimentarily observed by the dichroic behaviour of optical absorption spectra. Due to the molecular interaction, in the present LB films, the Soret band of porphyrins split into two bands; one is red-shifted J-like band, polarized in the film plane, and the other is blue-shifted H-like band polarized in out-of-plane direction. In particular, the shift of H-like out-of-plane band becomes larger for the edge-on case (that is, with HA) due to the stronger choromophore interaction. The dichroic ratio is very well reproduced by the orientation distribution determined by ESR. Figure 14 shows schematically the orientation of porphyrins with and without HA. Arrows indicate the transition dipole moment.

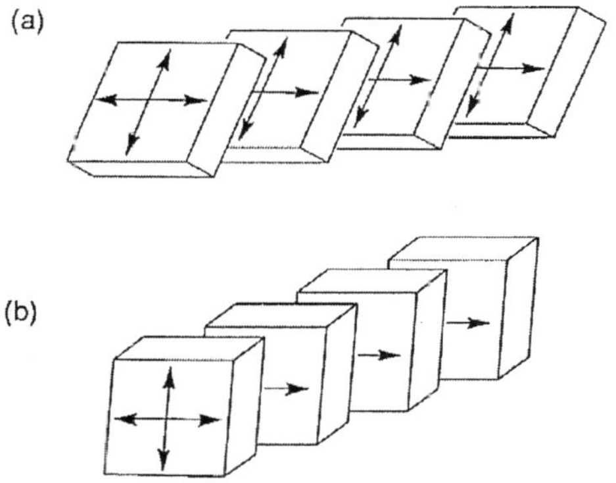

Fig. 14. Schematic view of the orientation of PM molecules in LB film (a) without hexatriacontane and (b) with hexatriacontane

The above results show that a long-chain alkane, HA, can control the orientation of functional molecules, porphyrin in LB films. The mechanism to change the orientation of porphyrin by HA would be presumably like that the HA act in a certain coordinating 'space' between porphyrins. The x-dependence of the area per C_{20} molecule in the mixed films deduced from the ΠA-isotherms in fact suggests that the amount of HA necessary to cause the orientation change of porphyrins is less than the concentration of HA of about x=0.5 (Fig. 15). That is, further increase of the HA concentration indicates that the excess amount of HA is expelled from the 'space'.

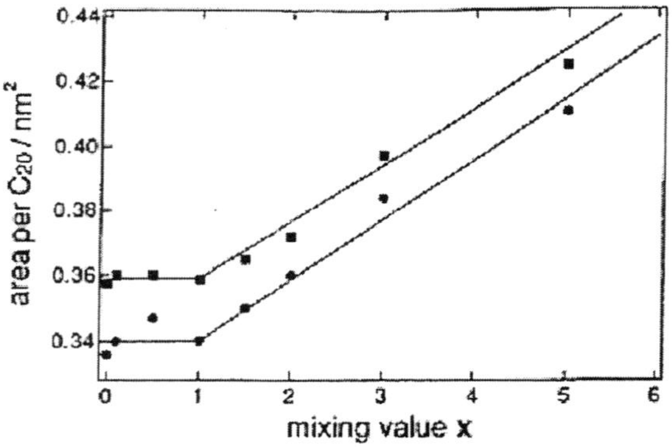

Fig. 15. Limiting area (■) and the area at 25mNm^{-1} (●) per cadmium arachidate as a function of the mixing ratio of n-hexatriacontane, *x*.

The concentration of x=0.5 means that the one HA is effective in controlling three porphyrin molecules, which is the origin of the naming of the 'trigger' used for this type of the phenomena. In this connection, the selective observation of the orientation of HA by FTIR by using deuterium-labelled HA (HA-d) has been performed to see whether their orientation is significantly modified compared with those of fatty acid matrix mixed in the system [31]. As a result, the alkyl-chain of HA has been found to be significantly tilted from the normal direction of the substrate compared with fatty acids. In the mixed LB film with a mixing ratio PM/C_{20}/HA-d=0/10/0.5, that is the case of fatty acid and alkane without PM, the tilt angles γ of the alkyl chain for HA-d and C_{20} with respect to the film normal were estimated at 12° and 13°, respectively. The introduction of PM with the mixing ratio 1.5/10/0.5 changed γ of HA-d to 37°, whereas that of C_{20} stay unchanged. It was also found that HA-d was oriented with its C-C-C plane almost perpendicular to the film surface.

Figure 16 shows the schematic view of the orientation of HA-d in the presence of PM. These facts strongly suggest that the action of trigger is effective through the specific molecular interaction between the functional molecules. Then it would be an interesting problem whether the trigger mechanism works for functional molecules other than monomer porphyrin. In fact the method has been found to be effective for more bulky molecules such as dimer-type porphyrin.

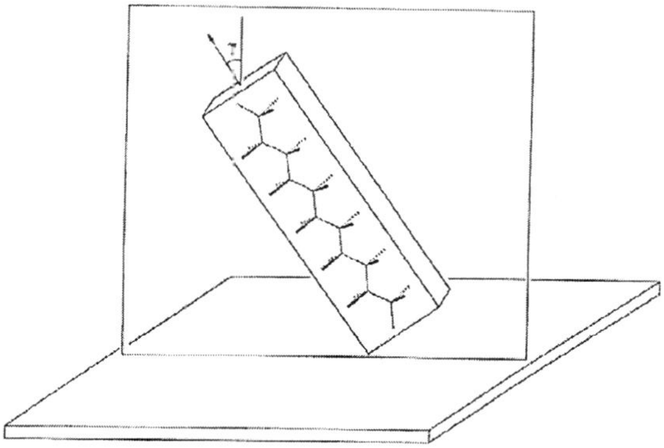

Fig. 16. Schematic view of the orientation of HA-d in the mixed LB film with the mixing ratio PM/C_{20}/HA-d=1.5/10/0.5. The C-C-C plane is almost perpendicular to the film surface.

3.2.2. Dimer-type porphyrin with trigger

Figure 17 shows a dimer-type Cu-porphyrin (hereafter abbreviated as PD) where the two Cu-porphyrins are conjugated. The orientation of PD was investigated by using ESR for the mixed LB films with the molar mixing ratio PD/C_{20}/HA=1.5/10/x in the range of $0 \leq x \leq 5$. $0 < x \leq 2$ was effective in the orientation control. Without HA, Pd had an orientation tilted with the most probable angle of $\theta \sim 60°$ between the plane normal of the molecule and the film normal, while the macrocycle plane of the porphyrin was oriented more vertically ($\theta \sim 75°$) with respect to the film surface for the LB film with the mixing ratio x=2.0. Moreover, ESR spectra clearly indicate that two unpaired electron spins of dicopper derivatives are interacting through a Heisenberg exchange interaction on two copper nuclei in a molecule. These features are discussed in this section.

The hyperfine structures of weakly interacting two unpaired electron spins critically depend on the relative magnitude of the Heisenberg exchange interaction J between two electron spins, compared with the magnitude of hyperfine coupling A [21]. When J is much smaller than A, the hyperfine structure becomes that of independent electron spin. On the contrary, if J>>A, the hyperfine structure becomes that of a spin interacting with two equivalent nuclear spins, the line spacing being a half of the coupling constant A. Thus in the case of Cu^{2+} dimer with $^{63,65}Cu$ nuclear spin of I=3/2, the hyperfine splitting becomes a 1:2:3:4:3:2:1 septet with the whole spectral span unchanged compared with the 1:1:1:1 quartet of monomer, as was observed in the case of PM.

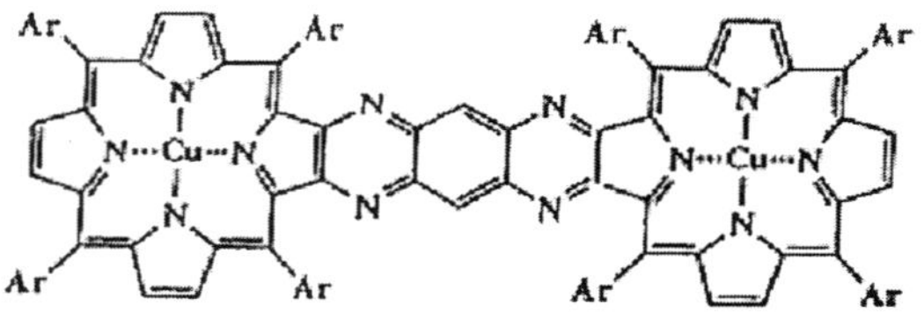

Fig. 17. Chemical structure of dimer-type porphyrin.

The full curves in Fig.18 show the orientation dependence of the X-band ESR spectra of the mixed LB films without HA. The external magnetic field is perpendicular and parallel to the film surface for the upper and lower curves, respectively. The spectra show anisotropy where septet copper hyperfine structures are weakly resolved for the lower curve. This shows that the plane normal of macrocycle points intermediate between the substrate normal and the film surface but slightly more inclined to the latter. No in-plane anisotropy was observed in this case and the case with HA discussed below, showing the molecular orientation is randomly distributed in the film plane.

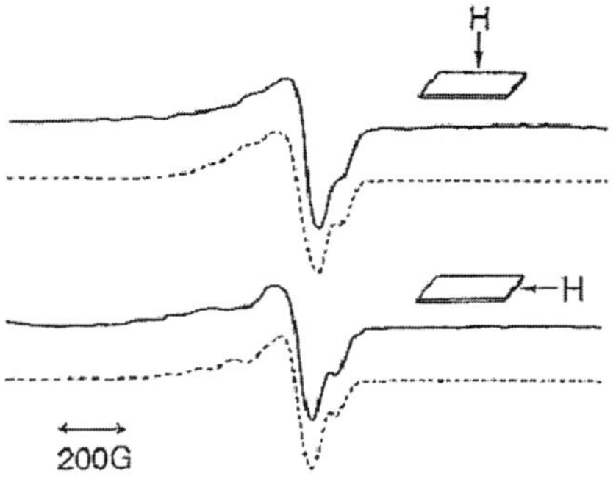

Fig. 18. Orientation dependence of the X band ESR spectra of a PD film without hexatriacontane at room temperature (solid line). The external magnetic field is perpendicular and parallel to the film surface for the upper and lower curves, respectively. Dashed line shows the calculated spectra.

The ESR spectra of the LB films containing HA are markedly different from those for the films without HA, as shown by the full curves in Fig. 19 for x=2.0. The direction of the external magnetic field is similar to those in Fig. 18 for the upper and lower curves. The hyperfine structure is not observed for the upper curve while the septet splitting owing to the parallel hyperfine component of the macrocycle is clearly resolved for the lower curve. This clear out-of-plane anisotropy shows that the plane normal of macrocycle is rather parallel to the film plane. The results also confirms that the porphyrins really exist as dimers with the case of J>>A. This situation can be further confirmed by ESR spectrum simulation.

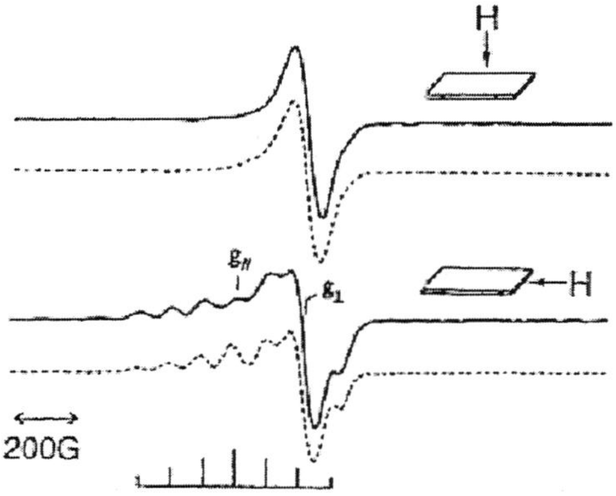

Fig. 19. Orientation dependence of the X band ESR spectra of a PD film with n-hexatriacontane (HA) with a molar ratio, PD:HA=2:3 (solid line) The external magnetic field is perpendicular and parallel to the film surface for the upper and lower curves, respectively. Dashed line shows the calculated spectra.

The broken lines in Figs. 18 and 19 show the calculated spectra according to the procedure applied to the monomer LB films cases discussed in the preceding section by taking the above septet splitting of the dimer into account in the present case. The simulation generally reproduces the observed results fairly well. The validity of the results have been further confirmed by the K-band ESR and simulation, as discussed in the case of merocyanine LB films in section 3.1.1. From these simulations, the most probable angle between the plane normal of the macrocycle and the substrate normal has been obtained as 58° for films without HA and 73° with HA. The former value agrees well with that of the monomer film while the latter value is slightly smaller than the corresponding monomer value of 80°. From the results of UV/Vis absorption spectroscopy, it has been shown that the long axes of the dimer are parallel to the substrate. Although the detailed mechanism of HA action in controlling the porphyrin is not completely clear, the above mentioned results of FTIR of HA-d in the case of PM LB films certainly shows the strong interaction between the HA and porphyrin molecules. The above difference between the orientations of macrocycle in the presence of HA between the monomer and dimer case would be related to the difference of coordinating space for inserting HA molecules due to the presence of the conjugated part of the dimer. Such complexation would be an interesting problem, for example, to study with molecular modelling techniques.

3.3. *Other examples*

As typical cases of concentrated spins, ESR studies of charge-transfer salts are briefly discussed in this section. These systems provide ultrathin conducting layers with molecular thickness and may be the possible candidate of molecular wires. Charge-transfer salts are either be semiconductors with low-dimensional magnetic properties or the systems with metallic conduction associated with disordered sites [18-20]. Unpaired electron spins arise from the mixed valence states of constituent molecules. Among them, 1:2 complex of long-chain pyridinium and TCNQ systems, which exhibit interesting magnetic and conducting properties, are discussed as typical examples. Figure 20 shows the chemical structures of two long-chain pyridinium-TCNQ salts, which are abbreviated as $C_{22}PT$ and APT(8-12), respectively [32,33]. The former possesses the ordinary long chain with 22 carbon atoms, while the latter includes the azobenzene unit in the long chain part.

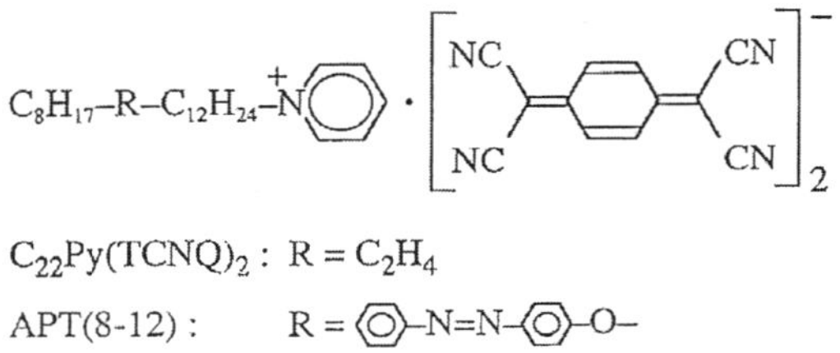

Fig. 20. Chemical structure of $C_{22}PT$ and APT(8-12)

In these systems, TCNQ possesses one unpaired electron per two molecules. The room temperature conductivity of the order of 0.01 S/cm suggests the formation of columnar structures of TCNQ molecules. From the spin susceptibility measurements of $C_{22}PT$, the system has been found to be the typical cases of Random-Exchange Heisenberg

Antiferromagnetic Chain (REHAC) system [12, 34]. REHAC system is characterized by linear chain antiferromagnet with the exchange integral J associated with the site of weak exchange of εJ, ($0 \leq \varepsilon < 1$) due to structural disorder with the concentration of *c* and hence the concentration of site of the regular exchange of (1-c). Temperature dependence of spin susceptibility χ shows a clear crossover from the regular antiferromagnetic chain behaviour at higher temperature to Curie-like behaviour at low temperatures characteristic of cluster magnet generated by cutting the regular chain by weak exchange. From the analysis, the exchange integral of J=700K was obtained with the defect concentration of *c*=0.07. This shows the regular columnar structure of TCNQ is formed accompanied by small amount of defect sites.

Similar behaviour of susceptibility was found also in APT system. The difference between the $C_{22}PT$ system was that the exchange interaction of the regular chain was alternating as J_1 and J_2 along the chain, that is, J_1=500K with J_2/J_1=0.7. The defect concentration was 0.03 in the as grown film where the azobenzene takes the *trans* form. The prominent feature of APT system is the photoisomerisation effect. When the *trans* form is converted to *cis* form by UV irradiation, conductivity increases. This change is reversible, that is, *cis* to *trans* back transformation by the other UV irradiation of different wavelength brings about the conductivity decrease. Despite of the increase of conductivity upon *trans* to *cis* isomerisation, the defect concentration was found to increase from 0.03 to 0.04. The overall mechanism of conductivity change may be governed by various factors and is beyond the scope of the present discussion.

From the ESR point of view, these systems provide typical example of spectrum simulation to deduce both the magnetic and structural parameters based on the ESR theory of quasi-one dimensional antiferromagnets [35,36] and such analysis has been successfully performed. This procedure requires more steps than the one of isolated spin cases discussed in the section 3.1, thus the only the result of APT case is presented here. Figure 21 shows the orientation dependence of the X-band ESR spectra of an APT LB film at room temperature with $\theta = 0°$, 45° and 90° (thick line) where θ denotes the angle between the external magnetic field and the plane normal. Thin line shows the simulated spectra. g_x, g_y and g_z denotes the g tensor components of the coordinate axes fixed to the one-dimensional TCNQ columnar structure. One of the prominent features of the quasi-one-dimensional antiferromagnetic chain in ESR is its typical dependence of the ESR linewidth with respect to the chain direction. ESR width is determined by the exchange-narrowed dipolar width and shows the angular dependence of

$3\cos^2\phi-1$, where ϕ denotes the angle between the chain direction and the external magnetic field. That is, the linewidth minimum occurs at the magic angle (ϕ=54°). In Fig. 21, the arrows in the middle spectrum show the structure corresponding to the linewidth minimum. Such linewidth minimum can act as spectral turning point as shown here.

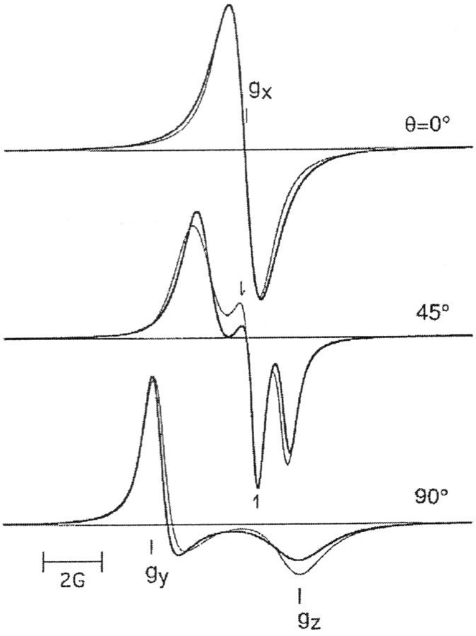

Fig. 21. Orientation dependence of the X band ESR spectra of an LB film of APT(8-12) at room temperature with θ = 0°, 45° and 90° (thick line). θ denotes the angle between the external magnetic field and the plane normal. Thin line shows the simulated spectra. The arrows in the middle spectrum show the structure corresponding to the linewidth minimum of the one-dimensional spin chain.

As shown by thin line, simulated spectra reproduce the observed curves fairly well. In the in-plane spectrum (θ=90°) the simultaneous observation of g_y and g_z shows the distribution of TCNQ columnar axis in the plane (two-dimensional powder spectrum). Lorentzian line shape of the top curve indicates that there is a finite interchain exchange interaction from the frame work of ESR theory of quasi-one-dimensional system. The analysis of line shape yields the interchain exchange integral of J'=0.1K. As for the structural information, the long axis of TCNQ molecule is nearly parallel to the plane normal and makes and angle of about 26°. It can be also deduced that the orientation distribution of the molecular plane is quite small (δ~5°), showing high orientational order existing in the system. As for the corresponding values in

$C_{22}PT$ system, J'=0.7K and the angle between the long axis of TCNQ and the plane normal of 22°. This difference may be reasonably ascribed to the difference of long-chain part, resulting in the modification of the structures. Further studies of LB films of APT analogues with different positions of azobenzene in long chain part can be found in the literature [20], which reports the in-situ observation of the effect of *cis-trans* isomerization of azobenzene by using ESR. These studies indicate that high orientational order existing in LB films that can be modified with the introduction of photoisomerised groups such as azobenzene.

Besides the above examples, other important cases of LB films of charge transfer salts are the systems with considerably high conductivity. In these cases, conduction electron spin resonance can be observed from which the Pauli spin susceptibility of conduction electrons yields the information of state density at the Fermi level [19, 37].

ESR studies of other charge transfer salts can be found in the literature [38, 39] . As for the systems with other kinds of paramagnetic species discussed in the present chapter, spin labels are another kinds of spins often incorporated in LB films [40-43]. Metallic ion other than Cu^{2+} or Mn^{2+}, such as Fe^{2+}, Co^{2+} or Ni^{2+} were also studied in stearate LB films [44]. In addition to the spins contained in the ground state, photogenerated spins, the example of which has been discussed in the section 3.1.3, obviously provide another class of spins, for which ESR can play fundamental role in their characterization.

4. CONCLUDING REMARKS

ESR spectroscopy has been shown to be useful in elucidating molecular orientation and electronic states of functional molecules in LB films. We have discussed isolated radical molecules of dyes, porphyrins bearing paramagnetic ions, strongly interacting spins in the mixed valence state in organic charge transfer salts, as typical examples of ESR application. Two control methods of molecular orientation have been presented based on the ESR analysis, that is, the flow orientation for in-plane orientation control by using semi-macroscopic phenomena, and the trigger molecule method which is based on the microscopic molecular interactions between the host and added trigger molecules. ESR spectroscopy of strongly interacting spins in the mixed valence states of charge transfer salts also provide the structural information in addition to the information concerning the electronic states. LESR spectroscopy would further extend the area of ESR application by detecting the radicals generated by photoinduced charge transfer in molecular aggregates. Not like in single crystals, the orientation distribution in LB films results in the appearance of structures either associated with different g values or hyperfine structures, broadening the ESR spectra etc. Even in this case, the emphasis is made on the importance of quantitative analysis, which is useful in revealing the essential features of the system studied. One of the functional molecules in recent molecular electronics are conjugated polymers which exhibit high conductivity as well as electroluminescence [45, 46]. In these systems, nonlinear excitations such as solitons and polarons are attracting attention and are observed by ESR and electron-nuclear double-resonance (ENDOR) techniques [28, 29, 47]. These nonlinear excitations may be also studied in LB films. Cluster molecules such as fullerens are also utilized in fabricated molecular assemblies. If we consider that there are increasing numbers of new organic systems developed by organizing functional molecules including in nanotechnology area, there would be more cases where ESR and related spectroscopies can play essential roles in the microscopic analyses.

5. ACKNOWLEDGMENT

The present author is grateful to M. Sugi, K. Ikegami, M. Matsumoto, R. Azumi and H. Tachibana for valuable discussions. S. Yasui is thanked for the synthesis of merocyanien dyes.

6. REFERENCES

1. H. Kuhn, D. Möbius and H. Bücher, in “Techniques of Chemistry,” Vol.1, A. Weissberger and B.W. Rossiter, Ed., Wiley, New York, (1973), Part IIIB
2. S. Kuroda, Colloids Surf., 72 (1993) 127.
3. J. Messier and G. Marc, J. Phys. (Paris), 32 (1971) 799.
4. P.A. Chollet, J. Phys. C7 (1974) 4127.
5. M. Pomerantz, Solid State Commun., 27 (1978) 1413.
6. A.F. Janzen, J.R. Bolton and M.J. Stillman, J. Am. Chem. Soc., 101 (1979) 6337.
7. J. Cunningham, E.E. Polymeropoulos, D. Möbius and F.Baer, Magnetic Resonance in Colloid and Interface Sciences, D. Reidel, Dordrecht, 1980, p. 603.
8. M. Vandevyver and A. Barraud, J. Mol. Electron., 4 (1988) 207.
9. S. Kuroda, K. Ikegami, M. Sugi and S. Iizima, Solid State Commun., 58 (1986) 493.
10. S. Kuroda, K. Ikegami, K. Saito, M. Saito and M. Sugi, Thin Solid Films, 159 (1988) 285.

11. S. Kuroda, K. Ikegami, Y. Tabe, K. Saito, M. Saito and M. Sugi, Phys. Rev. B, 43 (1991) 2531.
12. S. Kuroda, K. Ikegami and M. Sugi, Mol. Cryst. Liq. Cryst., 190 (1990) 111.
13. R. Azumi, M. Matsumoto, Y. Kawabata, S. Kuroda, M. Sugi, L.G. King and M.J. Crossley, J. Am. Chem. Soc., 114 (1992) 10662.
14. R. Azumi, M. Matsumoto, Y. Kawabata, S. Kuroda, M. Sugi, L.G. King and M.J. Crossley, J. Phys. Chem., 97 (1993) 12862.
15. R. Azumi, M. Matsumoto, S. Kuroda and M.J. Crossley, Langmuir, 11 (1995) 4056.
16. S. Kuroda, K. Ikegami, Y. Tabe, K. Saito, M. Saito and M. Sugi, Thin Solid Films, 210/211 (1992) 512.
17. S. Kuroda, K. Murata and K. Ikegami, Thin Solid Films, 284/285 (1996) 615.
18. K. Ikegami, S. Kuroda, M. Sugi, T. Nakamura, H. Tachibana, M. Matsumoto and Y. Kawabata, J. Phys. Soc. Jpn., 61 (1992) 2930.
19. K. Ikegami, S. Kuroda, T. Nakamura, G. Yunome, M. Matsumoto S. Horiuchi and G. Saito, Phys. Rev. B, 49 (1994) 10806.
20. K. Ikegami, S. Kuroda, H. Tachibana and M. Matsumoto, J. Phys. Soc. Jpn., 65 (1996) 237.
21. See for example, J.E. Wertz and J.R. Bolton, Electron Spin Resonance, McGraw-Hill, New York, 1972.
22. S. Kuroda, M. Sugi and S. Iizima, Thin Solid Films, 99 (1983) 21.
23. M. Sugi and S. Iizima, Thin Solid Films, 68 (1980) 199.
24. S. Kuroda, M. Sugi and S. Iizima, Thin Solid Films, 133 (1985) 189.
25. V. Czikkley, H.D. Försterling and H. Kuhn, Chem. Phys. Lett., 6 (1970) 11, 207.
26. N. Minari, K. Ikegami, S. Kuroda, K. Saito, M. Saito and M. Sugi, J. Phys. Soc. Jpan., 58 (1989) 222.
27. S. Kuroda, Colloids Surf., A 198-200 (2002) 735
28. S. Kuroda, K. Murata, Y. Shimoi, S. Abe, T. Noguchi and T. Ohnishi, ed. K. Kajimura and S. Kuroda, in Materials and Measurements in Molecular Electronics, Springer Proceedings in Physics, 81 (1996) 256.
29. S. Kuroda, K. Marumoto, H. Ito, N.C. Greenham, R.H. Friend, Y. Shimoi and S. Abe, Chem. Phys. Lett., 325 (2000) 183.
30. N. Kato, K. Saito, H. Aida and Y. Uesu, Chem. Phys. Lett., 312 (1999) 115.
31. R. Azumi, M. Matsumoto, S. Kuroda and M.J. Crossley, Langmuir, 11 (1995) 4495

32. M. Matsumoto, H. Tachibana and T. Nakamura, in "Organic Conductors", J.P. Frages, Ed., Marcel Dekker, New York (1994) p.759.
33. H. Tachibana, T. Nakamura, M. Matsumoto, H. Komizu, E. Manda, H. Niino, A. Yabe and Y. Kawabata, J. Am. Chem. Soc., 111 (1989) 3080
34. Z.G. Soos and S.R. Bondeson, Solid State Commun., 35 (1980) 11.
35. R.E. Dietz, F.R. Merritt, R. Dingle, D. Hone, B.G. Silbernagel and P.M. Richards, Phys. Rev. Lett., 26 (1971) 1186.
36. M.J. Hennesy, C.D. McElwee and P.M. Richards, Phys. Rev. B, 7 (1973) 930.
37. H. Ohnuki, M. Izumi, Y. Ishizaki, M. Nagata, T. Imakubo and R. Kato, Thin Solid Films, 327-329 (1998) 458.
38. J. Richard, M. Vandevyver, P. Lesieur, A. Ruaudel-Teixier, A. Barraud, R. Bozio and C. Pecile, J. Chem. Phys., 86 (1987) 2428.
39. J.P. Morand, R. Lapouyade, P. Delhaes, M. Vandevyver, J. Richard and A. Barraud, Synth. Met., 27 (1988) B569.
40. G. Martini, F. Bonosi, M.F. Ottaviani and G. Gabrielli, Thin Solid Films, 178 (1989) 279.
41. M. Romanelli, F. Bonosi, X. Chen and L. Kevan, Langmuir, 7 (1991) 1039.
42. K. Suga, Y. Iwamoto, H. Shibata, N. Yamada and Y. Fujihira, Thin Solid Films, 327-329 (1998) 837.
43. D. Zhang, Y. Xu, L. Ding Y. Liu, D. Zhu, Chem. Phys. Lett., 304 (1999) 236.
44. Y. Ando, T. Hiroike, T. Miyashita and T. Miyazaki, Thin Solid Films, 278 (1996) 144.
45. A.J. Heeger, S. Kivelson, J.R. Schrieffer and W.P. Su, Rev. Mod. Phys., 60 (1988) 781.
46. R.H. Friend, R.W. Gymer, A.B. Holmes, J.H. Burroughes, R.N. Marks, C. Taliani, D.D.C. Bradely, D.A. Dos Santos, J.L. Bredas, M. Lögdlund, W.R. Salaneck, Nature 376 (1999) 121.
47. S. Kuroda, Int. J. Mod. Phys., 9 (1995) 221.

CHAPTER 7
BIOTIN-STREPTAVIDIN SENSOR SURFACE: A VERSATILE PLATFORM FOR PERFORMING DNA HYBRIDIZATION INTERACTIONS

Dev Kambhampati [a, b] and Wolfgang Knoll [a]

[a] Max Planck Institut für Polymer Forschung, Ackermannweg 10, D55128 Mainz, Germany.
[b] Xomics Surface Technologies, Kraus Strasse 14, D89077 Ulm, Germany.

The properties of the biotin-streptavidin sensor system such as surface topology, regeneration of probes and non-specific interaction effects are investigated in detail by using DNA-DNA hybridization interactions as the test system. Surface Plasmon Resonance (SPR) and SPR enhanced fluorescence spectroscopy (SPRFS) detection schemes were widely employed to monitor the effect of probe/target length, mismatches and temperature on the hybridization kinetics. 1:1 Langmuir kinetic models were found to be sufficient to describe the hybridization interactions.

1. INTRODUCTION

The design and assembly of bio-functionalized surface architectures [1] is one of the most challenging aspects of materials science, with potential implications in diverse areas such as the construction of molecular machines [2], nanostructured materials [3], sensor development [4,5] and in the fundamental studies of a wide variety of biomolecular recognition events [6-9]. Numerous surface formats have been proposed for sensoric applications using a wide range of substrates such as glass, gold and plastics. While rapid advances continue in the fields of sensor surface chemistry and detection technologies, the biotin-streptavidin technology [10, 11] that was discovered over two decades ago still continues to play a pivotal role in most aspects of molecular biology and biosensor development. The elegance of this technology can be traced to the extremely high affinity between biotin and streptavidin (Kd ~ 1013 M^{-1}), which is one of the strongest non-covalent receptor-ligand interactions in nature. Rapid strides in research and technology to functionalize the biomolecules (nucleic acids, proteins, etc.) with biotin has led to a phenomenal growth in their biotechnology applications, ranging from affinity targeting, flow cytometry, epitope mapping, and immunodiagnostics to the areas of drug delivery and medical imaging.

Sensors based on the biotin-streptavidin system hold great promise in investigating the intricacies of various biomolecular interaction events. Self-assembly techniques [12-14] have been widely used to construct such sensor architectures since it is possible to exquisitely tailor the overall properties of the sensor architecture by individually controlling the chemical nature of the building blocks. Based on the pioneering work of Knoll and co-workers [15], numerous biotin-streptavidin sensors have been proposed on gold and glass substrates using mixed biotinylated self assembled monolayers [16-19]. Using such sensor systems, the kinetics of numerous nucleic acids, proteins and saccharide biomolecular interactions have been reported using real-time detection technologies such as surface plasmon resonance (SPR) [20-22] and SPR-enhanced fluorescence spectroscopy (SPRFS) [23].

Recently, the field of nucleic acid sensors has seen a surge in interest due to the deciphering of the initial code of the Human Genome. Understanding the properties of DNA hybridization interactions has become critical in various genotyping, gene expression, pharmacogenomic and diagnostic studies. High quality data is required for discriminating effectively between the various mismatches and the interaction behavior of DNA strands with similar properties. To address the above challenges, our group [24-26] has been actively developing novel experimental protocols to gather high content information based on kinetic studies. Numerous nucleic acid hybridization interactions have been quantitatively analyzed using the kinetic rate (k_{on} and k_{off}) and affinity constants, K_A.

Here, we extend our studies in this field by noting the practical aspects of this sensor technology pertaining to surface topology, sensor surface regeneration for multiple usage, and non-specific interaction analysis. Kinetic studies of DNA-DNA hybridization interactions was performed by varying parameters such as mismatches, probe/target length and temperature to highlight the versatile nature of the biotin-streptavidin sensor system. Real-time detection technologies such as SPR and SPRFS were used to monitor such interactions.

2. EXPERIMENTAL METHODS

2.1. Materials

Biotinylated DNA probes and MR-121 labeled fluorescent DNA targets were procured from Boehringer Mannheim. Cy5 labeled DNA targets (T2, T3, T4) were obtained from MWG Biotech. The sensor surface regeneration solutions, viz. sodium hydroxide (50 mM, Biacore AB) and HEPES buffer (pH 4.5, Aldrich) were used directly as received. PBS Buffer (10 mM phosphate, 2.7 mM KCl, 150 mM NaCl, pH 7.4, Sigma) was used to dissolve the nucleic acid solutions and was also used as the hybridization buffer.

Nucleic Acid Probe Sequences:

DNA Probe (15 T spacer): 5'- T15 – TGT ACA TCA CAA CTA – 3' (P1)

DNA Probe (30 T spacer): 5'- T30 – TGT ACA TCA CAA CTA – 3' (P2)

Nucleic Acid Target Sequences:

15mer Mismatch 1- 3' – ACA TGC AGT GTT GAT – MR 121 – 5' (T1)

25mer Mismatch 1- 3' – T5 ACA TGC AGT GTT GAT T5 – Cy5 – 5' (T2)

45mer Mismatch 1- 3' – T15 ACA TGC AGT GTT GAT T15 – Cy5 – 5' (T3)

75mer Mismatch 1- 3' – T30 ACA TGC AGT GTT GAT T30 – Cy5 – 5' (T4)

45mer Mismatch 0- 3' – T15 ACA TGT AGT GTT GAT T15 – Cy5 – 5' (T5)

75mer Mismatch 0- 3' – T30 ACA TGT AGT GTT GAT T30 – Cy5 – 5' (T6)

15mer Mismatch 0- 3' – ACA TGT AGT GTT GAT – MR 121 – 5' (T7)

2.2. Surface Architecture

A mixed self-assembled monolayer (SAM) architecture of biotinylated and hydroxy terminated thiols (Fig 1a) in the ratio of 1: 9 was used to assemble the streptavidin sensor layer on gold substrates (47nm). The typical concentration of the mixed thiol solution was 5 $*10^{-4}$ M. Streptavidin solutions with concentrations ranging between sub-micro to a few micro M were found to be sufficient to assemble a well-defined monolayer. The schematic of the biotin-streptavidin sensor system is shown in Fig. 1b. Biotinylated probes with varying concentrations were subsequently immobilized on top of the streptavidin surface and were used in hybridization studies with fluorescent labeled targets in solution.

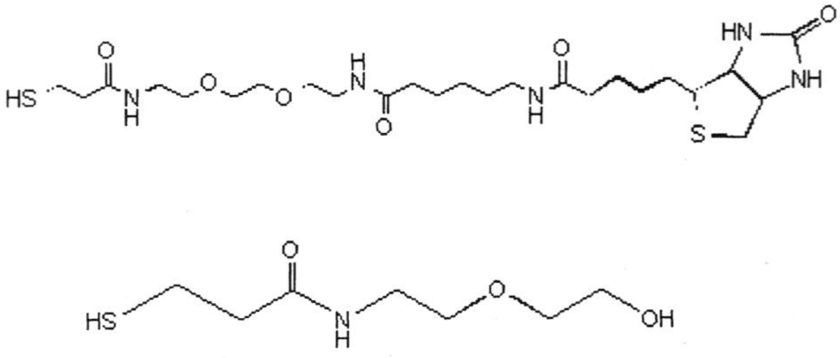

Fig. 1 (A) Structural formula of the biotinylated and hydroxy terminated thiols that were used to generate a stable streptavidin monolayer.

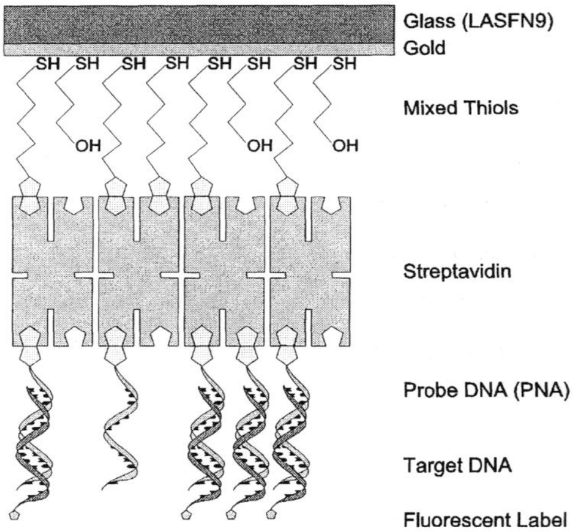

Fig. 1 (B) Schematic of the biotin-streptavidin sensor architecture.

Commercially available dextran surfaces conjugated with streptavidin (Biacore) were also used to monitor certain DNA hybridization interactions. The schematic of this sensor architecture is shown in Fig. 2. The subsequent hybridization analysis is similar to the above mentioned mixed SAM architecture.

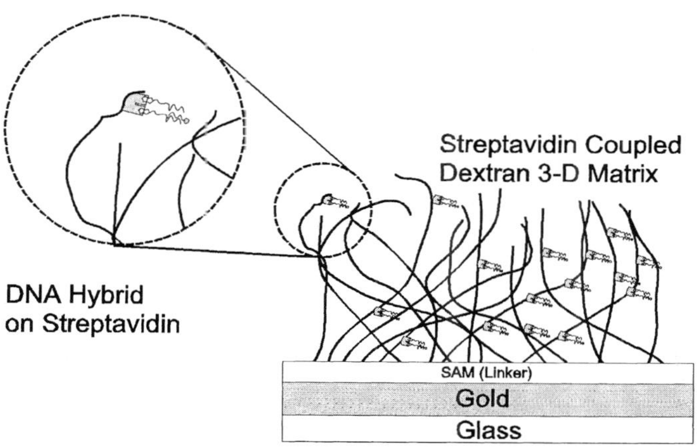

Fig. 2. Schematic of the streptavidin conjugated dextran sensor surface.

2.3. Detection Techniques

Surface Plasmon Resonance detection technology was used to monitor the hybridization interactions in real-time. For measurements conducted on the dextran surface, SPR detection mode (BIACORE 3000) was sufficient to detect the hybridization response. It is important to note that the surface density of the probes in the case of the streptavidin derivatized dextran chip is much higher than in the mixed SAM architecture, resulting in higher refractive index shifts. Also, a microfluidic liquid handling system was used in conjunction with the detection optics, resulting in better flow and exchange of liquids during the hybridization interaction.

In the case of the mixed SAM architecture, the resonance shifts for the hybridization interactions (probe-target interactions) are small, and as a result, plain SPR alone is not sufficient to accurately detect these interactions. To improve the signal-to-noise ratio, fluorescent labeling of the DNA targets was used. These labels lead to an amplification in the detection signal and thus, offer an interesting way of monitoring such low resonance shift interactions. These interactions were monitored by using the SPRFS detection scheme. A research grade cuvette (flow-through cell) was used for assembling the SAM and performing the hybridization studies. The cuvette allows for continuos flow only during the rinsing step (or desorption step).

2.4. Kinetic Models

A quantitative analysis of the DNA hybridization interactions was carried out by using 1:1 Langmuir kinetic models. The specific equations for determining the rate (k_{on}, k_{off}) and affinity constants (K_A) are described below:

$$R(t) = R_{max} (1\text{-}exp(-t/\tau)) \quad (1)$$

$$1/\tau = k_{on}\, c_o + k_{off} \quad (2)$$

$$R(t) = R_{max} \exp(-k_{off}\, t) \quad (3)$$

$$K_A = k_{on} / k_{off} \quad (4)$$

Equation 1 describes the association process, where R(t) indicates the SPR (or SPRFS) response at a given time and R_{max} indicates the final equilibrium response. In the case of the SPR response, the change in refractive index is measured either in the so-called refractive index values (RU, BIACORE) or in approximate units (a.u.). In the SPRFS measurements, the fluorescence intensity values are measured and are used in determining the above rate and affinity constant values. The association (k_{on}) and dissociation (k_{off}) rate constants are related to each other via an alternative time constant, τ (Eq. 2). Equation 3 described the dissociation process, and the koff parameter is obtained by fitting the dissociation segment of the hybridization interaction. Based on the k_{on} and koff values, the overall affinity constant, K_A, of the interaction is determined by using Eq. 4.

3. RESULTS AND DISCUSSIONS

3.1. Sensor Surface Topology

Earlier reports [15] on biotin-streptavidin architectures have shown that the surface coverage of streptavidin on biotinylated SAMs is around 53 % (from refractometric measurements). Even in model studies employing diffraction detection techniques, the surface coverage of the two-dimensional streptavidin crystals on a biotinylated surface was found to be about 66 %. Ellipsometric measurements on streptavidin monolayer assembled on biotinylated lipid layers assembled at the air-water interface indicate a total surface coverage of ~ 65%. Here, we performed AFM studies to check for the surface topology and to determine the distribution of the streptavidin molecules on the biotinylated SAM architecture assembled on gold coated mica substrates.

Tapping mode AFM studies were formed at every step of the sensor buildup, starting from the mixed thiol architecture (Fig. 3a) to the formation of the streptavidin surface (Fig. 3b). While the biotinylated mixed SAM architecture looks homogeneous on the mica terraces, the surface architecture changes dramatically after the streptavidin immobilization step. Since streptavidin is a bulky molecule, the above increase in surface roughness and heterogeneity is not surprising. Several pockets of streptavidin appear to cluster together and the underlying thiol layer is also visible in certain regions. These images provide further support to the above observations regarding the relatively low packing density of streptavidin on SAM surfaces. The distribution of the biotinylated DNA probes is shown in Fig 3c. The probes are uniformly distributed over the streptavidin surface. No further noticeable changes in the surface topology were observed upon subsequent hybridization with the targets. Later, we will show that in spite of the surface heterogeneity and roughness of the streptavidin layer, which can result in different orientations of the DNA probes, the hybridization interactions were found to adhere to the predictions of a simple 1:1 Langmuir kinetic model.

3.2. Non-specific Interaction Analysis

If the targets interact non-specifically with the sensor surface, then the entire hybridization response will be compromised. Thus, it is important to minimize the non-specific interactions between the sensor surface and the various biomolecules. Although the hydroxy terminated thiols act as the passivation layer and minimize non-specific interactions with the sensor substrate, complete elimination of such interactions is not possible. Also, since DNA is charged, certain electrostatic interactions with the sensor surface is inevitable and thus, it is important to get an estimate of this non-specific contribution to the final hybridization response. One method of determining the extent of non-specific interactions is by injecting the fluorescent labeled DNA targets on the probe free streptavidin surface. This test study was performed by injecting a 1μM solution of T1 targets on the streptavidin sensor surface (Fig. 4).

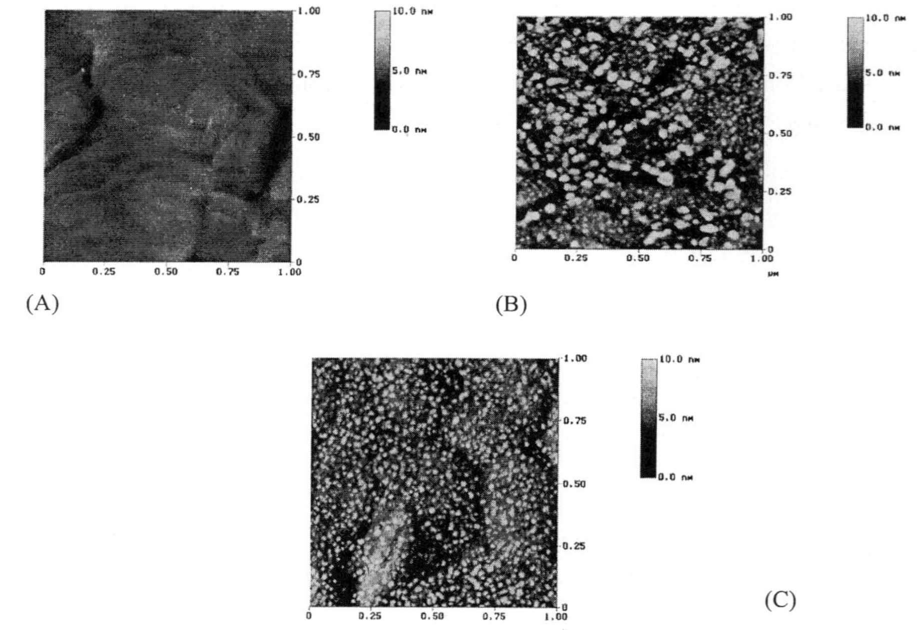

Fig. 3. Surface topology of the mixed self assembled monolayer (A), streptavidin (B) and DNA probes (C) on gold coated mica substrates.

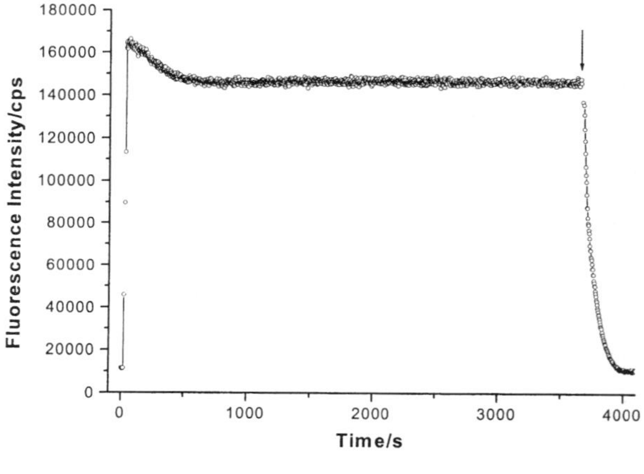

Fig. 4. Non-specific interaction studies using the T1 target. Since the fluorescence signal returns to the background level upon rinsing, no non-specific interaction of the target is expected on the streptavidin surface.

The initial rise in the fluorescence signal is attributed to the targets that are all lying within the exponentially decaying surface plasmon field, and as a result, are excited by it. Upon rising with buffer, the entire fluorescent signal is lost indicating that there is no non-specific interaction of the targets on the streptavidin surface. Even if some targets interacted non-specifically with the thiol layer, it is highly likely that their signal will be quenched due to their close proximity to the gold surface.

3.3. Regeneration of Sensor Surface

Some of the drawbacks of most biomolecular interaction studies is the limited availability of biological samples (DNA, Proteins, etc.) and their cost. Thus, from a practical point of view, it is important that the surface immobilized probes be regenerated so that they can be used in multiple experiments. Numerous regeneration agents such as acids (HCl, HEPES buffer), bases (NaOH), DMSO, Urea, etc, have been used in regeneration efforts.

These regeneration agents disrupt the probe-target hybrid interactions, leading to the dissociation of the targets. Once the targets are dissociated, the fluid cell is rinsed with hybridization buffer in order to flush out the targets. Thereafter, the probes can be re-used for conducting additional sets of hybridization experiments. Now, although regeneration agents are efficient in dissociating the targets, they also invariably affect the sensor architecture by causing damage to the underlying layers. To further illustrate this case, hybridization studies were performed between P1 probes (5μM) and mismatch-1 targets (T1, 1μM). Once the first hybridization interaction was conducted (Fig. 5), the hybrids were dissociated using HEPES buffer (pH 4.5). Once the fluid cell was flushed copiously with PBS buffer, a fresh solution of the same target was subsequently injected to perform the second experiment using the same set of probes. The second hybridization interaction displayed lower equilibrium fluorescence intensity level that was about 15 % lower than that observed during the first hybridization experiment. Some of the above loss can be attributed to the desorption of the biotinylated probes with time, although most of this loss arises from the desorption (erosion) of the underlying sensor layers.

To address the nature of such losses, a 50mM NaOH solution was used as a test regeneration agent to check whether this solution affects only the topmost sensor layers or the entire sensor assembly in general. The effect of NaOH on the biotinylated ($5* 10^{-4}$ M) mixed SAM architecture is shown in Fig. 6a. The NaOH injections (indicated by normal arrows) were followed by PBS rinsing (indicated by block arrows) during the test measurements. It is clear from the above six injections that the SAM architecture loses material constantly upon exposure to the basic regeneration solution. To check whether this basic treatment affects the natural binding characteristics of biotin, a streptavidin solution ($5 * 10^{-6}$ M) was injected over the biotinylated SAM surface. As expected under normal conditions, a rapid interaction of streptavidin (Fig. 6b) with the biotinylated surface was observed, indicating that the regeneration solution does not affect the binding properties of the biotinylated SAM architecture. The subsequent treatment of the streptavidin layer with the NaOH solution leads to a steady drop in the sensor thickness during each injection cycle. This drop, however, seems to be slightly lower than that observed in the case of the thiol architecture. Since it was earlier shown that the thiol layer is also affected by the basic treatment, it is difficult to determine what fraction of the streptavidin layer loss is purely coming from streptavidin or from the biotin-streptavidin combination in general.

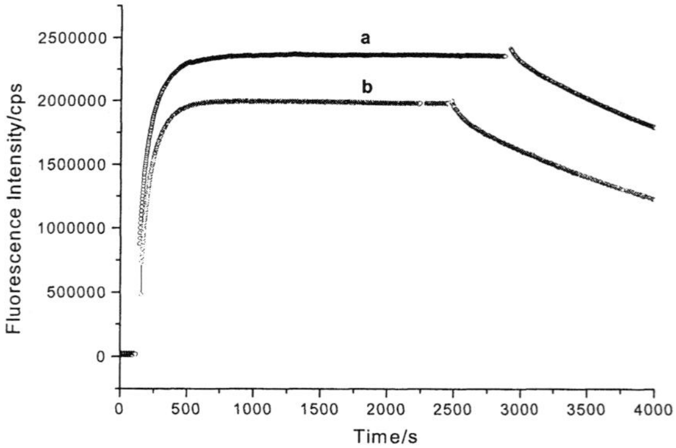

Fig. 5. Effect of regeneration agents (in this case, HEPES buffer, pH 4.5) on the hybridization interactions. Regeneration agents can result in loss of sensor architecture, leading to the drop in the final fluorescence intensity values between the first (a) and subsequent (b) hybridization interactions.

Similar to the underlying biotin-streptavidin interactions, the above basic treatment of streptavidin does not affect the its interaction with the biotinylated probes (P1, 1 μM). However, once the probes (Fig. 6c) are immobilized, subsequent injections of NaOH leads to a significant drop in the SPR signal (corresponding to the loss of sensor material), indicating that the nucleic acid probes are affected slightly more than the underlying biotin-streptavidin sensor assembly. As explained above, some of the above losses can be attributed to the erosion of the underlying layers.

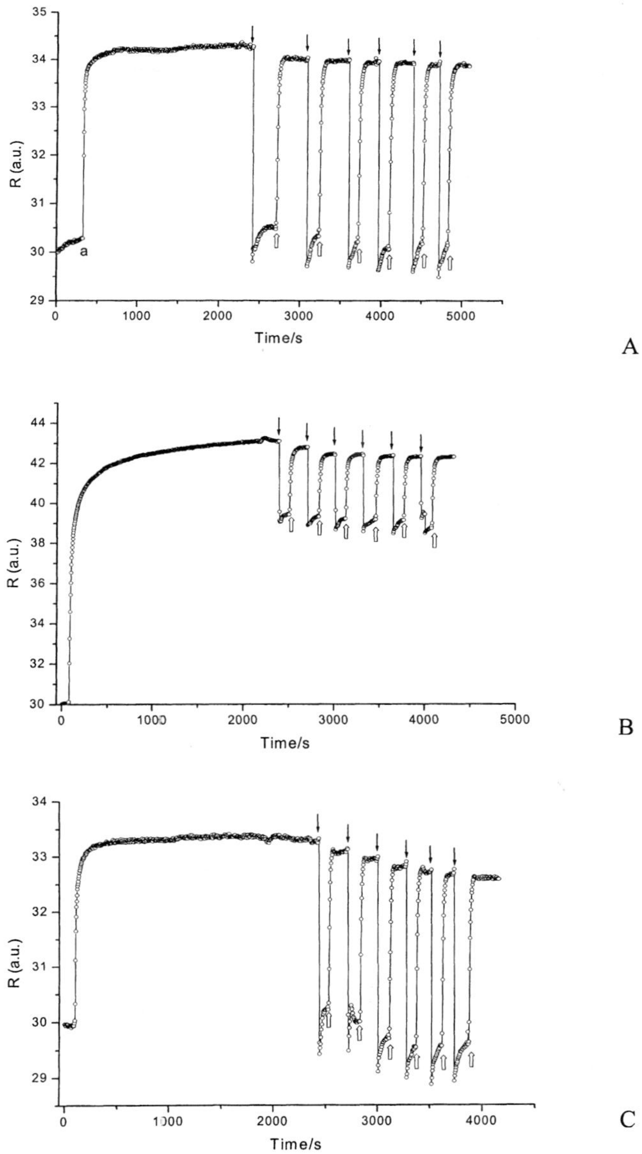

Fig. 6. Effect of NaOH regeneration solution on the mixed self assembled monolayer (A), streptavidin (B) and the DNA probe (C) layers. The short arrows indicate points of NaOH injections while the block arrows indicates the shift to PBS buffer rinsing.

It is clear from the above results that proper care should be taken while using regeneration solutions in order to minimize the loss of the sensor architecture. Such losses can directly interfere with the hybridization kinetics and the absolute values of the final equilibrium levels. Some of the above losses can be minimized by reducing the time of exposure (1-2 minutes is normally recommended) to these regeneration agents. For weaker interactions, a one-minute injection of 0.1 M HCl would suffice.

3.4. DNA Hybridization Interactions using P1 probes (15 T spacer and 15mer recognition sequence)

In our earlier studies [24, 25], we have reported extensively on the interaction of P1 probes with 15mer targets. Here, we will focus primarily on longer targets and describe the effect of target length on the hybridization kinetics and also check if the Langmuir 1:1 kinetic model holds true or not when global analysis experiments are conducted using long DNA targets. Unless otherwise mentioned, all hybridization interactions were conducted at 22°C.

The interaction of mismatch 1 targets of varying lengths (T2-T4, 100nm) with the P1 probes (100nM) is shown in Fig. 7. All the targets displayed fast interaction kinetics with the probes, and such interactions could be simulated using the 1:1 Langmuir kinetic model. Using equations 1,3 and 4, the kinetic rate and affinity constants were calculated (Table 1a) for the above interactions. From this table, it is clearly evident that as the length of the target increases, there is a steady drop in the overall affinity constant values. This trend is to be expected since the actual recognition sequence amongst all these targets is still a 15mer, and as the target length increases they face increasing steric and conformational restrictions while locating the complementary recognition on the surface immobilized smaller DNA probes. This effect is more pronounced on longer targets which is evident from the drop in the kon values. The stability of these hybrids is also lowered due to the non-interacting T's which are present at both ends of the targets. Thus, during the rinsing step, faster koff values are observed leading to an overall drop in the affinity constants.

Another set of experiments were performed by using mismatch 0 targets (T5-T6, 100nm). These studies are important to determine whether good mismatch discrimination can be observed in the case of the long strands, and also to verify if the longer mismatch 0 targets behave similar to the trend observed in the case of the mismatch 1 targets. The interactions of the T5 (45mer) and T6 (75mer) mismatch 0 targets with the P1 probes is shown in Fig. 8.

Table 1. Kinetic parameters for mismatch 1 (a) and mismatch 0 (b) target interactions with P1 probes. A drop in the affinity constant values with increasing target length is clearly observed here.

P1 Probe	Target	k_{on} (M^{-1} s^{-1})	k_{off} (s^{-1})	K_A (M^{-1})
	25	27869.87	$5 * 10^{-4}$	$5.57 * 10^{7}$
Mismatch 1	45	7949.981	$8 * 10^{-4}$	$9.94 * 10^{6}$
(a)	75	10474.89	$1.4 * 10^{-3}$	$7.48 * 10^{6}$
Mismatch 0	45	27533.33	$5.8 * 10^{-4}$	$4.74 * 10^{7}$
(b)	75	23545.57	$8.1 * 10^{-4}$	$2.91 * 10^{7}$

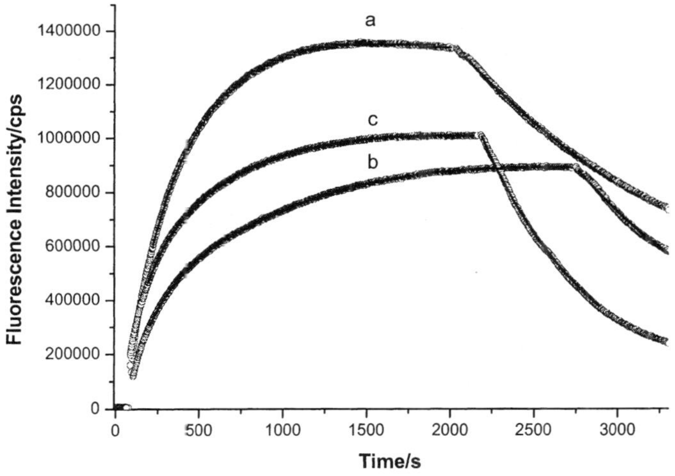

Fig. 7. Mismatch 1 hybridization kinetics between the long strand targets and the surface immobilized P1 probes. Here, a,b & c indicate the hybridization responses for the T2, T3 & T4 targets respectively.

The kinetic rate and affinity constants for these interactions are shown in Table 1b. It is interesting to note that unlike in the case of 15-mer mismatch 0 [24] hybridization interactions which displayed very low koff rates (data not shown), the above interactions displayed faster desorption rate constants. Also, as the length of the targets increases, we observe a similar drop in the k_{on} rates and an increase in the k_{off} rates of the hybrids, as experienced in the case of the mismatch 1 system. The affinity constant also decreases with increasing length of the targets. The final fluorescence intensity level of the 75mer target is higher than the 45mer target, and this can be attributed to the location of the fluorophore, which in the case of the 75mer might be located slightly away from the gold surface (and as a result, experiencing slightly lower quenching effect).

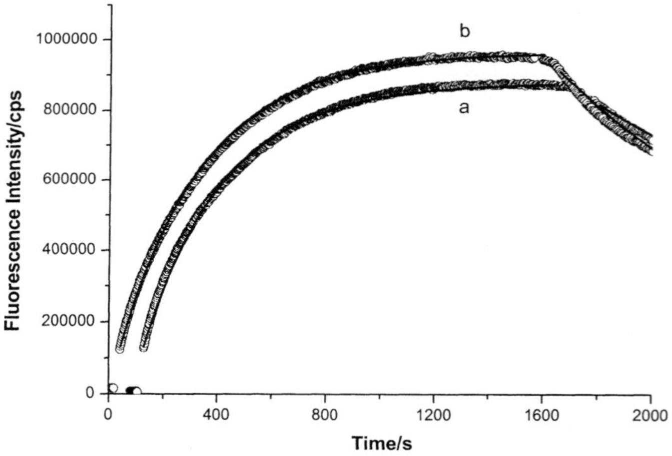

Fig. 8. Mismatch 0 hybridization kinetics between the long strand targets and the surface immobilized P1 probes. Here, a & b indicate the hybridization responses for the T5 & T6 targets.

It is also interesting to note (cf. Figs. 7, 8 and Table 1) that the affinity constant and the final intensity level of the 25mer (T2) mismatch 1 target is higher than both the 45- and 75-mer mismatch 0 targets. From this table, it is also clear that for longer targets, the differences in the affinity constant values, which is essential for mismatch discrimination, between the mismatch

0 & 1 targets also reduces substantially. This trend is in stark contrast to the 15mer hybridization interactions, where a few orders of magnitude differences were observed between the mismatch 0 & 1 targets.

Another rigorous test of the 1:1 kinetic model can be performed by conducting global analysis experiments using long targets (75mer, T4) with varying concentrations. For such experiments, a 1μM solution of the P1 probes was used and the concentration of the targets was varied between 50nm- 1μM. The interaction of these targets with the probes is shown in Fig. 9a. The τ analysis (Eq. 3) of the individual hybridization responses (Fig. 9b) is shown in Table 2. The k_{off} values displayed in the same table were calculated from the dissociation segments of the above responses. The k_{on} rate constant was determined from the slope of Fig. 9b, and was calculated as 2665.7 M^{-1} s^{-1}. From Table 2, the average k_{off} value was found to be 1.18 * 10^{-3} s^{-1}. By using Eq. 4, the overall affinity constant for this interaction was calculated as 2.26 * 10^{6} M^{-1}. Now, this value is lower than that observed in the case of the individual P1-T4 (Table 1a) interactions that were performed by using 100nm probe and target concentrations. The reduction in the affinity constant at higher probe concentrations could result from the close packing of the probes, which might impose certain steric restrictions for the longer targets in locating the recognition sequence. From a practical point of view, higher surface probe coverages are required while performing global analysis experiments since these experiments are performed over a long time interval (in this particular case, the total duration of the experiment was six times longer than in the individual experiments). Even if some probes desorb from the surface, there is still plenty of them left to conduct good kinetic analysis. If lower probe concentrations (e.g., 100nM) are used, the steady desorption of the probes with time might end up compromising the kinetic results.

Tab. 2. Kinetic parameters for global analysis experiments using the P1-T4 hybrid system. The k_{off} value was obtained by fitting the dissociation curve while τ was calculated from the association segment of the kinetic curve.

Concentration (nM)	τ	k_{off} (s^{-1})
50	505	1 *10^{-3}
100	467	1.1 *10^{-3}
250	340	1.1 *10^{-3}
500	300	1.3 *10^{-3}
1000	222	1.4 *10^{-3}

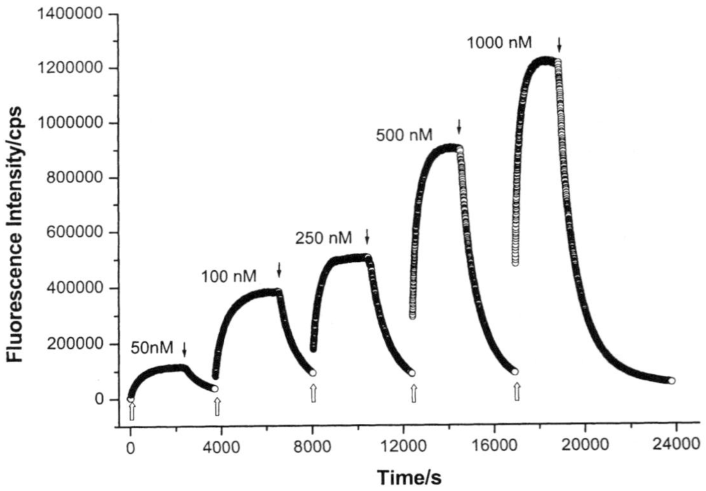

A

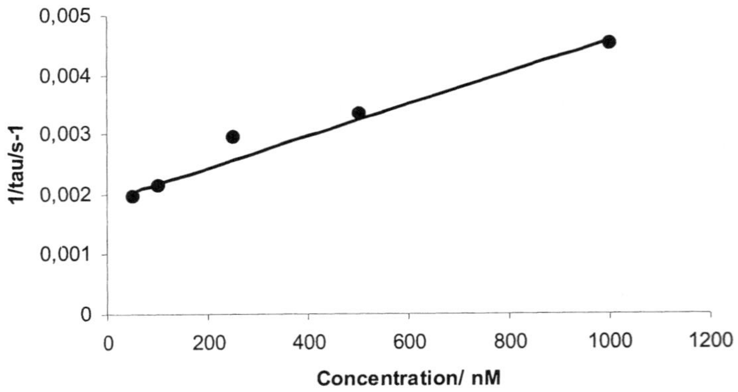

B

Fig. 9. (A) Global analysis experiments using P1 probes and T4 targets. The normal arrows indicate the start of buffer rinsing while the block arrows indicate the points of target injection. (B) Determination of the kinetic rate constants by using the 1:1 Langmuir kinetic model. The association rate constant, k_{on}, can be calculated from the slope of the curve, while the intercept gives the k_{off} value.

3.5. DNA Hybridization Interactions using P2 probes (30 T spacer and 15mer recognition sequence)

The effect of increasing the spacer length on the hybridization kinetics is reported here. The recognition sequence (15mer), however, stays identical to the P1 probes. Except for this change, the rest of the experimental protocols are identical to that of the above sub-section. The interaction of the long targets (T2-T4, 100nM) with the P2 probes (100nM) is shown in Fig. 10. These interactions are marked by rapid hybridization kinetics, with target responses reaching fluorescence intensity levels close to saturation within the first 10-20 minutes. The kinetic rate and affinity constants are displayed in Table 3a. It is clear from the kinetic data that although there is a decreasing trend in the affinity constant values with increasing target length, the differences aren't that significant. These affinity constants are comparable (or slightly higher) to the P1 hybridization interactions. Even the k_{off} rates for the long targets are lower those observed in the case of the P1 probes. These results indicate that increasing the spacer length results in increased flexibility for the longer targets to zip into the smaller recognition sequence found on the P2 probes, and ultimately results in hybrids of higher stability. Even if some of the probes are immobilized within close proximity to each other, the additional spacer length provides the opportunity for them to move away from the interface, and thereby avoiding some of the steric hindrance effects that are normally present within the vicinity of the probe conjugated biotin units. Another interesting point to note is that the final fluorescence intensity levels of these hybrids is much higher that obtained in the P1 system, even though the affinity constants obtained by using these two kinds of probes fall within a close range. Since the P2 probes have the flexibility to stretch farther from the interface than the P1 probes, the resulting P2 hybrids experience a slightly lower level of quenching from the metal surface than those observed in the case of the shorter P1 probes. This effect can explain why higher fluorescence intensity levels are observed in the P2 hybrid system.

The interaction of mismatch 0 targets (T5-T6, 100nm) with the P2 probes (100nM) is shown in Fig. 11. The kinetic rate constants for these interactions are displayed in Table 3b. The drop in the affinity constants with increasing length of the targets follows the same trend as in the case of the P1 probes. One of the striking differences, however, is the final intensity level (I_{max}) that is achieved for the mismatch 0 hybrids. The I_{max} values for the T5 –P2 hybrids is three times higher than that observed in the case of P1 system.

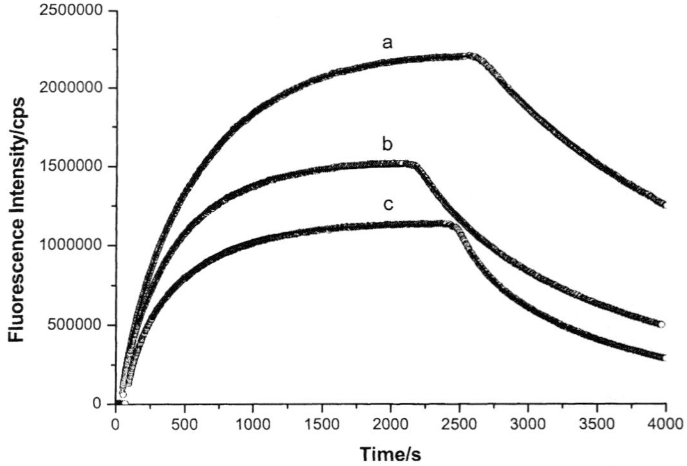

Fig. 10. Mismatch 1 hybridization kinetics between the long strand targets and the surface immobilized P2 probes. Here, a,b & c indicate the hybridization responses for the T2, T3 & T4 targets respectively.

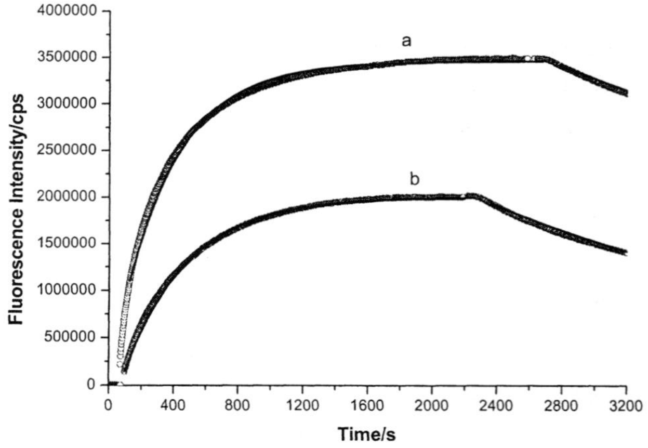

Fig. 11. Mismatch 0 hybridization kinetics between the long strand targets and the surface immobilized P2 probes. Here, a & b indicate the hybridization responses for the T5 & T6 targets.

Table 3. Kinetic parameters for mismatch 1 (a) and mismatch 0 (b) target interactions with P2 probes. A drop in the affinity constant values with increasing target length is clearly observed here.

P2 Probe	**Target**	k_{on} (M^{-1} s^{-1})	k_{off} (s^{-1})	K_A (M^{-1})
	25	13348.48	$4.2 * 10^{-4}$	$3.18 * 10^{7}$
Mismatch 1	45	17380.67	$7.8 * 10^{-4}$	$2.23 * 10^{7}$
(a)	75	14890.24	$9.5 * 10^{-4}$	$1.57 * 10^{7}$
Mismatch 0	45	26735.63	$2 * 10^{-4}$	$1.34 * 10^{8}$
(b)	75	21616.46	$3.7 * 10^{-4}$	$5.84 * 10^{7}$

It is possible that such hybrids have a higher hybridization efficiency, resulting in higher number of fluorescent labeled targets interacting with the surface immobilized probes. Since the fluorescence intensity signals depend upon the number of labels (fluorescent dye) that are excited by the evanescent field, the higher the hybridization efficiency, the higher is the resulting I_{max} value. Since the quenching phenomena on gold is still poorly understood, an accurate estimate of the hybridization efficiency is not possible. As a result, the above interpretation is speculative at this point. Even for the same set of the P2 probes, it is clearly evident from Figs. 10 & 11 that the I_{max} intensity levels for the mismatch 0 hybrids is much higher (as high as a factor of two for the 45mer hybrids) than their corresponding mismatch 1 hybrids. Since the differences between the affinity constant values between the mismatch 0 & 1 hybrids is not high, the clear differences in the I_{max} values lends credence to our above interpretation regarding the hybridization efficiencies of these interactions.

Global analysis experiments were performed (similar to the P1 system) by using a 1μM solution of the P2 probe and varying concentrations of the T4 targets (50nm- 1μM). The hybridization interactions using the above set of targets are shown in Fig. 12a. The results of the τ and k_{off} analysis for the individual plots are displayed in Table 4. From the slope of Fig. 12b, the k_{on} rate constant was calculated as 2883.4 M^{-1} s^{-1}, while the average k_{off} rate, from Table 4, was found to be $7.94 * 10^{-4}$ M^{-1}. Using the above rate constants, the overall affinity constant of this interaction was calculated as 3.63 * 106 M^{-1}. Similar to the P1 analysis, the above affinity constant value is lower than that observed at 100 nM probe concentration (Table 3a).

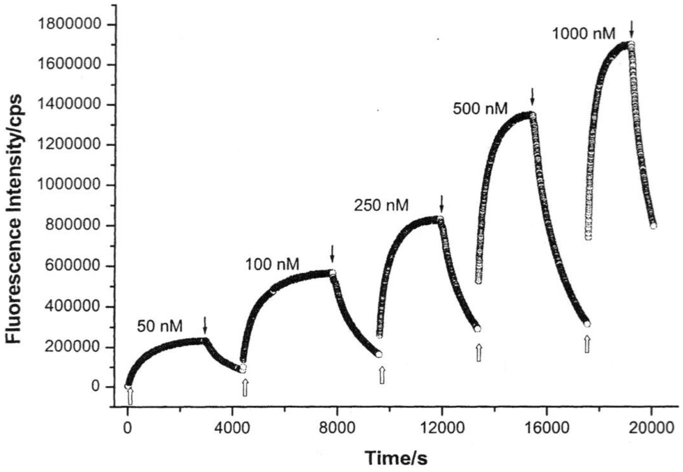

(A)

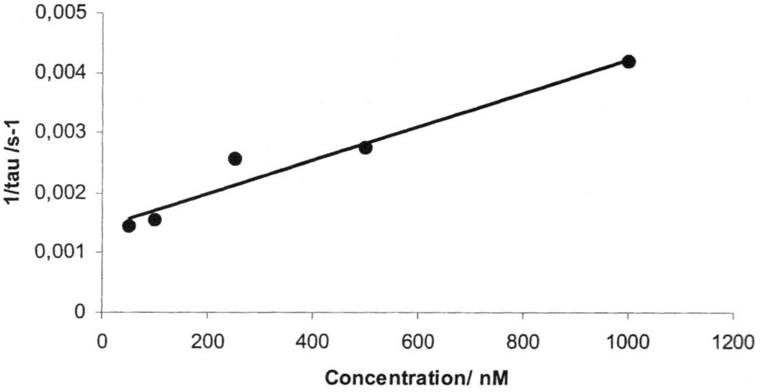

(B)

Fig. 12. (A) Global analysis experiments using P2 probes and T4 targets. The normal arrows indicate the start of buffer rinsing while the block arrows indicate the points of target injection. (B) Determination of the kinetic rate constants by using the 1:1 Langmuir kinetic model. The association rate constant, kon, can be calculated from the slope of the curve, while the intercept gives the koff value.

Table 4. Kinetic parameters for global analysis experiments using the P2-T4 hybrid system. The koff value was obtained by fitting the dissociation curve while τ was calculated from the association segment of the kinetic curve.

Concentration (nM)	τ	k_{off} (s^{-1})
50	690	$8 * 10^{-4}$
100	645	$7.9 * 10^{-4}$
250	390	$7.7 * 10^{-4}$
500	365	$8 * 10^{-4}$
1000	240	$8.1 * 10^{-4}$

3.6. Melting' Analysis of Hybridization Interactions

Nucleic acid hybrid stability is most widely judged by its melting temperature [27-29]. The melting temperature of the hybrid is defined as the temperature at which half of the hybrids split into single strands, and it is usually measured by monitoring changes in absorbance during UV-VIS spectrophotometer measurements or by the use of calorimeters, where changes in thermodynamic parameters (free energy, etc.) are recorded. According to these measurements, a fully complementary hybrid will have a higher melting temperature than hybrids containing mismatches, and this result is plainly based on the stability of the interaction. Now, all of the above measurements are solution-based and thus, are not directly applicable to biochip formats. The stability of the hybrids on sensor surfaces is a function of sensor geometry and since the probes are anchored to the surface, these molecules experience additional spatial restrictions with respect to their physical orientation. The restricted orientation of these molecules can interfere with their interactions with the targets, leading to the determination of affinity constants that are quite different from those calculated from solution based measurements.

Here, we address the fundamental question of what melting means on sensoric surfaces, by investigating the effect of temperature on hybridization interactions that are at equilibrium. Specifically, the effect of temperature ramp rates and the nature of the nucleic acid targets (i.e., wild type or targets containing mismatches) on hybrid stability are reported. Temperature is an important physical parameter that can be used to discriminate between wild type and strands containing mismatches. Here, we will explain the 'melting concept' within the Langmuir framework, where the variation in the affinity constants with temperature results in the observed drop/gain in fluorescence intensities.

The hybridization of 15mer mismatch1 targets (T1, 100nM) with the surface immobilized probes (P1, 200nM) at room temperature (22°C) is shown in Fig. 13. As expected, this interaction is quite fast and once it has reached equilibrium, the temperature of the heating element was increased to 45°C in a single step. Raising the temperature results in a rapid loss of fluorescence, which indicates that the hybrids start dissociating from the catchers and begin to move away from the interface. The refractive index of the solvent also changes during this temperature step, which is clearly visible in the SPR response. The drop in SPR reflectivity value causes a change in the plasmon excitation level, which should in principle increase the fluorescence intensity since we are approaching closer to the resonance situation. However, the temperature effect seems to dominate the entire event (which is in this case happens to be the dissociation process) and the additional gain that is to be expected from the SPR dip, does not manifest itself into the final measured fluorescence intensity.

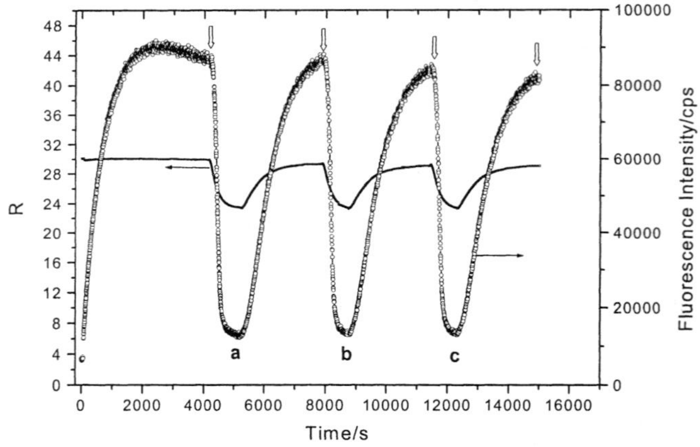

Fig. 13. Effect of temperature on P1-T1 hybrid stability. The vertical arrows indicate the points of heating, where dissociation of the hybrids occur. Once the temperature is cooled down to the initial temperature (indicated by the points a,b & c), re-hybridization starts to occur. The SPR responses correspond to the bulk refractive index shifts.

The entire hybridization event is reversible, as indicated by the re-hybridization that occurs when the temperature is brought back to the room temperature. As the temperature starts dropping, the targets start approaching the sensor surface and begin to interact with the probes.

This interaction causes the fluorescence signal to increase, although the rate of increase in fluorescence is slightly different from that observed during the initial binding event. This behavior is to be expected since re-binding occurs amidst falling temperatures as opposed to the initial binding interaction that was conducted at a fixed temperature. The recovery of fluorescence intensity during the re-hybridization process is almost 100% during the first cycle and subsequently starts dropping. The drop in fluorescence recovery could be attributed to the desorption of the probes and certain materials from the underlying sensor architecture (biotin & streptavidin) with time. It is also possible that the rapid increase in temperature followed by slow cooling cycles can introduce certain thermal shocks on the sensor architecture, leading to lowered stability of the interface. Nevertheless, the re-hybridization process is preserved even after 3 heating cycles spanning a period of more than four hours The SPR changes that are seen in the above figure illustrate simply the effect of temperature changes on the refractive index of the bulk solvent, and does not correspond to hybridization events. Since the signal-to-noise ratio of the SPR response is poor, the only way to make meaningful analysis of the above studies is to focus on the fluorescence intensity changes that occur while varying the temperature. This experiment illustrates the point that melting experiments can be conducted on the sensor architecture, although the results that are obtained should be analyzed by keeping the sensor stability and behavior in perspective. The above experiment involved a one step temperature shock treatment and was conducted to show the effect of temperature on the hybridization dynamics.

The analysis of the above experiment can be readily carried out by using the Langmuir isotherm (Fig. 14). When sufficient amount of target species are present, hybridization occurring at a fixed temperature (Fig. 14A, at T_1) results in the complete saturation of the surface species, and usually follows the isotherm shown in Fig. 14 C. After the hybridization has reached equilibrium, increasing the temperature (Fig. 14 B) results in shifts in the affinities between the catchers and targets. Thus, as the temperature is increased from T_1 to T_2, a drop in the affinity constant is observed, and the resulting Langmuir isotherm shifts to higher concentrations to reach full surface coverage. Since, the concentration of the targets within the fluid cell is fixed, the above temperature shift obviously results in a complete drop in fluorescence. When the temperature is cooled from T_2 to T_1, a gain in fluorescence intensity is observed, which is in-line with the increasing affinity constants. At temperature T_1, and in the absence of any hysteresis, the affinity constant should be identical to the starting condition, and the fluorescence intensity level should also match the initial final value. In the absence of any matrix effects, the above process can be repeated over several temperature cycles, and should result in identical binding/dissociation during the cooling and the heating steps respectively. Again, the above analysis is based purely on the kinetic model. Under normal experimental conditions, the sensor matrix and the immobilized catcher stabilities are affected by the above temperature changes, and thus, we do not observe complete recovery of the final fluorescence intensity. This recovery level keeps dropping with increasing number of temperature cycles, since a corresponding loss in the sensor architecture occurs (each time), during the temperature treatment.

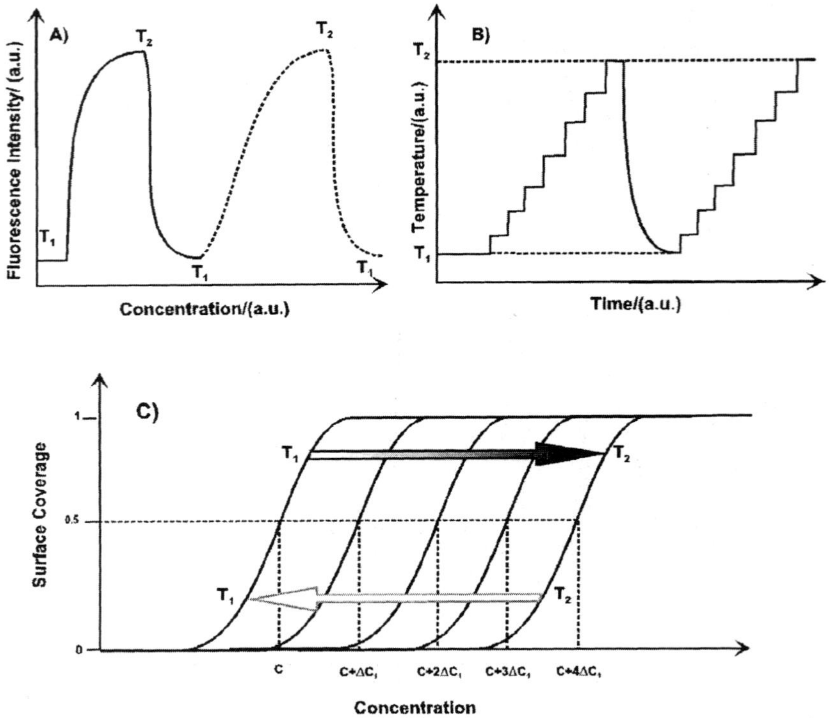

Fig. 14. Effect of temperature on DNA-DNA hybrid stability. (A) Here, hybridization is conducted at temperature, T_1, and once the interaction has reached equilibrium, the temperature is raised to a higher value, T_2. At this temperature, the hybrids start dissociating, resulting in a drop of fluorescence. Once the temperature is brought back to T_1, re-hybridization starts to occur. The temperature program is indicated in B. The shifts in the Langmuir isotherm curves, due to changes in the affinity constants, when the temperature is increased from T_1 to T_2 and back to T_1, is shown in C.

To determine the exact temperature when half of the fluorescence intensity is lost (which corresponds to our definition of the melting temperature), it is important to increase the temperature in a systematic way. This study was carried out by using the same sets of probes (P1, 1μM) and targets (T1, 100nM). The hybridization of this mismatch target with the probes (conducted at 30°C) is shown in Fig. 15a.

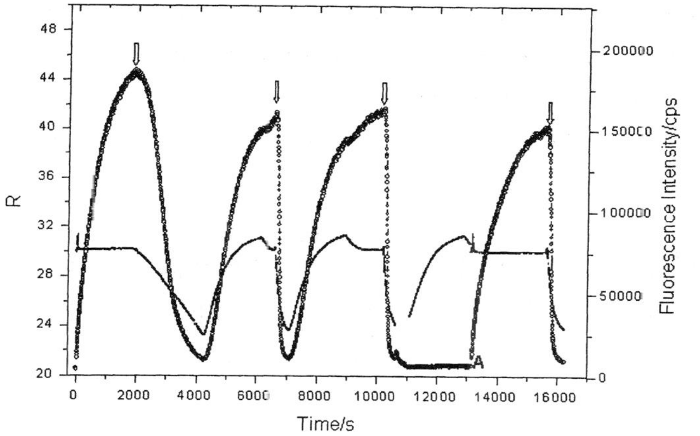

15(A)

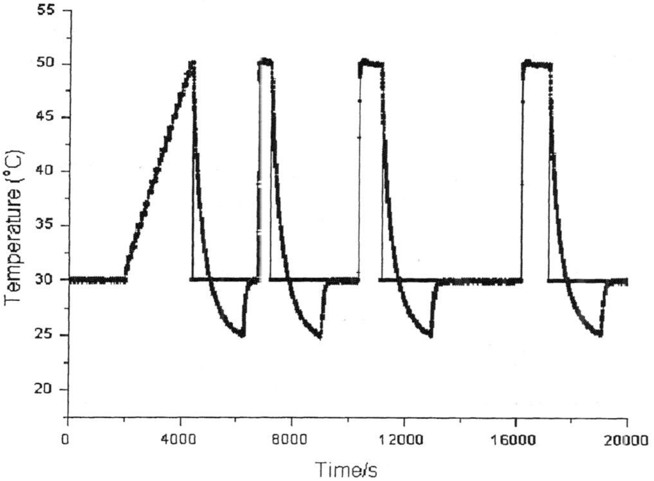

15(B)

Fig. 15. (A) Effect of multiple heating/cooling cycles on the P1-T1 hybrid system. It is clear from this figure that the fluorescence recovery keeps dropping after every heating cycle, indicating that some of the sensor matrix is also eroding during this temperature treatment. 'A' indicates the point where fresh target was injected. (B) Temperature program. During this analysis, the temperature was raised in small increments and maintained at each level for a certain period of time before the next ramp step.

Once the interaction reached equilibrium, the temperature of the heating element was increased in small steps of 1°C, and maintained there for 2 minutes. During this experiment, the temperature of the heating element was increased till 50°C over a 40 minute period, which corresponds to 42°C within the fluid cell. Since the thermal conductivities of glass and the heating element are different, the temperature within the fluid cell is lower than that of the heating element. The exact profile of the temperatures within the fluid cell was determined from temperature calibration experiments (ref to Fig 15b).

The gradual increase in temperature is accompanied by a slow but steady loss in fluorescence. Based on this measurement, the melting temperature (corresponding to the level where half the final fluorescence signal is lost) of the mismatch-1 hybrid was determined as 35°C. Once the temperature was cooled down to 30°C, re-hybridization starts to occur. The fluorescence recovery is not 100 % since some of the streptavidin sensor surface is lost during the heating step. The temperatures that were used during this experiment approached near the maximum stability temperature values (40-45°C) of the biotin-streptavidin architecture (i.e., the SAM layer), and thus, it is highly likely that prolonged exposure to these temperature conditions could cause some of the interface material to erode. The next two heating cycles were shock treatments, where the temperature of the heating element was raised to 50°C in a single step. This heating step was conducted to check the behavior of the interface under gradual and shock temperature treatments. The subsequent recovery of the fluorescence signal upon cooling is quite good. Thus, it is tempting to speculate that the interface is affected more during prolonged periods of slow but gradual heating rather than during short but rapid heating cycles. Subsequent injection of fresh probes does not result in the recovery of the fluorescence intensity levels to the starting level, indicating once again, that the underlying sensor surface is irreversibly damaged.

The above analysis was based on mismatch 1 hybridization studies. Here, we put to test the above 'melting analysis' approach in mismatch discrimination studies by using mismatch 0 targets (T7, 100nM). The concentration of the surface immobilized P1 probes was 1μM. The hybridization interactions (Fig. 16) were conducted at 30°C, similar to the mismatch-1 situation. Since the affinity constant of the wild type targets is higher than the mismatch-1 targets, we observe faster hybridization kinetics while using these targets. Once the hybridization reached equilibrium, heating was conducted in small steps (similar to Fig. 16b).

Since the stability of these hybrids is much higher than in the mismatch-1 situation, heating to higher temperatures is required for the complete dissociation of the hybrids. The final temperature of the heating element was 70°C, which corresponds to the fluid cell temperature of 58°C. The above heating cycle was conducted over a period of 80 minutes. This prolonged heating time coupled with higher temperatures results in a poor recovery of the re-hybridization fluorescence signal. It is obvious that this heat treatment leads to a strong loss of the sensor surface. Subsequent heating cycles were conducted in single steps, which all resulted in further loss of fluorescence recovery. An accurate value of the melting temperature of this interaction can not be calculated, since beyond 40°C, the hybrid desorption is also accompanied by a steady loss of the streptavidin surface. All we can say, qualitatively, is that these mismatch-0 targets have a higher melting temperature than the mismatch-1 targets.

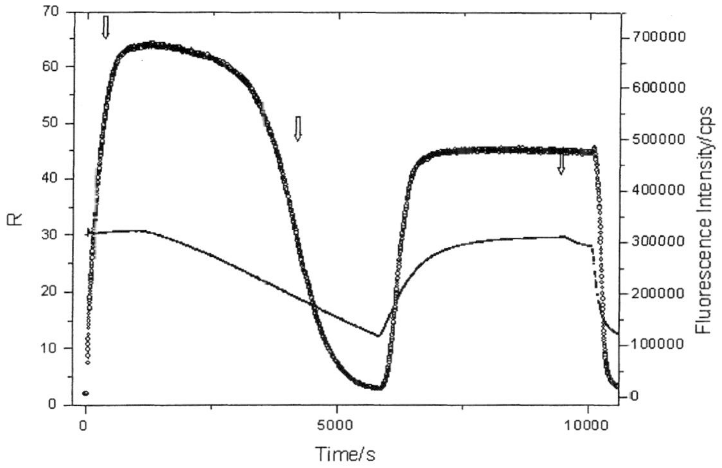

Fig. 16. Effect of temperature on P1-T7 hybridization interactions. Since these hybrids are more stable than the P1-T1 hybrids, elevated temperatures are required to completely dissociate these hybrids. The negative consequence of this elevated temperature treatment was a substantial damage of the underlying sensor architecture.

Thus, from the above experiments, it is clear that DNA melting on sensoric surfaces takes an entirely different meaning under the Langmuir picture, and the results that are obtained from these temperature studies should be analyzed by keeping the sensor architecture in mind. Based on the sensor architecture and the orientation of the catchers and the targets, different melting profiles can occur, unlike in the case of solution based experiments. Now, even repeated melting studies conducted on the same sensor architecture can lead to slight variations in the determination of the melting temperature, since certain loss of the architecture material is to be expected. If the melting temperature of the hybrids is high, chemical reagents (e.g. formamide) can be used to lower the melting temperature of the hybrids. In spite of the above challenges, applying temperature to discriminate between the various strands is highly desirable among the currently used biochip formats that are facing increasing signal resolution problems imposed by high packing densities and detection of absolute fluorescence intensity levels.

3.7. Effect of Temperature on Hybridization Interactions

Most of the SPRFS measurements reported in earlier sub-sections was conducted at room temperature (22°C). In this section, we will discuss the effect of temperature on DNA hybridization kinetics by performing such interactions at 10, 20, 25, 30 and 40°C respectively. Since temperature is known to influence the kinetic rate constants, it is interesting to investigate what trends can be observed when such measurements are conducted on the streptavidin surface. The dextran surface derivatized with streptavidin was used during these studies. These interactions were monitored within the BIACORE instrument using conventional SPR detection schemes. The hybridization studies were conducted by using P1 probes and the mismatch 1 targets (T1, with varying concentrations). The probes were immobilized at 25°C with the final SPR response of about 1024 RU, a value that is near surface saturation. The target concentration was varied in the following steps: 100, 250, 500, 750 and 1000nM respectively, while conducting the DNA hybridization studies. The flow rate that was used during the target injection was 20μl/min. All the hybridization responses were background subtracted from the reference cell in order to minimize artifacts arising from the sensor matrix. A 50mM NaOH solution (one minute injection of 20 μl) was used for probe regeneration.

The hybridization kinetic studies performed at 10°C and the corresponding Langmuir analysis are shown in Fig. 17 a and 17 b. All the above interactions could be simulated by using the simple Langmuir 1:1 model. Using Fig. 17b, k_{on} was determined from the slope and the k_{off} from the intercept. Since the contribution of k_{off} to τ is small, k_{off} was determined from the individual desorption curves (Table 5). From the slope, the k_{on} was calculated as 516.11 M^{-1} s^{-1}

and from Table 5, the average k_{off} was found to be 1.8 10^{-4} s^{-1}. Using these values, the affinity constant, K_A, for this interaction at 10°C was calculated as 3.23 * 10^6 M^{-1}. The same set of analysis was conducted at 20 (Fig. 18 a and 18 b, Table 6), 25 (Fig. 19 a and 19 b, Table 7), 30 (Fig. 20 a and 20 b, Table 8) and 40°C, respectively. The rate constants (k_{on} and k_{off}) along with affinity constants at various temperatures are shown in Table 9.

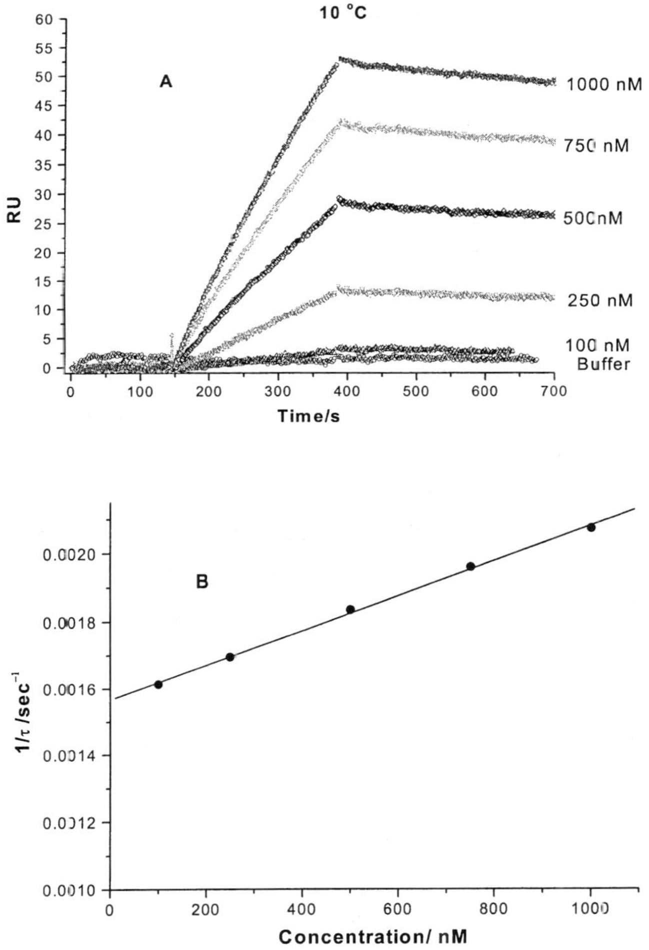

Fig. 17. (A) P1-T1 hybridization interactions conducted at 10°C. The buffer response is shown simply to indicate that solvent artifacts are absent on this surface. (B) Langmuir 1:1 kinetic analysis for determining k_{on} and k_{off}.

Table 5. Kinetic parameters obtained from the 1:1 Langmuir simulations of P1-T1 hybridization studies performed at 10°C.

Concentration (nM)	τ / s	k_{off} / s^{-1}
100	620	0.00016
250	590	0.00016
500	545	0.00017
750	510	0.00015
1000	482	0.00016

Tab. 6. Kinetic parameters obtained from the 1:1 Langmuir simulations of P1-T1 hybridization studies performed at 20°C.

Concentration (nM)	τ / s	k_{off} / s^{-1}
100	290	0.00029
250	260	0.00025
500	220	0.00023
750	190	0.00017
1000	170	0.00021

Tab. 7. Kinetic parameters obtained from the 1:1 Langmuir simulations of P1-T1 hybridization studies performed at 25°C.

Concentration (nM)	τ / s	k_{off} / s^{-1}
100	265	0.00087
250	189	0.00077
500	129	0.00075
750	99	0.00075
1000	80	0.00078

No binding was observed at 40°C for all target concentrations. This result is not surprising since the melting temperature of the mismatch-1 target is 35 C, which was determined from SPRFS measurements (refer to sub-section 3.6) conducted on 2-d biotin streptavidin architectures. So, for temperatures approaching near or above the melting temperature, there is a strong thermodynamic barrier that has to be overcome before hybridization can occur. It is worth noting that as the temperature is increased from 10°C to 30°C, there is a substantial increase in the target binding response, with curves reaching close to saturation rapidly for targets of higher concentration.

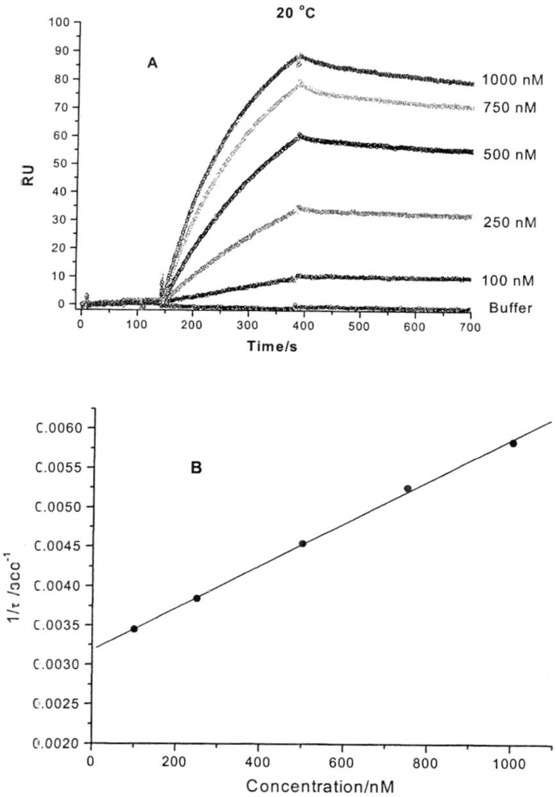

Fig. 18. (A) P1-T1 hybridization interactions conducted at 20°C. (B) Langmuir 1:1 kinetic analysis for determining k_{on} and k_{off}.

From Table 9, it is evident that the affinity constant increases steadily from 10°C to 25°C and then drops off at 30°C. One of the possible reasons for this drop can be attributed to the abrupt increase in the desorption rate constant, k_{off}, at this temperature. Again, since this temperature is approaching closer to the melting temperature of the mismatch-1 hybrid, it is likely that increased thermal stresses could start destabilizing the hybrid. Beyond the melting temperature, no binding is expected. The hybridization studies conducted at 40°C proves this point.

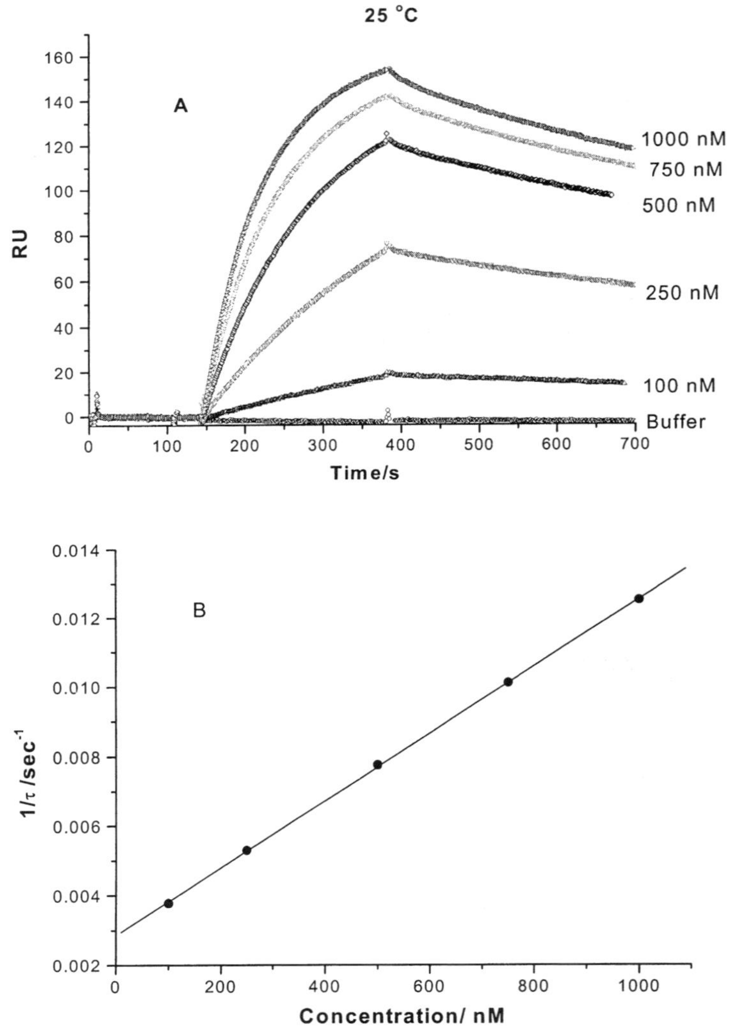

Fig. 19. (A) P1-T1 hybridization interactions conducted at 25°C. (B) Langmuir 1:1 kinetic analysis for determining k_{on} and k_{off}.

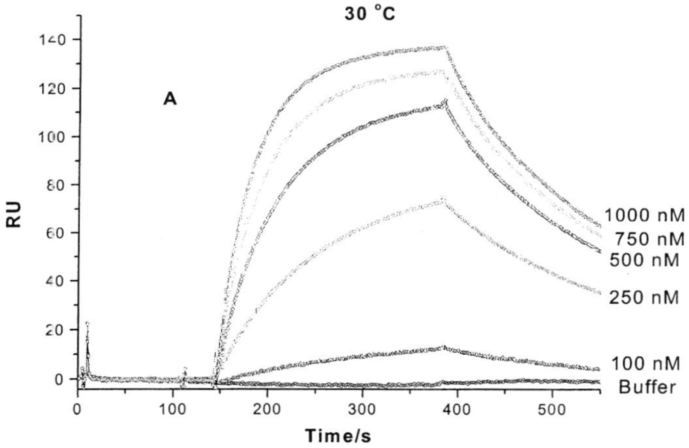

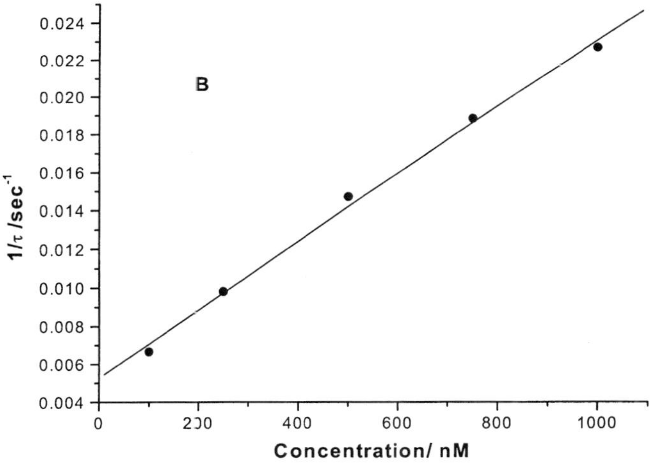

Fig. 20. (A) P1-T1 hybridization interactions conducted at 30 oC. (B) Langmuir 1:1 kinetic analysis for determining kon and koff.

Table 8. Kinetic parameters obtained from the 1:1 Langmuir simulations of P1-T1 hybridization studies performed at 30°C.

Concentration (nM)	τ / s	k_{off} / s^{-1}
100	150	0.0070
250	102	0.0043
500	68	0.0043
750	53	0.0042
1000	44	0.0042

Tab. 9. Effect of temperature on the affinity constant for P1-T1 hybridization interactions.

Temperature (°C)	k_{on} (M^{-1} s^{-1})	k_{off} (s^{-1})	K_A (M^{-1})
10	516.11	$1.6 * 10^{-4}$	$3.23 * 10^{6}$
20	2701.5	$2.3 * 10^{-4}$	$1.17 * 10^{7}$
25	9671.1	$7.8 * 10^{-4}$	$1.24 * 10^{7}$
30	17838	$4.25 * 10^{-3}$	$4.19 * 10^{6}$

4. CONCLUSION

This study has clearly highlighted the applicability of the biotin-streptavidin sensor surfaces in performing high quality DNA interaction studies. Self-assembled monolayers are expected to play an increasingly important role in sensor design since it is possible to effectively tune the surface properties and minimize non-specific interactions. The hybridization data that is characterized with high signal-to-noise ratios clearly demonstrates the integrity of the sensor architecture. The effect of interfacial properties on hybridization kinetics, especially those pertaining to spacer length, surface topology, and its stability under the influence of both chemical and physical stimulus, was also elucidated in detail. The use of temperature as a tool in mismatch discrimination was also clearly demonstrated along with its effect on the kinetic rate and affinity constants. Additionally, a new concept of monitoring 'DNA melting' studies within the context of the Langmuir kinetic model was also introduced. This approach is expected to play an important role in SNP (single nucleotide polymorphism) typing studies [30, 31] and also in the implementation of rational drug design based on sound pharmacogenomic investigations.

5. ACKNOWLEDGMENTS

The authors would like to express their appreciation to collaborating members at Boehringer Mannheim and Biacore AB for their help. In addition, D.K. would like to thank Jürgen Worm, Bernhard Menges, Volker Scheumann, Gretl Dworschak, Thorsten Lieberman & Thomas Neumann for their support during the past years.

6. REFERENCES

1. Ulman, A., Self-assembled monolayers, An Introduction to Ultrathin Organic Films: From Langmuir-Blodgett to Self- Assembly, Ed. A. Ulman, 237-301, Academic Press, San Diego, 1991.
2. Smith, S., S., A self-assembling nanoscale camshaft: Implications for nanoscale materials and devices constructed from proteins and nucleic acids, Nano Letters, Vol. 1(2), 51-56, 2001.
3. Niemeyer, C.M., Adler, M., Pignataro, B., Lenhert, S., Gao, S., Chi, L., Fuchs, H. and Blohm, D., Self-assembly of DNA-streptavidin nanostructures and their use as reagents in immuno-PCR, Nucleic Acids Research, Vol. 27, 4553-4561, 1999.
4. Southern, E., Mir, K. and Shchepinov, M., Molecular interactions on microarrays, Nature Genetics Supplement, Vol. 21, 5-9, 1999.
5. Jordan, C.E., Frutos, A.G., Thiel, A.J. and Corn, R.M., Surface plasmon resonance imaging measurements of DNA hybridization adsorption and streptavidin/DNA multilayer formation at chemically modified gold surfaces, Analytical Chemistry, Vol. 69, No. 24, 4939-4947, 1997.
6. Malmqvist, M. and Karlsson, R., Biomolecular interaction analysis: affinity biosensor technologies for functional analysis of proteins, Current Opinion in Chemical Biology, Vol. 1, 378-383, 1997.

7. Jonsson, U., Fagerstam, L., Ivarsson, B., Johnsson, B., Karlsson, R., Lundh, K., Loefas, S., Persson, B., Roos, H., Ronnberg, I., Sjolander, S., Stenberg, E., Stahlberg, R., Urbaniczky, C., Ostlin, H. and Malmqvist, M., Real-time biospecific interaction analysis using surface plasmon resonance and a sensor chip technology, Biotechniques, Vol. 11, No. 5, 620-, 1991.

8. Weisser, M., Tovar, G., Mittler-Neher, S., Knoll, W., Brosinger, F., Freimuth, H., Lacher, M. and Ehrfeld, W., Specific bio-recognition reactions observed with an integrated Mach-Zehnder interferometer, Biosensors and Bioelectronics, Vol. 14, No. 4, 405-411, 1999.

9. Jung, L.S., Nelson, K.E., Stayton, P.S. and Campbell, C.T., Binding and dissociation kinetics of wild-type and mutant streptavidins on mixed biotin-containing alkylthiolate monolayers, Langmuir, Vol. 16, No. 24, 9421-9432, 2000.

10. Bayer, E.A. and Wilchek, M., Avidin-Biotin Technology, Methods in Enzymology, Vol. 184,1990, Academic Press, London.

11. Wilchek, M. and Bayer, E., A., Foreword and introduction to the book (strept) avidin-biotin system, Biomolecular Engineering, Vol. 16, 1-4, 1999.

12. Tomalia, D.A., Wang, Z.G. and Tirell, M., Experimental self-assembly: The many facets of self.assembly, Current Opinion in Colloid & Interface Science, Vol. 4, 3-5, 1999.

13. Brunsveld, L., Folmer, B.J.B. and Meijer, E.W., Supramolecular polymers, MRS Bulletin, Vol. 25, No. 4, 49-53, April 2000.

14. Xia, Y. and Whitesides, G.M., Soft lithography, Angew. Chem. Int. Ed. Vol. 37, 550-575, 1998.

15. Knoll, W., Liley, M., Piscevic, D., Spinke, J., & Tarlov, M., J., Supramolecular architectures for the functionalization of solid surfaces, Adv. Biophys., Vol. 34, 231-251, 1997.

16. Knoll, W., Optical characterization of organic thin films and interfaces with evanescent waves, MRS Bulletin, Vol. XVI, No. 7, 29-39, 1991.

17. Perez-Luna, V.V., O'Brien, M.J., Opperman, K.A., Hampton, P.D., Lopez, G.P., Klumb, L.A. and Stayton, P., S., Molecular recognition between genetically engineered streptavidin and surface bound biotin, J. Am. Chem. Soc., Vol. 121, 6469-6478, 1999.

18. Kambhampati, D.K., Jakob, T.A.M., Robertson, J.W., Cai, M., Pemberton, J.E. and Knoll, W., Novel silicon dioxide sol-gel films for potential biosensor applications: A surface plasmon resonance study, Langmuir, Vol. 17(4), 1169-1175, 2001.

19. Kambhampati, D.K. and Knoll, W., Surface-plasmon optical techniques, Current Opinion in Colloid & Interface Science, Vol. 4, 273-280, 1999.

20. Raether, H., Surface plasmons on smooth and rough surfaces and on gratings, Springer tracts in Modern Physics, Vol. 111, Springer Verlag, Berlin, 1988.

21. Kretschmann, E., Die Bestimmung optischer Konstanten von Metallen durch Anregung von Oberflächenplasmaschwingungen, Z. Phys., Vol. 241, 313-324, 1971.

22. Liebermann, T. and Knoll, W., Surface-plasmon field-enhanced fluorescence spectroscopy, Colloids & Surfaces A, Vol. 171, 115-130, 2000.

23. Lieberman, T., Knoll, W., Sluka, P. and Hermann, R., Complement hybridization from solution to surface-attached probe-oligonucleotides observed by surface plasmon-field enhanced fluorescence spectroscopy, Colloids & Surfaces A, Vol. 169, 337-350, 2000.

24. Kambhampati, D., Nielsen, P. E. and Knoll, W., Investigating the kinetics of DNA-DNA and PNA-DNA interactions using surface plasmon resonance-enhanced fluorescence spectroscopy, Biosensors & Bioelectronics, Vol. 16, 1109-1118, 2001.

25. Neumann, T. and Knoll, W., Mismatch discrimination in oligonucleotide hybridization reactions using single strand binding protein. A surface plasmon fluorescence study, Isr. J. Chem., Vol. 41, 69-78, 2001.

26. Blake, R.D. and Delcourt, S., G., Thermal stability of DNA, Nucleic Acids Research, Vol. 26, 3323-3332, 1998.

27. Pompizi, I., Häberli, A., and Leumann, C.J., Oligodeoxynucleotides containing conformationally constrained abasic sites: a UV and fluorescence spectroscopic

investigation on duplex stability and structure, Nucleic Acids Research, Vol. 28, 2702-2708, 2000.

28. Vallone, P.M. and Benight, A., S., Melting studies of short DNA hairpins containing the universal base 5- nitroindole, Nucleic Acids Research, Vol. 27, 3589-3596, 1999.

29. Howell, W.M., Jobs, M., Gyllensten, U. and Brookes, A.J., Dynamic Allele-Specific Hybridization: A new method for scoring single nucleotide polymorphism, Nature Biotech, Vol. 17, 87-88, 1999.

30. Prince, J.A., Feuk, L., Howell, W.M., Jobs, M., Emahazion, T. and Brookes, A.J., Robust and accurate single nucleotide polymorphism genotyping by dynamic allele-specific hybridization (DASH): Design criteria and assay validation, Genome Research, Vol. 11, 152-162, 2001.

CHAPTER 8

APPLICATIONS OF ORGANISED MOLECULAR FILMS TO ELECTRONIC AND OPTO-ELECTRONIC DEVICES

M. C. Petty

School of Engineering and Centre for Molecular and Nanoscale Electronics
University of Durham, DH1 3LE, UK

This article reviews recent research activities on possible applications of organised molecular films in electronic and opto-electronic devices. Although much of the work concerns the use of Langmuir-Blodgett (LB), self-assembled and electrostatically deposited layers, devices using thin films produced by the methods of spin coating and thermal evaporation are also discussed. The latter techniques may be more suited to commercialisation. The first part of the paper provides an overview of electrically insulating and conductive organic films. Semiconductive thin films can exhibit rectifying behaviour when contacted by suitable metal electrodes. It is also possible to fabricate organic field effect transistors. Such devices may find application in niche areas of electronics, e.g. plastic smart cards, where processing speed is not an important requirement. The field of organic light emitting displays is reviewed separately. Here, there has already been some degree of commercial success. Simple device structures based on a fluorescent organic material sandwiched between two electrodes are contrasted to those incorporating electron-transporting layers. The use of thin insulating layers (e.g. LB films) to increase the minority carrier injection is also discussed. There has been considerable interest in the exploitation of organic thin films in chemical sensors, particularly for environmental monitoring. Organic layers can conveniently be tailored to provide a particular response, either by chemical modification or by variation of the thin film deposition conditions. The combination of multiple sensors with pattern recognition software (i.e. electronic nose) can provide improved selectivity. The ability to orient the dipole moments in multilayer assemblies, for example by depositing alternate-layer LB films, can result in asymmetric molecular architectures with pyroelectric behaviour. These may find application in heat sensors or infrared detectors.

1. INTRODUCTION

The past 30 years has witnessed the emergence of *molecular electronics* as an important technology for the 21st Century [1, 2]. An increasing number of organic materials are now finding use in the electronics industrial sector. The liquid crystalline display is well established. Other application areas include piezoelectric and pyroelectric plastics (polyvinylidene fluoride), gas-sensitive polymers (polypyrrole) and electroluminescent displays tris(8-hydroxyquinoline) aluminium.

Certain organic compounds can possess semiconductive, and even metallic, properties [3, 4]. It is therefore possible to fabricate diodes and field effect transistors (FETs) using these compounds [5, 6]. Organic light emitting displays may be based on a fluorescent organic material sandwiched between electrodes of high and low work functions [7, 8]. Work on such devices represents a burgeoning field of both academic and commercial endeavour. However, over the next 20 years, it is unlikely that organic materials will displace silicon as the dominant material for fast signal processing. The charge carrier mobility (velocity per unit field) is low in organic materials. Silicon technology is firmly established and Moore's law, namely that device dimensions halve every two to three years, will probably describe developments over at least the next decade. The Semiconductor Industry Association bases a Microelectronics Roadmap on the future of CMOS technology [9]. Its prediction is for silicon-based devices with a 35 nm minimum feature size and 10^8 transistors per cm^2 by the year 2012.

There has been considerable interest in the exploitation of organic materials in chemical sensors, particularly for environmental monitoring [10-12]. Organic thin films can conveniently be tailored to provide a particular response, either by chemical modification or by variation of the thin film deposition conditions. The combination of multiple sensors with pattern recognition software as a means to provide improved selectivity (i.e. electronic nose) is currently attracting much interest. The ability to orient the dipole moments in multilayer assemblies can result in asymmetric molecular architectures with pyroelectric behaviour [13]. Such layers could form the basis of infrared or heat detectors.

The use of thin film technologies that enable layers of molecular dimensions to be built-up on solid surfaces can provide a valuable insight into the physical processes at work in devices such as those described above, e.g. by elucidating the dependence of the device characteristics on the thickness and molecular orientation in the thin film structures. The methods could also be exploited in their commercial development. In this article, we review progress on the application of organised organic films to a variety of electronic and opto-electronic devices. Most of the work will focus on structures fabricated using the Langmuir-Blodgett (LB) technique, self-assembly and layer-by-layer electrostatic deposition [14, 15]. However, studies using simpler thin film deposition methods, such as thermal evaporation and spin coating, are also included [16].

The first section discusses the electrical (d.c. and a.c. conductivities) of multilayer assemblies. The applications of such structures to organic diodes and transistors, light emitting displays, and sensors are then described.

2. ELECTRICAL CONDUCTIVITY OF MOLECULAR ASSEMBLIES

2.1. D.C. Conductivity

The d.c. behaviour of simple (e.g. fatty acid) Langmuir-Blodgett (LB) assemblies has been reviewed previously [17]. Organised assemblies of organic molecules usually exhibit highly antisotropic electrical conductivities. Different experimental procedures are used to measure the conductivities in the in-plane and out-of-plane directions. For example, Langmuir-Blodgett multilayer films containing polypyrrole mixed with palmitic acid exhibit room temperature values of the in-plane and out-of-plane dc conductivities of 20 S cm^{-1} and 3 x 10^{-6} S cm^{-1} (i.e. $\sigma_{||}/\sigma_{\perp} \approx 10^7$), respectively [18]. Particular problems arise from the procedures used to establish electrical contacts and from the measurement conditions. Quantum mechanical tunnelling is noted in multilayer films of a few layers in thickness. Above this thickness, a variety of conduction mechanisms, including space-charge injection, Schottky emission and Poole-Frenkel conduction, have been reported. In many cases, the d.c. conductivity appears to be dominated by defects rather than by the intrinsic properties of the organic film.

Semiconductive multilayer films often exhibit Ohmic electrical behaviour (i.e. current $\propto$ applied voltage), which is thermally activated, following an exponential dependence of conductivity σ on temperature T of the form [19]

$$\sigma = \sigma_0 \exp\left(-\frac{\Delta E}{k_B T}\right) \tag{1}$$

where σ_0 is a constant, ΔE is the thermal activation energy and k_B is the Boltzmann constant. Following doping, room temperature, in-plane d.c. conductivities of between 10^{-5} S cm^{-1} and 10 S cm^{-1} and ΔE values in the range 0.1 – 1 eV are frequently reported [19].

Since such thin films could find applications in electronic devices, more recent work has focused on the effect of film preparation details and of subsequent processing (e.g. doping) on the conductivity. For example, LB layers deposited from (*N*-octadecylpyridimium)-Ni(dmit)$_2$ (where H_2dmit =4,5-dimercapto-1,3-dithione-2-thione) have been shown to exhibit structural and electrical behaviour that depends on the time that the floating film remains on the subphase surface [20]. This material does not spread easily at the air/water interface, corroborating results with similar compounds [21]. Only if the floating film is left in its uncompressed state on the water surface for many hours will true molecular films be obtained. LB layers built-up from such a film do not show significant in-plane electrical conductivity, either in their as-deposited state or after iodine doping. It has been suggested that the molecular organisation in these films prevents the organometallic moieties from packing closely together [20]. In contrast, LB films deposited from 'fresh' floating films exhibit stable in-plane conductivity values up to 0.2 S cm^{-1}. These films are two-phase structures with micron-sized organometallic particles embedded in a multilayer matrix. Fig. 1 shows the area per complex for the floating film, measured at a surface pressure of 30 mN m^{-1}, versus the time that the floating film was left in its uncompressed state on the subphase surface. Electrical data are also given. These were recorded for a film that was built-up over five LB deposition cycles (up and down movements of the substrate). This resulted in a film nine layers in thickness for deposition from a 'fresh' floating film (Y-type deposition) and five layers in thickness in the case of deposition from an 'aged' floating film (Z-type deposition). Measurements were taken, following iodine doping, in vacuum and in the in-plane direction using silver paste electrodes.

The electrical stability of the metallic contacts (Au, C and Ag) after iodine doping, important for device applications, has also been examined and the results are shown in Fig. 2 [20]. Measurements taken within about 10 days (the closely spaced experimental data points in the Fig.) were made with the LB films in vacuum (.10^{-2} mbar); subsequent measurements were taken with the LB layers exposed to air. Immediately after iodine-doping, the in-plane conductivity measured for both carbon- and gold-contacted LB films increased to 10^{-2} – 10^{-4} S cm^{-1} (Fig. 2(a)).

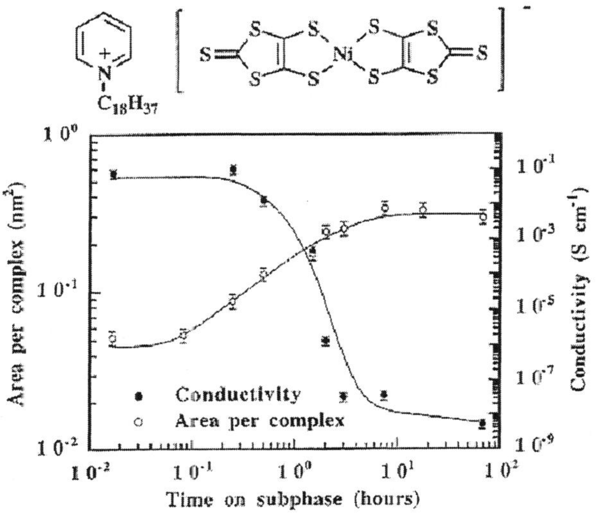

Fig. 1. In-plane current (stabilised values) versus time in the uncompressed state on the subphase surface for LB films (five deposition cycles) of a $Ni(dmit)_2$ complex. LB films were deposited onto glass and doped for 2000 s with iodine. Silver paste electrodes; applied voltage 100 V. The area per complex at 30 mN m^{-1} is also shown. The molecular formula of the dmit complex is given above. After Pearson et al. [20].

The conductivity then continued to increase over a period of a few minutes to reach a maximum value of about 0.2 S cm^{-1}. This effect is probably related to the removal of excess iodine from the film and the formation of a mixed-valence compound. The final conductivity was reasonably stable for several months for films stored in air and at room temperature. Silver-contacted LB films showed similar electrical conductivity immediately after iodine doping (Fig. 2(b)). However, a deterioration in the current monitored in these films was noted a few hours after the maximum conductivity had been achieved. After three months, the conductivity had fallen to 10^{-3} –10^{-4} S cm^{-1}. If fresh silver contacts were then made to the same film, the current was restored to its original value for a few hours before the current again was observed to fall (Fig. 2(b)). This experiment suggests that the deteriorating electrical characteristics resulted from the metallic contact regions rather than from a change in the properties of the LB film.

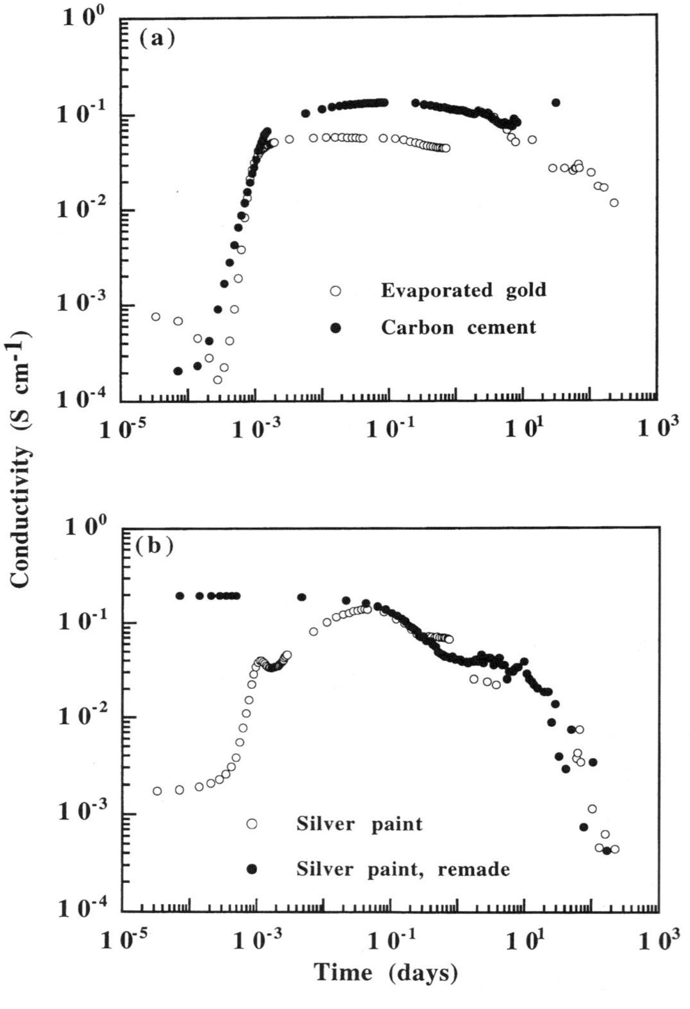

Fig. 2 Contact stability data for LB films of the $Ni(dmit)_2$ complex shown in Fig. 1 with different metallic electrodes. The floating film was left for 10 min in its uncompressed state on the subphase surface. The LB films were deposited onto glass and doped in iodine vapour for 2000 s. In-plane measurements were taken at an applied voltage of 10 V. (a) Au and C electrodes; (b) Ag electrodes. After Pearson et al. [20].

Post-deposition processing can be used to fabricate novel organic nanostructures. An interesting example is that of conductive polypyrrole (PPy) dispersed in a fatty acid matrix. Fig. 3 illustrates a method to produce such a two-phase network [18, 22].

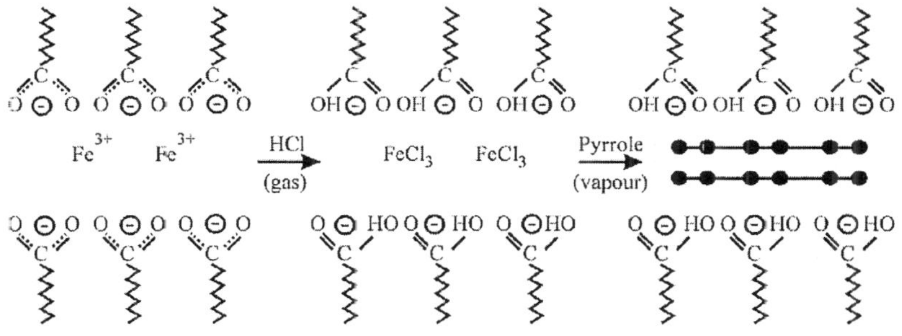

Fig. 3 Idealised solid-state reactions of ferric stearate multilayers with HCl and pyrrole to form a conductive polymer film within the fatty acid matrix [22].

First, monolayers of the iron salt of a long-chain fatty acid (e.g. ferric palmitate) are assembled on an appropriate substrate. The multilayer film is then exposed to saturated HCl vapour at room temperature for several minutes. During this process, a chemical reaction transforms the fatty acid salt into layer of ferric chloride separated by layers of fatty acid. In the third and final step, the film is exposed to pyrrole vapour and a reaction occurs between the pyrrole and the ferric chloride producing polypyrrole distributed within the multilayer assembly.

An analysis of both the in-plane and out-of-plane d.c. conductivity versus temperature of the final multilayer assembly was undertaken with the variable-range hopping (VRH) model [23] and using an approach used by other workers for PPy [24]. In this model, the most frequent hopping process may not be to the nearest neighbour site. For example, it can be more convenient for the charge carrier to hop to a site at a longer distance if the energy barrier is smaller than that associated with the nearest neighbour site. The best fit to the VRH model, shown in Fig. 4, was found using an equation corresponding to three-dimensional hopping

$$\sigma = \sigma_0 \exp[-(T_0 / T)^{1/4}] \qquad (2)$$

where σ_0 and T_0 are constants.

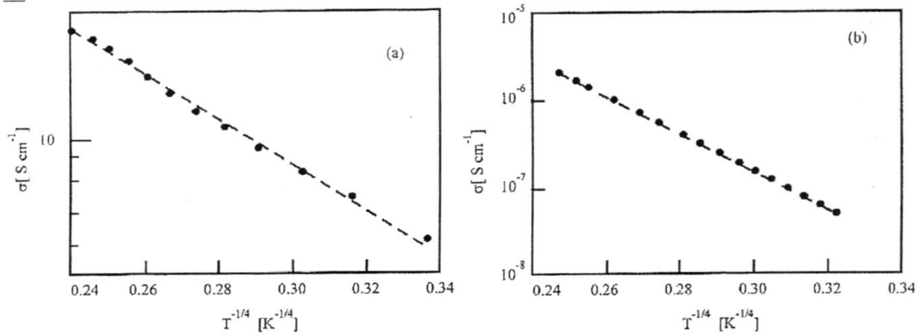

Fig. 4 Conductivity versus $T^{-1/4}$ for a polypyrrole plus palmitic acid LB film. The broken lines are the best fit with equation (2). (a) In-plane direction; LB film 21 monolayers thick; measurement voltage 1 V. (b) Out-of-plane direction; LB film 115 monolayers thick; measurement voltage 20 mV. After Casalini et al. [18].

It should also be noted that acceptable fits to the d.c. conductivity data were also obtained using the model, known as fluctuation-induced tunnelling, of Sheng et al. [25]. Here, the conductivity is assumed to be dominated by tunnelling of electrons across small potential barriers that separate conductive regions (20 nm). Such a model might be expected to apply to films described here (which are islands of conductive polymer separated by insulating fatty acid molecules). The d.c. conductivity is given by

$$\sigma = \sigma_0 \exp[-T_1/(T + T_2)] \tag{3}$$

where σ_0, T_1 and T_2 are constants. For $T \ll T_2$, the conductivity is temperature independent and equation (3) reduces to the form expected for quantum mechanical tunnelling through a parabolic barrier. However, at high temperatures the electrical conductivity is thermally activated with an energy k_BT_1. The Sheng model includes three unknowns, rather than two in the VRH model and is consequently rather easier to fit to the experimental data.

2.2 A.C. Conductivity

Direct current conduction and high frequency phenomena are essentially separate and, for the most part, independent quantities. However, at low frequencies, both can contribute to a measured conductance. The macroscopic effect observed when a dielectric (insulating) material is placed in an electric field is the creation of a second electric field that tends to oppose the original field. Microscopically, this polarisation arises from two main contributions: induced and orientational polarisation [26]. The former is present when the applied electric field causes a displacement of the electronic cloud relative to the nuclei in each atom (electronic polarisation) or when there is a displacement of the atomic nuclei relative to one another (atomic polarisation). Orientational polarisation is related to the orientation of molecules or to part of them. Chemical bonding between atoms of different electronegativity in a molecule can create an asymmetric charge distribution (i.e. so that the centres of positive and negative charge are not coincident) forming a permanent dipole moment (polar molecules). Such moments experience a torque in an applied electric field that tends to orient them in the field direction (molecular polarisation).

The relative permittivity of a material $\varepsilon(\omega)$ can be defined as a complex quantity

$$\varepsilon(\omega) = \varepsilon'(\omega) - j\varepsilon''(\omega) \quad (4)$$

The response may also be described in terms of a complex conductivity

$$\sigma(\omega) = \sigma'(\omega) + j\sigma''(\omega) \quad (5)$$

where

$$\sigma'(\omega) = \omega\varepsilon_0\varepsilon''(\omega); \quad \sigma''(\omega) = \omega\varepsilon_0\varepsilon'(\omega) \quad (6)$$

and ε_0 is the permittivity of free space. In materials showing a measurable d.c. conductivity σ_{dc}, $\varepsilon(\omega)$ is associated only with the a.c. losses

$$\sigma'(\omega) = \sigma(0) + \omega\varepsilon_0\varepsilon''(\omega) = \sigma_{dc} + \omega\varepsilon_0\varepsilon''(\omega) \quad (7)$$

The first model for dielectric relaxation was presented by Debye [26] for the interpretation of the dielectric response of polar liquids. Debye supposed the existence of a system of N non-interacting dipoles in a viscous medium. He neglected dipole-dipole interactions and any other interactions dependent on the molecular arrangement, considering instead an average interaction in the form of a friction with the medium.

In a static field, the dipoles will have a slight preferential orientation parallel to the field. A sudden removal of the field will cause an exponential decay of this ordered state due to the randomising of the Brownian movement. The relaxation time τ measures the time required to reduce the order to e^{-1} of its original value. Debye's result, in the time domain, corresponds to having a step relaxation function F(t) given by

$$F(t) = e^{-t/\tau} \tag{8}$$

In the frequency domain, the relationship is usually found in the form

$$\varepsilon(\omega) = \varepsilon_\infty + (\varepsilon_S - \varepsilon_\infty)\frac{1}{1 + j\omega\tau} \tag{9}$$

where ε_∞ and ε_S are the high frequency and low frequency (static) values, respectively, of the permittivity. The Debye equation has been found adequate to describe the orientational polarisation for a large number of simple condensed materials (e.g. solvents). However, for more complex systems, such as macromolecules, deviations are observed. These take the form of a broader and asymmetric peak in the $\varepsilon^{//}$ versus frequency curve, or in a depressed and asymmetric semicircle in a plot of $\varepsilon^{/}$ versus $\varepsilon^{//}$ (Cole-Cole plot).

Electrical admittance, or impedance, spectroscopy may be used to study the dynamics of bound or mobile charge in the bulk or interfacial regions of any kind of solid or liquid material [27, 28]. This is now proving a useful method to investigate the electrical properties of thin organic layers, such as self-assembled or LB films [14]. The general approach is to apply a known voltage to the electrodes and observe the response. The admittance is measured in the frequency domain by monitoring the phase shift and amplitude (or real and imaginary parts) of the resulting current. The procedure is then repeated at different frequencies to provide an admittance spectrum.

For the polypyrrole/fatty acid multilayer structure discussed above, in the in-plane direction, the imaginary part of the relative permittivity $\varepsilon^{//} = \sigma_{ac}/(\omega\varepsilon_0)$ was dominated by the dc conductivity. No information could be found from $\varepsilon^{/}$ because of the very small capacitance of the electrodes ($\approx 10^{-17}$ F). The equivalent circuit used for the interpretation of the out-of-plane direction includes only the LB film and the electrodes. Contributions from the oxide on the metal electrode and the connecting leads were negligible. The LB film was modelled as a capacitance C_{LB} in parallel with a conductance G_{LB}, both frequency dependent. The electrodes were assumed to contribute a resistance r in series with the sample. For the frequencies under consideration, the following conditions were satisfied

$$rG_{LB} << 1; \quad \omega r C_{LB} << 1; \quad G_{LB} >> (\omega C_{LB})^2 r \tag{10}$$

The measured conductance $G(\omega)$ and capacitance $C(\omega)$ were, to a good approximation, equal to G_{LB} and C_{LB}, respectively.

In the out-of-plane direction, the permittivity was in good agreement with a Cole-Davidson function [29], including a term related to the dc conductivity, namely

$$\varepsilon(\omega) = \varepsilon_\infty + (\varepsilon_S - \varepsilon_\infty)\frac{1}{(1+\omega\tau)^\beta} - j\frac{\sigma_{dc}}{\omega\varepsilon_0} \tag{11}$$

β is a number in the range $0 \le \beta \le 1$ called the shape parameter. The Cole-Davidson relaxation function is asymmetrical with respect to the characteristic frequency $\omega_0 = 1/\tau$ and has two power law relationships for $\omega << 1/\tau$ and $\omega >> 1/\tau$ with exponents 1 and β, respectively, with $\varepsilon'(\omega)$ tending to ε_S and ε_∞, respectively.

The analysis of the experimental data over the temperature range 160-300 K, was based on the fitting both of the real and imaginary parts of the permittivity. It should be noted that the analysis for $\varepsilon'(\omega)$ was more accurate because $\varepsilon''(\omega)$ was dominated by the d.c. conductivity. For temperatures lower than 160 K, τ became large and the relaxation was outside the measurement range. The data and best fits for the real part of the permittivity are shown in Fig. 5.

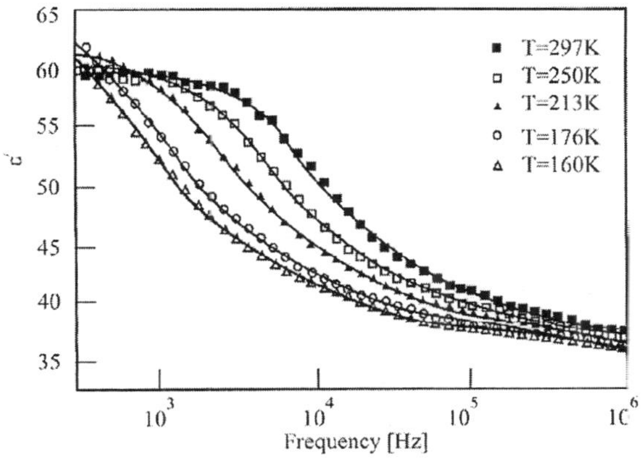

Fig. 5. The real part of the permittivity ε' as a function of frequency for a 115 monolayers thick polypyrrole/palmitic acid LB film for five different temperatures. Measurement in the out-of-plane direction. The full lines are the best fits using equation (11). After Casalini et al. [18].

The high-frequency permittivity ε_∞ is approximately constant over the temperature range. In contrast the static permittivity ε_S increases with decreasing temperature, as expected. The shape parameter β decreased with decreasing temperature (β = 0. 43 at 297 K and 0. 39 at 160 K) whereas the relaxation time was found to increase with decreasing temperature ($\tau = 2.9 \times 10^{-5}$ s at 297 K and 4.6×10^{-4} s at 160 K).

3. APPLICATIONS OF ORGANIC MULTILAYER ASSEMBLIES

3.1. Organic Diodes and Transistors

Since the discovery of semiconducting behaviour in organic materials, there has been a considerable research effort aimed at exploiting these properties in electronic and optoelectronic devices. Organic semiconductors can have significant advantages over their inorganic counterparts. For example, thin layers of polymers can easily be made by low-cost methods such as spin coating. High temperature deposition from vapour reactants is generally needed for inorganic semiconductors. Synthetic organic chemistry also offers the possibility of designing new materials with different bandgaps. Of course, the electrical properties of the organic devices cannot be directly compared to those based on single crystal silicon and gallium arsenide. The mobilities (carrier velocity per unit electric field) of the charge carriers in organic field effect transistors are low and similar to those found in amorphous silicon. Nevertheless, the simple fabrication techniques for polymers have attracted several companies to work on polymer transistor applications such as data storage and thin film device arrays to address liquid crystal displays [30].

Semiconducting LB films have been used similarly to inorganic semiconductors (e.g. Si, GaAs) in metal/semiconductor/metal structures. Perhaps the simplest example is that of a diode. Here, the LB film is sandwiched between metals of different work functions. In the ideal case, an n-type semiconductor should make an Ohmic contact to a low work function metal and a rectifying Schottky barrier to a high work function metal [31]. The reverse will be true for a p-type semiconductor.

Figure 6 shows the electronic potential energy band structure for a Schottky barrier formed by sandwiching a phthalocyanine LB film between aluminium and indium-tin-oxide electrodes

[32]. The LB material is a p-type semiconductor and the aluminium/LB film interface provides the rectifying junction. The current versus voltage characteristics of this device, shown in Fig. 7, are asymmetrical, exhibiting significant rectification. Under illumination, the structure also acts as a photovoltaic cell. Although the open circuit voltage produced by such devices is of the same order as those from silicon solar cells, the short-circuit current density is many orders of magnitude less.

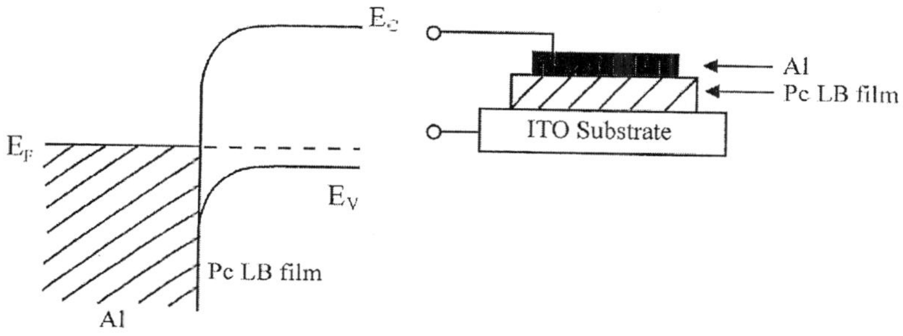

Fig. 6 Energy band structure of a rectifying contact between an unsubstituted phthalocyanine LB film and an aluminium contact. The electrical contact between the indium-tin-oxide (ITO) substrate and the LB film is Ohmic.

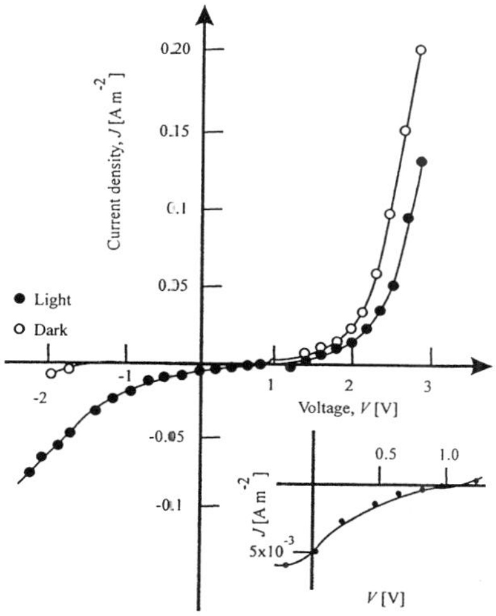

Fig. 7 Current density versus voltage characteristics in the dark (o) and light (•) for an ITO/30 LB layers of phthalocyanine/Al structure at room temperature. The light characteristic is shown in an expanded form in the inset. After Hua et al. [32].

Of particular interest is the possibility of observing molecular rectification using monolayer or multilayer films. This follows the prediction of Aviram and Ratner that an asymmetric organic molecule containing a donor and an acceptor group separated by a short σ-bonded bridge (allowing tunnelling) should exhibit diode characteristics [33]. There have been many attempts to obtain this effect in LB systems [14]. Asymmetric current versus voltage behaviour has certainly been recorded for many LB film metal/insulator/metal structures. Some of these results are open to several interpretations as a result of the asymmetry of the electrode configuration. However, the demonstration that the rectification is affected by a change in the film structure (by bleaching) [34] and a recent report of rectifying behaviour with symmetrical gold electrodes [35] suggest that 'true' molecular rectification is achievable.

Organic thin films have also been used as the semiconducting layer in FET devices [6]. Here, the organic material should possess a reasonable in-plane conductivity: suitable candidates are amphiphilic charge-transfer compounds and conductive polymers. For transistor applications, the organic semiconductor must possess several other important attributes. Charge must be readily injected from the source electrode and the carrier mobility should be high enough to allow useful quantities of source-drain current to flow. The semiconductor and other materials with which it is in contact must also withstand the operating conditions without thermal, electrochemical or photochemical degradation. Two performance parameters to be optimised in organic field effect transistors are the field effect mobility and on/off ratio [6]. The former is related to the absolute quantity of 'on' current that can be induced in the device and is defined for the regimes where current is linear with respect to source-drain voltage and 'saturated' (independent of source-drain voltage). The on/off ratio expresses the current relative to that that would flow in the absence of the gate field.

Many organic materials are p-type semiconductors, where the organic molecules would be considered electron donors, and the majority charge carriers are holes. In general, it is difficult to obtain carrier inversion by the application of an electric field to the gate electrode and organic FET devices operate in the accumulation mode. Carrier mobilities that have been obtained in transistor devices are listed in Table 1 [19, 36, 37]. A significant increase in carrier mobility has been reported over the last 10 years [36]. This has been brought about by both improvements in the material synthesis and in the thin film processing techniques.

Table 1: Field effect carrier mobilities for FETs based on organic semiconductors [19, 36, 37].

Material	Mobility [cm^2 V^{-1} s^{-1}]
Polythiophene	10^{-5}
Polyacetylene	10^{-4}
Phthalocyanine	10^{-4}-10^{-2}
Thiophene oligomers	10^{-4}-10^{-1}
Organometallic dmit complex	10^{-1}
Pentacene	10^{-3}-3
C_{60}	0.3

However, the data in Table 1 suggest that molecular crystalline solids have superior carrier mobilities to polymeric semiconductors. State-of-the-art organic thin film transistors possess characteristics similar to those of devices prepared from hydrogenated amorphous silicon, with mobilities around 1 cm^2 V^{-1} s^{-1} and on/off ratios greater than 10^6. With the use of high permittivity gate insulators, operating voltages as low as 5 V can be achieved [38]. Recently, FET devices using pentacene have been reported that can operate in both accumulation (p-type) and inversion (n-type) [37, 39]. The use of ambipolar thin film devices should lead to a significant simplification of complementary logic circuits.

For commercialisation, the LB technique is probably not the most appropriate to form the relatively thick and stable semiconductive layers required for these transistor devices. The processing method, in which single monolayer films are transferred from a water surface to a solid substrate, does not combine well with other processing technologies used in the semiconductor industry. Thermal evaporation, spin-coating and even screen-printing may be more practical for the fabrication of low-cost FETs [6, 36-41]. Thin film transistors based on organic semiconductors are likely to form key components of plastic circuitry for use as display drivers in portable computers and pagers, and as memory elements in transaction cards and identification tags.

3.2. Light Emitting Displays

Following the initial report of organic light emitting devices (OLEDs) incorporating the conjugated polymer poly(p-phenylene vinylene) (PPV) [7], there has been much academic and commercial interest in improving the device characteristics [42]. External quantum efficiencies of up to 8 % and colours covering the entire visible range have now been obtained [43]. Many techniques have been applied in attempts to optimise the performance of OLEDs for commercialisation. Either polymers or molecular compounds can be used as the emissive layer. Fig. 8 shows the structure of some compounds that have been used as the emissive layer in OLED structures. Polymers may also be doped with organic dyes (e.g. rubrene) to improve the device efficiency [44]. Hole and electron transporting materials, such as those shown in Figs. 9 and 10, can be sandwiched between the cathode and the emissive layer and between the anode and the emissive layer, respectively, to improve and balance the injection of carriers [45].

The use of low work function cathode materials such as calcium [46] or magnesium [47] has led to high efficiency devices. These metals are unstable on exposure to air, however. Another approach is to insert an inorganic insulating layer such as CaO [48], SiO_2 [49] or LiF [50] between the cathode and the emissive material. The anode electrode, usually indium tin oxide (ITO), can be treated (e.g. with an oxygen plasma [51]) to reduce the turn-on voltage and increase the device performance. ITO/polyaniline [52], ITO/poly-(3,4-ethylene dioxythiophene) [53] and ITO/phthalocyanine [54] bilayer anodes have also been shown to produce improved device performance.

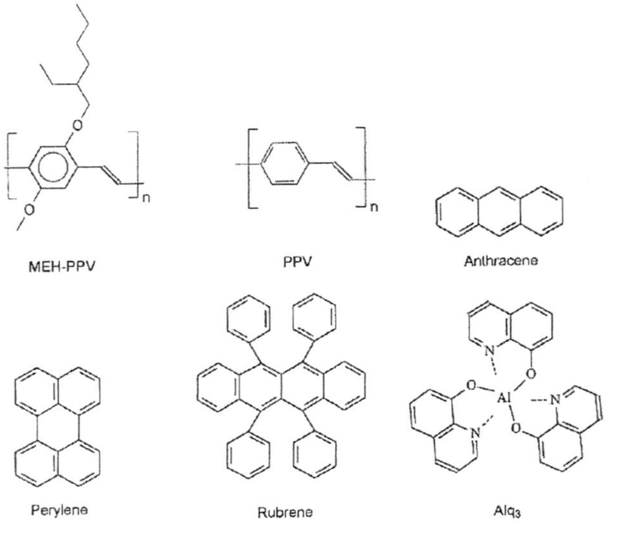

Fig. 8 Organic materials commonly used as the emissive layer in OLED structures.

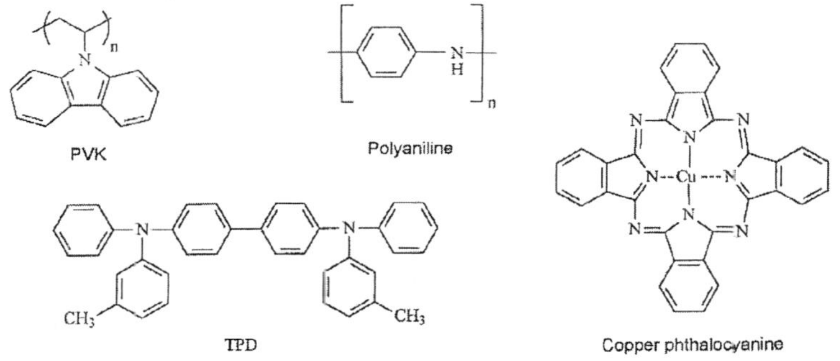

Fig. 9.Organic hole-transport materials.

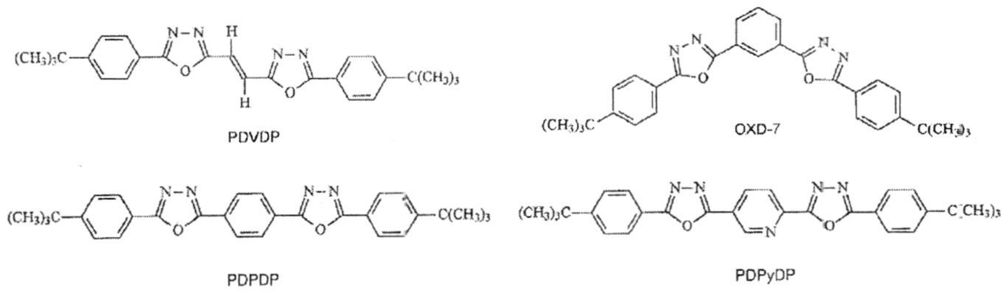

Fig. 10 Organic electron-transport materials.

The thin polymeric or oligomeric layers required for OLEDs are usually deposited by spin coating or thermal evaporation. The LB technique and layer-by-layer self-assembly offer alternative methods for building up ultra-thin organic films of nanometre dimensions. These approaches to thin film deposition allow the thickness requirements of OLEDs to be fully explored in the laboratory. They may also offer certain advantages for the fabrication of OLEDs containing ultrathin organic films and/or alternate-layer films.

One of the first applications of LB technology to produce the emissive layers for OLEDs exploited a substituted anthracene derivative [55]. Although blue electroluminescence was observed, the lifetime of these devices was very short (a few minutes). Poly(2-methoxy-5-(2'-ethylhexyloxy)-p-phenylenevinylene) (MEH-PPV) has been developed as a processible EL material [56]. This compound is insoluble in water and forms a condensed floating layer at the air/water interface. LB films may be built up and the behaviour of OLEDs based on this compound explored [57, 58]. Other LB materials that have been used for electroluminescence studies have included cyanine dyes [59], oligothiophenes [60-62] and tris(8-hydroxyquinoline)aluminium (Alq_3) [63].

The relatively ordered nature of LB arrays of molecules may also be exploited in light emitting structures. For example, the preferential alignment of the molecules can result in polarised light emission [61, 64]. Langmuir-Blodgett deposition can also be used to control the positions of luminescent species within metal mirror microcavities [65]. In this way, the position of the absorbing/emitting layers allows the optical mode structure of the cavities to be probed by both the emission and absorption of the chromophore.

Although the LB technique enables some elegant experiments to be undertaken on OLED structures, the quantum efficiencies of single-layer devices are generally lower than those for structures produced by spin coating. However, marked improvements can be achieved by the incorporation of an electron-transporting LB film between the emissive layer and the Al cathode [58] and/or by the insertion of an LB film 'spacer' layer between the emissive layer and the cathode [66, 67]. The latter approach has previously been applied successfully using GaP as the emissive material [68]. Fig. 11 shows the light output versus voltage characteristics for a series of ITO/MEH-PPV/arachidic acid (AA)/Al OLEDs deposited using the LB technique [66]. Insertion of the fatty acid layer both increases the light output at a fixed voltage and lowers the turn-on voltage to observe electroluminescence. This results in an improvement of the quantum efficiency by a factor of between two and three.

An insight into the underlying physics can be obtained by reference to the energy band structures in Fig. 12.

The schematic diagram in Fig. 12(a) represents the situation in forward bias for an ITO/MEH-PPV/Al device. The electron current injected from the Al is limited by a barrier of about 1.4 eV at the Al/polymer interface. It is only at a large applied bias, when electrons can tunnel through this barrier that the electron current becomes significant and EL can be observed. An important difference for the devices incorporating the AA layers is that the Fermi level is no longer 'tied' to the energy band structure of the polymer (Fig. 12(b)).

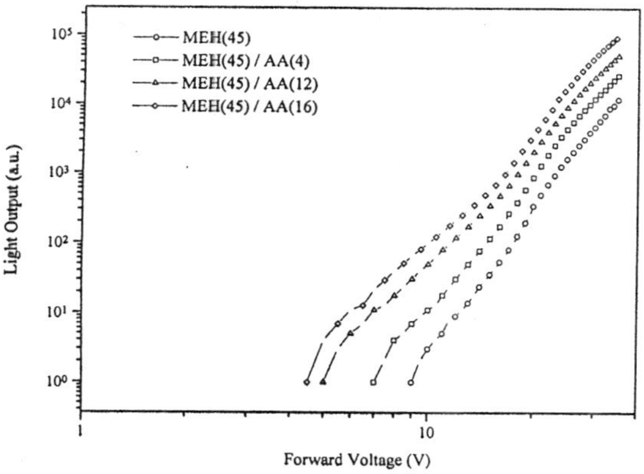

Fig. 11. Light output versus voltage for ITO/MEH-PPV/AA/Al devices containing different thicknesses of fatty acid. The number of LB deposition cycles is shown for each device structure. After Jung et al. [66].

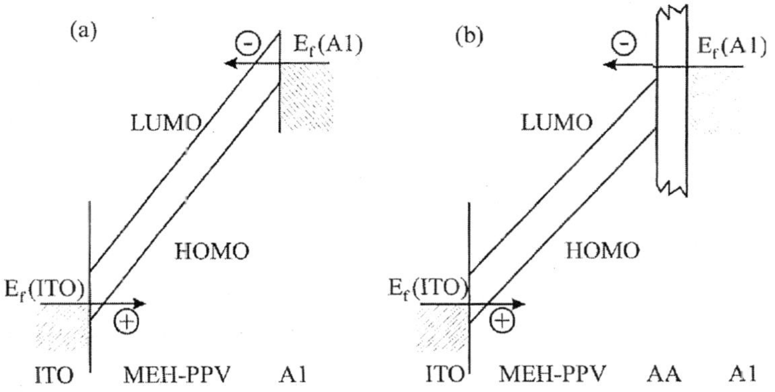

Fig. 12. Potential energy diagrams for (a) ITO/MEH-PPV/Al structure and (b) ITO/MEH-PPV/AA/Al structure. In both cases a positive voltage is applied to the ITO electrode. After Jung et al. [66].

In particular, as the applied voltage is increased, the Al Fermi level can move with respect to the LUMO level in the MEH-PPV aligning filled electron states in the metal with vacant states in the polymer. As long as the fatty acid layer is transparent to electrons, the electron current, and hence the EL efficiency increases. The precise mechanism of conduction through the AA film is unclear. The film thicknesses used in the work are too high to support simple quantum mechanical tunnelling. It is possible that fortuitous alignment of energy levels associated with localised states in the AA with the energy bands of the MEH-PPV could provide a path for the increased electron current [17].

The technique of using insulating LB layers to improve the efficiency of OLEDs seems generally applicable and it has also been used to enhance the operation of N,N'-diphenyl-N,N'-bis(3-methylphenyl)-1,1'-biphenyl-4,4'-diamine (TPD)/ tris(8-hydroxyquinoline) aluminium (Alq_3) bilayer devices in which both the emissive layer (Alq_3) and the hole-transport layer were deposited by thermal evaporation [69].

A further important parameter for organic light emitting devices is their operating lifetime. Suggested degradation mechanisms for OLEDs include: crystallisation of the amorphous organic films [70]; a reaction between the organic material and the electrode(s) [71]; a reaction of the top (outer) electrode with water and oxygen [72]; oxidation of the organic layer by indium-tin oxide [73]; and poor contact between the organic layer and the electrode [74]. The degradation is often associated with the formation of dark areas in the image of the electroluminescent output. To improve the lifetime, elaborate encapsulation and packaging schemes are used to exclude both water and oxygen [75]. Other methods include the use of a high T_g (glass transition temperature) hole transport material [76], a polymeric anode [77] and doping the emissive layer [78].

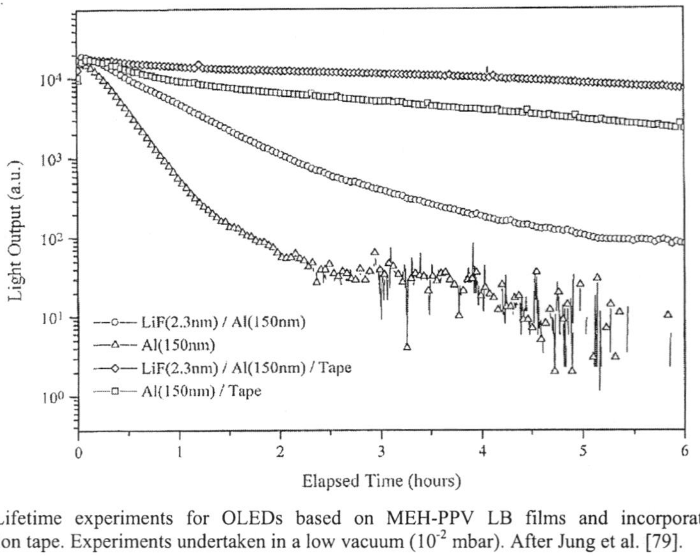

Fig. 13. Lifetime experiments for OLEDs based on MEH-PPV LB films and incorporating LiF and/or encapsulation tape. Experiments undertaken in a low vacuum (10^{-2} mbar). After Jung et al. [79].

Fig. 13 shows the results of some experiments to improve the lifetime of OLEDs based on MEH-PPV LB films [79]. In each case, the structure investigated incorporated an LB film that had been heat treated in vacuum at 35°C. In the first experiment, a LiF layer (≈ 2 nm) was thermally evaporated onto the annealed polymer layer to prevent any chemical reaction between this and the Al electrode. Simple device encapsulation using adhesive tape was also used to protect the aluminium electrode from direct contact with air and moisture. The figure shows the light output versus time behaviour for different OLED configurations when a constant voltage stress (20 V) was applied. Incorporation of the LiF buffer layer provides a useful increase in the device lifetime. However, a more significant improvement is obtained with the encapsulation tape. These data illustrate that the protection of the top electrode from moisture is a crucial factor if long operating times are to be achieved.

Currently, the most efficient OLEDs are fabricated using spin coating or thermal evaporation. In the case of devices using PPV and PPVs substituted with electron-donating groups, hole injection is relatively easier than electron injection under forward bias. A well-established approach to balanced charge injection is to use an additional layer of a material with a lower LUMO level than that of the corresponding PPV derivative beneath the cathode to complement the excess hole injection from the opposite electrode. This additional layer becomes an electron-injection layer. Molecular and polymeric 1,3,4-oxadiazole (OXD) derivatives are the most widely studied class of electron-injection and/or hole-blocking materials, mainly because of their electron deficiency, high photoluminescence quantum yield and good thermal stability, Fig. 10. In this context 1,3-bis(2-(4-*tert*-butylphenyl)-1, 3, 4-oxadiazol-5-yl) benzene (OXD-7) has been used in OLEDs with different emissive materials, and significant improvement in device performance has been observed [80]. Similarly, as pyridine is a moderately electron-deficient π system, poly(2, 5-pyridylene) has also acted as an electron-injection/hole-blocking materials for OLED fabrication [81].

More recent work has focused on a hybrid of pyridine and oxadiazole which was reasoned could lead to a further increase in the electron-injection [82]. This material, 2, 5-bis[2-(4-tert-butylphenyl)-1, 3, 4-oxadiazol-5-yl] pyridine (PDPyDP), is shown in Fig. 10, where its structure is contrasted with the phenylene and trans-vinylene analogues and OXD-7. These compounds were thermally evaporated onto spin-coated MEH-PPV layers and the electrical response of the resulting OLEDs studied. Fig. 14 shows the light output versus current density curves for four devices. Approximately linear behaviour was measured for all the device structures. The most efficient device was produced using the PDPyDP layer. This structure possessed an external quantum efficiency of about 0. 2 %, a factor of 50 greater than the value measured for the single layer MEH-PPV device.

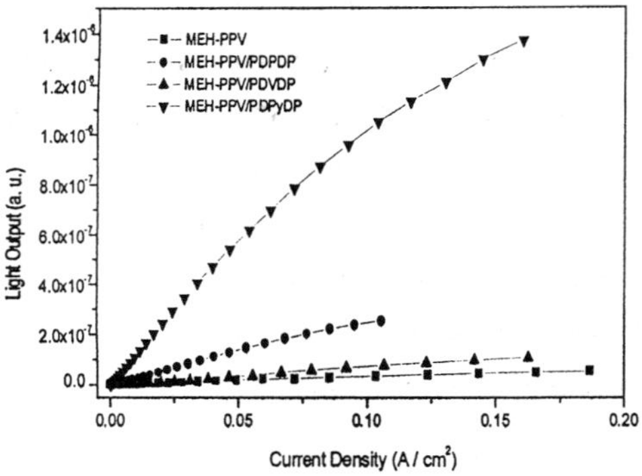

Fig. 14. Light output versus current density for the single-layer and bilayer OLEDs based on spin-coated MEH-PPV and incorporating the electron-injection materials shown in Fig. 10. After Wang et al. [82].

3.3. Gas Sensors

The development of effective devices for the identification and quantification of chemical and biochemical substances for process control and environmental monitoring is a growing need [10, 11]. Many current devices do not possess the specifications to conform to existing or forthcoming legislation; some systems are too bulky/expensive for use in the field [83] Inorganic materials such as the oxides of tin and zinc have been favoured as the sensing element [12]. However, one disadvantage of sensors based on metallic oxides is that they usually have to be operated at elevated temperature, limiting some applications. As an alternative, there has been considerable interest in trying to exploit the properties of organic materials. Many such substances, in particular phthalocyanine derivatives, are known to exhibit a high sensitivity to gases [84]. Lessons can also be taken from the biological world. For instance, one household carbon monoxide detector currently being marketed is designed to simulate the reaction between CO and haemoglobin. A significant advantage of organic compounds is that their sensitivity and selectivity can be tailored to a particular application by modifications in their chemical structure (e.g. a change of functional groups in the case of polymers). Moreover, thin film technologies, such as self-assembly or layer-by-layer electrostatic deposition, enable ultra-thin layer structures to be engineered at the molecular level.

Many different types of organic materials have been used for gas sensing. These include porphyrins, phthalocyanines, and insulating and conductive polymers [11, 12]. Phthalocyanines are an important group of synthetic charge-transfer complexes, offering good thermal stability together with chemical stability (resistance to weak acids and bleaches) [84]. These compounds are generally highly coloured and have been used extensively by the chemical industry as dyes. Porphyrins are related to phthalocyanines and comprise a hydrocarbon ring structure with a central metal ion and substituent organic groups around the periphery. Many of these compounds occur naturally, e.g. haemin and chlorphyll.

Electrochemical sensors (i.e. chemiresistors) have been widely investigated, also studies of mass and optical changes have been undertaken to gain a better understanding of the polymer/ gas interactions. The electrical properties of such materials can be greatly influenced by interactions with oxidising/reducing agents. Generally, the effects are not fully reversible for exposure to high concentrations of strong oxidising/reducing agents (e.g. NO_2, NH_3) but show reversible changes for exposure to organic vapours (e.g. ethanol, methanol). The interactions can be explained in terms of a charge transfer between the gas molecules and the polymer. However, other mechanisms, such as film swelling, can be important.

Non-conductive polymers (i.e. those not containing (π-bonds) have been used as mass sensors, thermal sensors, optical sensors and dielectric sensors. The detection principle is usually based on a bulk dissolution of the gas into the polymer film with a consequent change of a physical property such as mass, permittivity, refractive index or thickness. To monitor such changes, a transducer is required that is more complicated than a simple chemiresistor. However, the advantage of not requiring a chemical reaction between the gas and the polymer molecules can provide improved reversibility and stability. This class of polymer is more wide-ranging than that of conductive polymers, offering further scope for chemical modification to tailor the gas/polymer interaction. For example, amorphous elastomeric polymers represent an important category of sensor coatings. In the elastomeric phase (i.e. above the glass transition temperature) constant thermal motion of the polymer chains allows rapid vapour diffusion. Adsorption and desorption of the vapour leaves the material in the same state. The softness of elastomeric materials has an additional advantage for piezolectric sensors such as surface acoustic wave devices that are sensitive to changes in material stiffness. An example of such materials is the family of polysiloxanes characterised by a repeat unit (-RR$^{/}$-Si-O-) where R and R$^{/}$ are generic functional groups [85, 86]. By modifying the side chains, nonpolar, polar and polarisable polymers can be obtained.

There are many physical principles upon which sensing systems might be based. However, the main challenges in the development of new sensors are in the production of cheap, reproducible and reliable devices with adequate sensitivities and selectivities. In the following sections, some examples of the use of organic materials in gas sensing devices are given. The transduction techniques chosen are those that are specifically suited to organic compounds in the form of thin films.

3.3.1 Field Effect Devices

A simple chemiresistor sensor exploits the resistance change of a thin layer of a gas-sensitive material. For example, the conductivity of phthalocyanine chemresistors can be changed in the presence of oxidising or reducing gases [87]. This effect is analogous to the doping of an inorganic semiconductor, such as silicon, with acceptor or donor impurities. A problem associated with these devices is that the resistance of the organic layer is usually very high. Consequently, the current outputs are low (typically picoamperes) requiring elaborate detection electronics and careful shielding and guarding of components. This difficulty may be overcome by incorporating the organic sensing layer into a silicon field effect transistor. A schematic diagram of such a structure is shown in Fig. 15 [88, 89].

This device is a p-channel metal-oxide-semiconductor field effect transistor (MOSFET) in which the gate electrode is partly replaced by a gas-sensitive material. On application of a voltage to the gate, the metallic sections of the electrode rapidly charge to the applied voltage. Due to the high resistance of the sensing layer, there is a delay before this is uniformly charged. The time required for the complete charging of the gate, and hence the time needed to turn on the transistor, depends on the sheet resistance of the sensing layer.

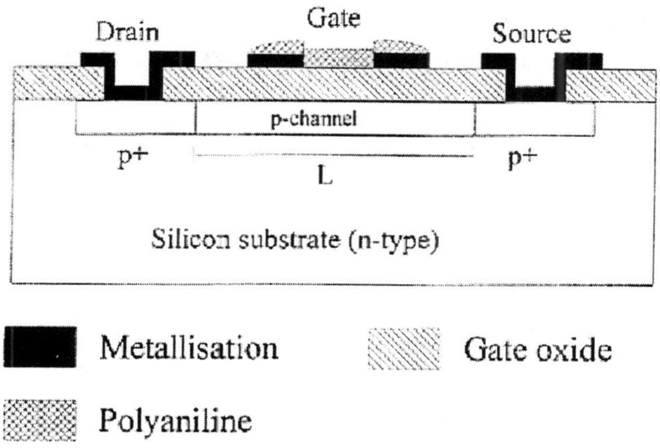

Fig. 15. Schematic cross-section of organic/silicon hybrid field effect transistor [88].

The d.c. operating characteristics for a device with 15 μm gate gap, and with polyaniline deposited by spin coating within this hole, are shown in Fig. 16 [89]. These curves are of the general form for an entirely metallic gate electrode device. The d.c. response for a fixed gate bias has three distinct regions as the applied drain bias V_{dd} is increased from 0 to −12 V. These are: (i) for low magnitude bias the drain current I_{ds} varies linearly as the negative bias is increased; (ii) as the negative bias is further increased the conducting channel induced by the gate voltage begins to 'pinch-off' at the drain end reducing the carrier (hole) density and leading to a saturated drain current; and finally (iii) further increase in drain bias leads to an additional depletion of carriers from the channel which prevents the current from increasing above the saturation value.

It is the transient response of the hybrid MOSFET that is important for gas sensing. Fig. 17 shows a set of such characteristics measured for a device incorporating a spin-coated phthalocyanine film [90]. The baseline response was established in a nitrogen ambient before every exposure to NO_2. Following the exposure, the supply of NO_2 was terminated and the baseline re-established. The complete response (i.e. until I_{ds} saturates) was not recorded since the most useful and characteristic data are accumulated in the initial part of the transient. The increase in the rate of the turn-on of I_{ds}, as the concentration of NO_2 increases, can be attributed to an increase in the conductivity of the phthalocyanine film within the gate electrode. The gas is chemisorbed on particular molecular sites and becomes negatively charged. This produces an increase in the hole concentration and, hence, in the conductivity of the phthalocyanine.

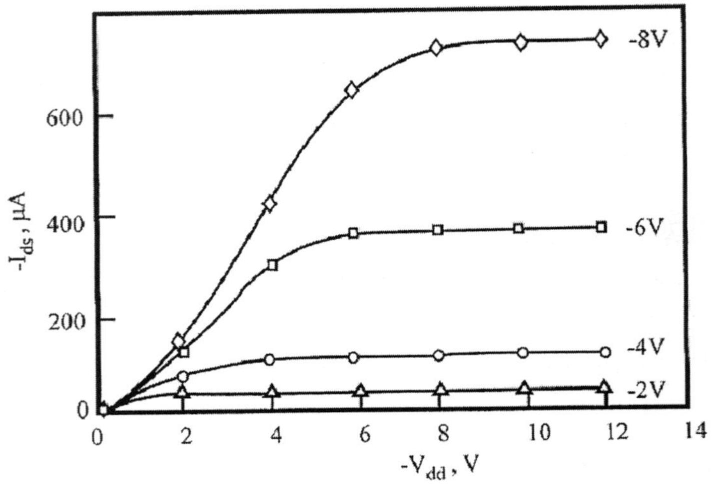

Fig. 16. Source-drain current I_{ds} versus drain voltage V_{ds} for 15 μm gate gap MOSFET, with polyaniline deposited in the gap, with varying gate voltages. After Barker et al. [89].

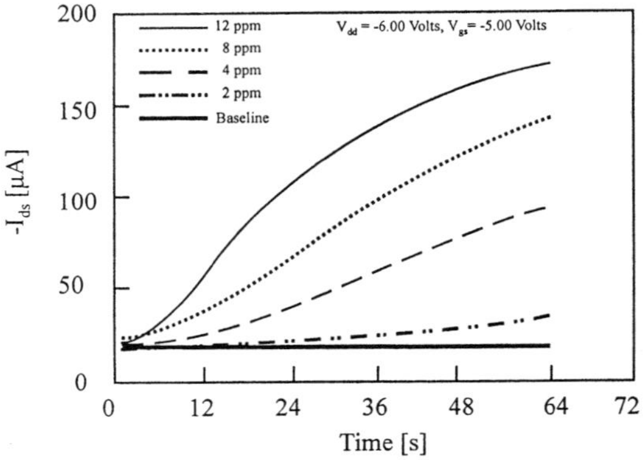

Fig. 17. Source-drain current versus time for a phthalocyanine-coated sensor device (35 μm gate gap) exposed to various NO_2 concentrations. After Barker et al. [90].

While gas sensing devices, such as that described above, may shows adequate sensitivity to some gases, the selectivity is generally poor. For example, the transistor device will respond to water vapour as well as to NO_2. To get around this difficulty, one approach that is being embraced enthusiastically is to use an array of sensing elements, rather than a single device. This is the method favoured by nature. The human olfactory system has many sensors, which are individually non-specific. Signals from these sensors are fed to the brain for processing. As a result, the olfactory system is a highly sensitive and selective detector of odours. The electronic nose is an attempt to mimic nature's olfactory system [91]. There are now several companies marketing such equipment. Individual sensors can be based on polymer films. Each element is treated in a slightly different way during deposition so that it responds uniquely on exposure to a particular gas or vapour. The pattern of resistance changes in the sensor array can then be used to fingerprint the vapour. An alternative technique to using an array of sensing elements is to make multiple measurements on one sample. The next two methods illustrate this approach.

3.3.2. Admittance Spectroscopy

Admittance spectroscopy has been described in section 2.2 and noted as a useful technique for the investigation of the dynamic electrical properties of organic thin films. The capacitance and conductance characteristics of thin organic layers can also be exploited for gas and vapour sensing. Discrimination between different vapours may be achieved by monitoring the complex admittance at several frequencies.

When exposed to different organic vapours at room temperature, transient effects were observed with rapid changes of both C and G that saturated after approximately 10 min. The inset in Fig. 18 shows the transient response of the capacitance to 3.3% ethanol measured at 1 kHz. The observed changes were reversible when the samples were re-exposed to nitrogen. Admittance changes were also noted for both benzene and acetonitrile with similar response and recovery times to those for ethanol.

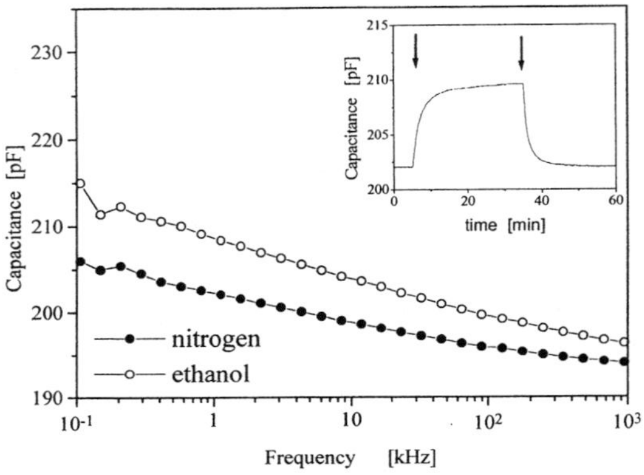

Fig. 18. Capacitance versus frequency for a co-ordination polymer LB film in nitrogen and exposed to ethanol vapour (3.3 %). The inset shows the transient behaviour measured at a fixed frequency (1 kHz) when the ethanol vapour was turned on and off (indicated by the arrows). After Casalini et al. [92].

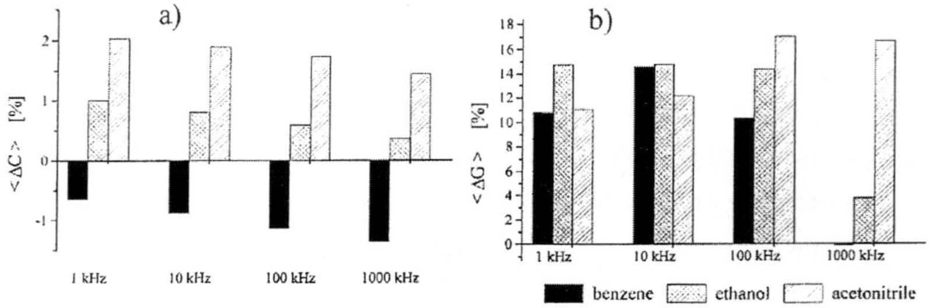

Fig. 19 Average changes in (a) the capacitance ΔC and (b) the conductance ΔG, measured at four frequencies, for three different vapours. After Casalini et al. [92].

The admittance of the device was measured in the organic vapours after 15 mins. exposure (the time taken to complete a frequency sweep was less than a minute). The capacitance data for the device exposed to ethanol are also given in Fig. 18. A frequency-dependent increase in capacitance is evident. Fig. 19 contrasts the average percentage changes in the capacitance C and the conductance G divided by the vapour concentration for the different vapours at four measurement frequencies. Three concentrations, in the range 3-7%, were generated for each vapour and the results (for each vapour) averaged. The figure shows that the capacitance increases when the device is exposed to ethanol (dipole moment 1.69 Debye) and acetonitrile (dipole moment 3.92 Debye) with a greater fractional increase for the acetonitrile. In contrast, a decrease in capacitance is observed with the non-polar benzene. This is almost certainly associated with the swelling of the film.

The changes in capacitance and conductance for the ethanol and acetonitrile may be the result of a change in the film permittivity due to the bulk dissolution of the vapour in the LB film. As the dielectric relaxations for the organic solvents are expected to occur at a very high frequency (beyond the maximum used in the work), this should produce changes in C and G that are constant over the frequency range measured, which is not evident in Fig. 19. An alternative explanation is that the admittance changes result from a variation in the polymer permittivity induced by the interaction between the polymer and the vapour molecules. A detailed discussion is presented elsewhere [93]. However, from the point of view of a practical sensor, it is clear that monitoring the admittance at different frequencies can provide a useful means to discriminate between the vapours studied.

3.3.3 Surface Plasmon Resonance

Surface plasma waves are collective oscillations of the free electrons at the boundary of a metal and a dielectric [94]. Fig. 20 shows one experimental arrangement, which involves the optical excitation of surface plasmons by means of evanescent electromagnetic waves.

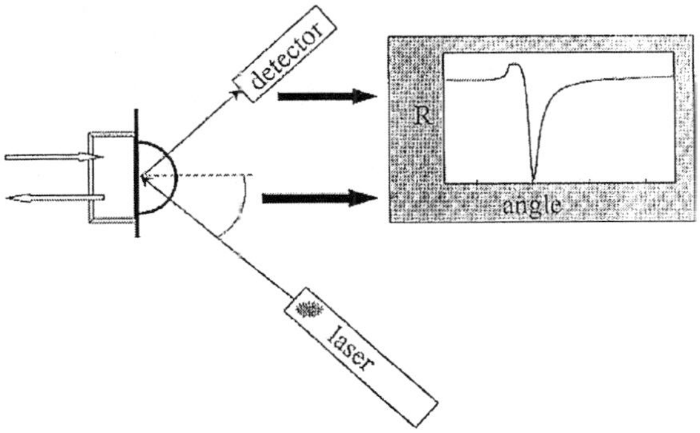

Fig. 20. Surface plasmon resonance technique for gas sensing.

Monochromatic light (*p*-polarised) enters a prism and is refracted onto the base and therefore onto a metal (e.g. silver) film, where it is reflected. The resulting evanescent field penetrates to the opposite side of this layer (i.e. the metal/air or metal/dielectric interface) if the metal is sufficiently thin. Changing the angle of incidence of the light alters the component of its wavevector parallel to the prism base. When this component matches the real part of the surface plasmon wavevector, then surface plasmons are excited; this process may be detected by a pronounced reduction in the intensity of the reflected beam. This phenomenon is known as surface plasmon resonance (SPR).

The position of the SPR minimum is very sensitive to alterations in the thickness or optical constants of any overcoating on the metal. For example, the SPR minimum shifts to higher angles as the overlayer thickness increases. A similar effect may be predicted if the refractive index of the overlayer is altered. The sensitivity of SPR is noteworthy, with claims that changes in refractive index of 10^{-5} may be monitored and, more generally, that the technique compares favourably with ellipsometry.

For sensing applications, changes in the optical properties of an 'active' layer in response to external ambients are detected. By choosing an angle of incidence close to the plasmon resonance angle, small changes in the resonance conditions (e.g. angle of resonance minimum) produce large changes in the reflected intensity [92].

The influence of a gas or vapour on the active thin film can be quite complex. For example, the interaction may lead to changes in the film thickness or in the real or imaginary components of permittivity (or all three). The usual method of SPR measurement involves monitoring the change in reflectivity versus time and cannot distinguish between these various effects. However, if a CCD camera is used as the detecting system, then more data can be collected. Fig. 21 illustrates how theoretical changes in the SPR curves can provide more detailed information in the form of CCD images [95]. In each case, the change to the SPR curve is shown on the left and the corresponding CCD image of the individual pixels on the right. The dark band in the centre of the SPR image (Fig. 21a(ii)) corresponds to the points where the incident light excites surface plasmons and is absorbed (the resonance minimum in Fig. 21b(i)). If the SPR curve moves to higher angles (shown by the dashed line (theoretical) in Fig. 21b(i)), the corresponding CCD image (Fig. 21b(ii)) will also move to the right (i.e. to higher angles). Fig. 21c illustrates the effect of a broadening, while Fig. 19d shows the effect of the SPR curve becoming less defined (loss of resonance depth).

Experimental results from a single LB layer of zinc tetrakis ((3,3-dimethyl-1-butoxy)carbonyl) phthalocyanine (ZnPcBC) deposited onto Ag-coated glass are presented in Fig. 22 [95].

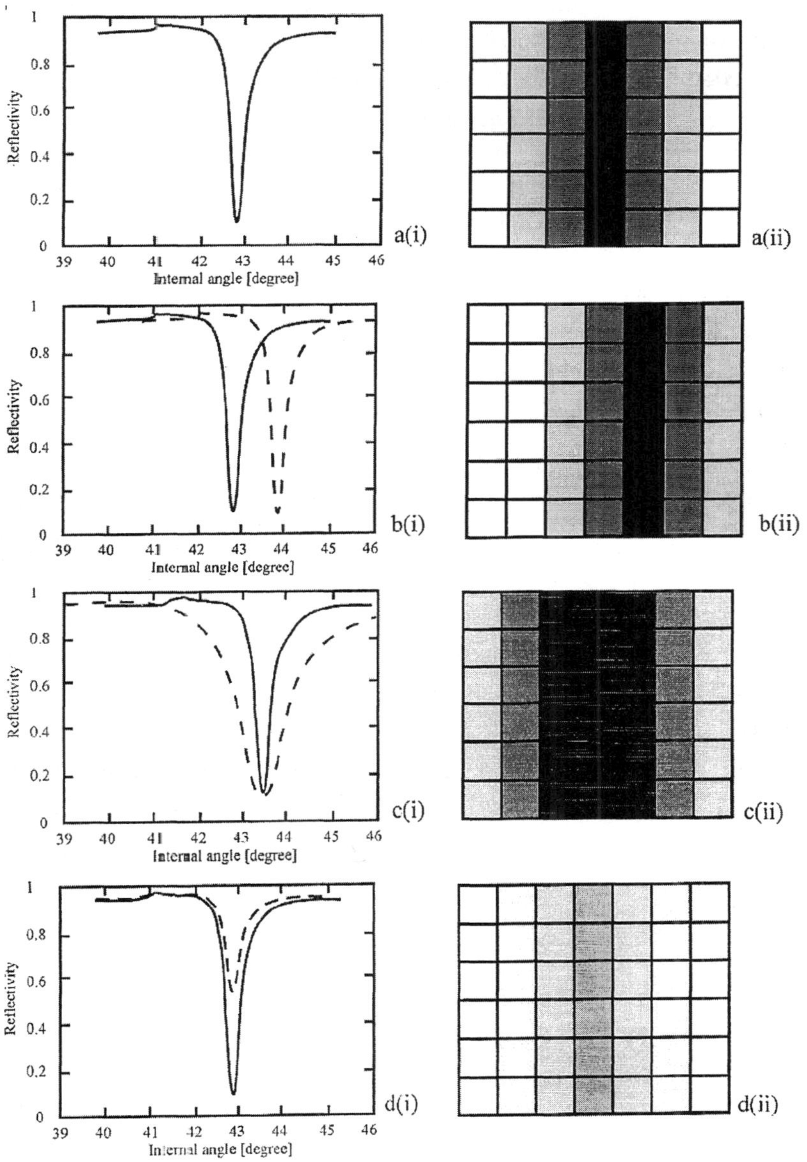

Fig. 21. (a) Theoretical SPR curve (i) and expected CCD image (ii) for a single LB layer deposited on Ag. (b) shift in SPR curve to higher angles. (c) broadening of SPR curve. (d) Loss of resonance depth [95].

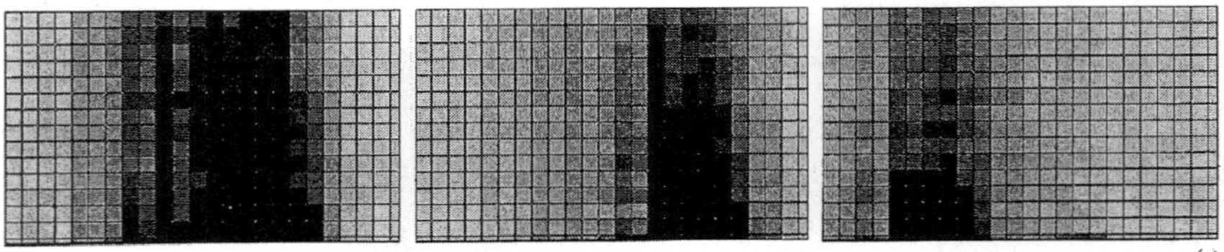

Fig. 22. (a) CCD image for one LB layer of ZnPcBC as-deposited. (b) After 50 min exposure to dry N_2. (c) Following 60 min exposure to 20 ppm NO_2 diluted in N_2. After Wilde et al. [95].

Fig. 22(a) shows the individual pixels in the CCD image of the as-deposited film following image processing using contrast and brightness filters provided in the software package. The sample was first exposed to a flow of dry nitrogen at a rate of 600 cc min^{-1}. After 50 minutes' exposure, the SPR point (shown in Fig. 22(b)) has shifted to higher angles by approximately five pixels and has become more defined (narrowed). This effect was not reversible within the timescale of the experiment (a few days). It was suggested that it is probably the result of the removal of trapped water in the film following LB deposition. The LB sample was then exposed to 20 ppm NO_2 at the same flow rate. After 60 minutes exposure (the exposure time was purposely chosen to be large so that the changes in the CCD images could be seen by eye) the SPR image (Fig. 22(c)) has shifted 10 pixels to the left and has started to lose intensity (loss of resonance). Following 24 hr. in N_2, the SPR image partially recovered to its original position in angle (i.e. before the introduction of NO_2). However, the reduction in the depth of resonance was permanent. These effects were attributed to a reaction between the ZnPcBC film and the NO_2 (reversible) and an oxidisation between the NO_2 and the silver base layer (irreversible). The latter may be eliminated by the use of a thin catalytic overlayer [96] or the use of a gold base layer.

3.4. Ion Sensors

The combination of molecular recognition and self-assembly processes offers a powerful route to the development of nanoscale systems that may have technological applications as sensors, devices and switches. A chemosensor is a molecular system for which the physiochemical properties change upon interaction with a chemical species in such a way as to produce a detectable signal. A device can, therefore, be based on molecular recognition phenomena where the specific binding of an analyte to a sensor material (the receptor) is transduced to the signalling unit, which then responds by undergoing a change in its electronic state. This is monitored by different spectroscopic, electrochemical or structural techniques. Most investigations based on this idea have concerned molecular recognition in solution, and signal transduction using luminescence [97] and electrochemical techniques [98] have been reviewed. However, the knowledge of these specific guest-host interactions and their application to sensing devices in thin films, especially thin solid films, is less well developed.

Much work has been undertaken on the development of sensing devices based on natural and synthetic ionophores incorporated into lipid membranes and organic films. Different types of response exploit different functional properties of the ionophore molecules; normally changes in either the electrical behaviour (e.g. ion-selective electrodes) or optical properties (optrodes) are used. There are several features that are important to all sensor structures. One of these, a fast response, requires a rapid penetration of analyte through the sensing element. This suggests the use of thin films.

Ion-sensitive organic thin films can be deposited using the electrostatic layer-by-layer absorption technique that has been pioneered by Iler [99] and refined by Decher et al. [100, 101]. This involves the assembly of oppositely charged layers of organic materials and can provide a 1-2 nm growth step for a bilayer. A solid substrate with a negatively charged planar surface is first immersed in a solution containing the cationic polyelectrolyte, and a layer of polycation is adsorbed. A number of ionic groups remain exposed to the interface with the solution, and thus the surface charge is effectively reversed. This prevents further polyion adsorption. After rinsing in pure water, the substrate is immersed in the solution containing the anionic polyelectrolyte. Again, a layer is adsorbed and the surface charge is reversed. By repeating both steps, alternating multilayer assembles are obtained with precisely repeatable layer thickness.

The layer-by-layer deposition method has the advantage that electrostatic attraction between oppositely charges is the driving force for the multilayer build-up. In contrast to chemisorption techniques that require a reaction yield of 100% or lateral cross-linking after each deposition step, no covalent bonds need to be formed. Additionally, an advantage over LB deposition is that adsorption processes are independent of the substrate size and topology. Layer-by-layer adsorption studies have now been extended to functional polymers [102], biopolymers [103], small molecules [104] and inorganic materials [105]. The procedure provides a simple method for the formation of thin films of a wide range of materials without the need for complex and expensive equipment.

Fig. 23 shows the how SPR may be used to monitor the real-time assembly of a bilayer of poly(ethyleneimine) (PEI) and poly(ethylene-co-maleic acid) (PMAE) [106]. Following an initial scan of an uncoated gold substrate, water in the test cell was exchanged with PEI solution, causing an increase in the reflectivity by approximately 0.04. There are two contributions to the enhanced reflectivity: the formation of an adsorbed layer on the gold and an increase in the refractive index of the solution. The latter was estimated to be approximately 0.003, much less than the change seen in Fig. 23. The measured reflectivity change is therefore a result of the adsorption of the polycation onto the gold surface, the step-like shape of the characteristic indicating rapid adsorption kinetics. The adsorption of polyions is complete after less than 4 min. After this, the reflectivity remains constant while the polycation solution is circulated.

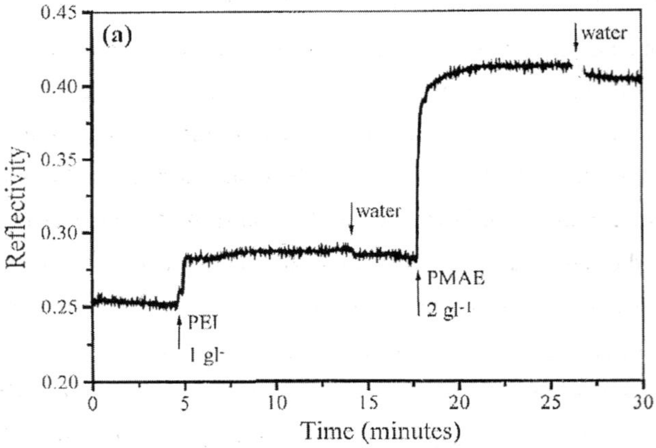

Fig. 23. Reflectivity (angle of incidence 56.0°) against time measured during adsorption of PEI and PMAE. After Pearson et al. [95].

On recirculating pure water, a slight decrease in the reflectivity is observed. This change was of the same order as that predicted for a refractive index decrease of the solution, suggesting that the PEI layer did not desorb. The water was then replaced by the PMAE solution, producing a steep increase in reflectivity, which was significantly greater than that noted for the adsorption of PEI onto the gold. After the initial rapid rise, the reflectivity continued to increase slowly to a maximum. This may result from (i) the distribution of molecular weights in the polymer, giving a rapid adsorption of low molar mass fractions, or (ii) rapid adsorption of the polymer followed by slow changes in the conformation of the adsorbed molecules (changing the thickness and refractive index of the adsorbed layer). It was suggested that the latter explanation was more likely [106]. Immersion in water again produced a slight decrease in reflectivity, perhaps a result of partial desorption of the film. Curve fitting of the SPR data for the bilayer adsorbed onto the gold surface provided an average thickness of 2.5 nm.

Sensing experiments were undertaken using the PEI/PMAE bilayer by exposing it to aqueous solutions of various ions. Exposure to copper ions produced a shift in the SPR curve to higher angles (by about 0.5°). The critical angle was unchanged so that any effects due to different refractive indices of the pure water and copper solution could be neglected. Therefore, the shift in the SPR curve suggests that copper ions are incorporated into the organic sensing film. Both chemical adsorption (reaction of copper ions with the maleic acid unit to form a chelate) and physical adsorption were probably responsible. The sensitivity to copper ions was found to be significantly greater than to either nickel or sodium. On immersion in hydrochloric acid (0. 01 M) for 10 min, the SPR curve returned almost to its original position in angle, indicating a reversible pH-dependent reaction. Fig. 24 shows the dynamic response to various copper ion concentrations [106]. In this experiment, the incident angle of the laser was fixed and changes in the intensity of the reflected beam were monitored. The measurement cell was immersed in solutions of hydrochloric acid and copper acetate. Between the change of copper acetate and hydrochloric acid (and vice versa) the cell was purged with water. On exposure to the copper acetate solution, the reflectivity increased and on exposure to the hydrochloric acid solution, the reflectivity decreased to its former value. The recovery time was less than 3 min for all copper concentrations. The results reveal that concentrations down to 10^{-6} M could be measured.

Prospective applications of supramolecular systems based on macrocyclic molecules have been described by Lehn [107]. In particular, the receptor and carrier characteristics of macrocyclic molecules, allowing the recognition and transport of metal and molecular ions, could lead to the eventual creation of supramolecular ionic devices. These structures may be analogous to biological systems, which process information using ions. One approach, therefore, is to understand biological function in terms of structural relationships and then to replicate this in synthetic products with simpler and more compact superstructures [108].

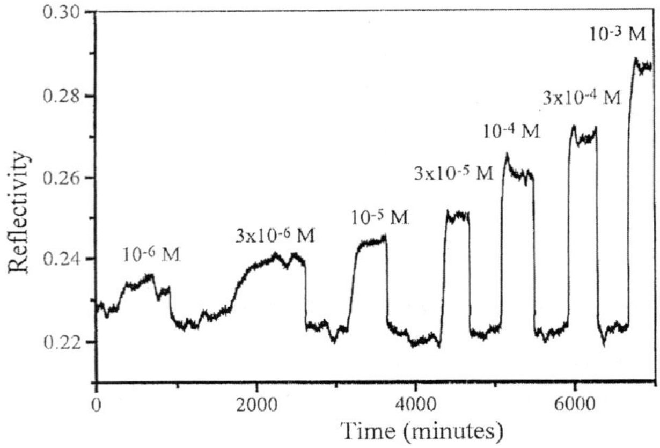

Fig. 24. Reflectivity (angle of incidence 56.5 °) versus time for a PEI/PMAE polyelectrolyte bilayer exposed to increasing concentrations of copper acetate. The bilayer sample was immersed in hydrochloric acid and purged with pure water between the different copper ion concentrations. After Pearson et al. [95].

The importance of membrane transport in biology has stimulated an intense study of this phenomenon using many different artificial systems [109]. Macrocyclic structures incorporated in membranes may be used both as carrier molecules and as transmembrane channels [107, 108]. There has been some work on the development of sensing devices using LB layers incorporating the natural ionophore valinomycin. This molecule is composed of three identical segments linked to form a 36-membered ring. A schematic diagram is shown in Fig. 25. There are six amide groups and six ester groups around the ring. Nine isopropyl and three methyl side groups are attached to the ring. Valinomycin forms complexes with many alkali and alkaline earth cations. It is highly selective for K^+ and Rb^+ over Li^+, Na^+, Mg^{2+} or Ca^{2+} [110]. The size of the valinomycin internal cavity is too large to selectively complex K^+ if it only binds in a two-dimensional sense. The valinomycin-K^+ complex exhibits a so-called 'tennis-ball-seam' conformation. The cation is located in the centre of the macrocycle while the side groups reside on the outside [110]. This produces a hydrophobic exterior that facilitates the incorporation of the complex into a lipid membrane. Six hydrogen bonds formed between the amide carbonyl groups and the >N-H hydrogens additionally stabilise the appropriate conformation of the complex [111].

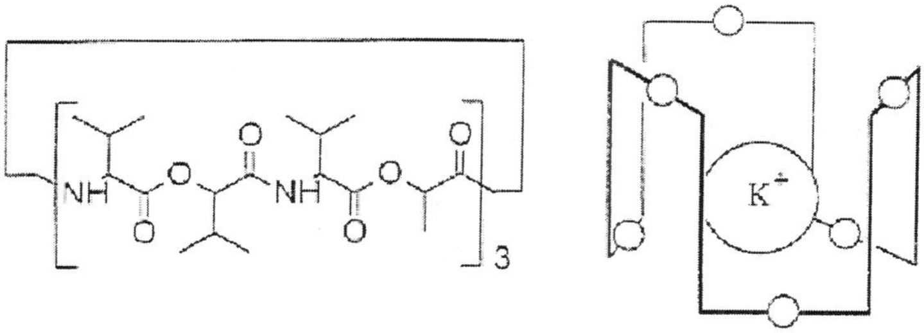

Fig. 25. Schematic representation of valinomycin and its complex with a potassium ion.

Valinomycin is known to form a condensed monolayer at the air/water interface [112, 113]. The surface pressure versus area isotherm is of an expanded type. Provided that the monolayer is not compressed beyond a molecular area of 2 nm^2, the floating film can be expanded and recompressed. The surface pressure versus area and the surface potential versus area isotherms were found to be insensitive to the presence of K^+ cations in the aqueous subphase with concentrations up to 0.1 M [114]. Above this concentration, changes in the isotherms became evident.

The floating monolayer may be transferred to solid substrates using the LB technique. An improvement in the film deposition is obtained by mixing the ionophore with a fatty acid or phospholipid. Infrared spectroscopy has been used to study the interactions of the LB films with K^+ ions. Using the attenuated total reflection (ATR) technique with samples of pure valinomycin up to 100 layers in thickness, dipped on a 0.05 M KCl subphase, no evidence for the valinomycin-K^+ complex was observed [115]. This suggests that no complexation occurs in a pure valinomycin film floating at the air/water interface. In contrast, complex formation with K^+ has been found for mixed valinomycin-arachidic acid [115] and mixed valinomycin-didodecyl phosphate monolayers [116]. Complexation of valinomycin with potassium cations, as a result of LB immersion in a salt solution, has also been studied by ATR-IR spectroscopy [115, 117, 118]. In the case of a 9 mol. % valinomycin-arachidic acid LB film of 14 layers thickness, significant complexation occurred after 1 min. immersion in saturated KCl solution [115]. Structural changes in the arachidic acid matrix after 1 min. immersion were also inferred from the IR spectra. It was suggested that these reflect the migration of valinomycin molecules towards the polar head groups upon complexation. The connection with K^+ transport is obvious and demonstrates the close resemblance of the mixed LB film system with the biological membrane. However an important distinction is that dissociation of the valinomycin-K^+ complex did not appear to occur in the LB films when immersed in pure water. Reversible complexation of K^+ by valinomycin on consecutive immersion of the LB film in KCl solution and in pure water has been found for valinomycin films in which the ionophore is mixed with phosphatidic acid [118]. However, both the valinomycin and phospholipid were found to partly disappear from the film on immersion. This has been attributed to the more flexible (liquid-like) nature of the phospholipid matrix. A simple fatty acid or fatty acid salt will hold the ionophore very rigidly and this does not allow the dissociation to take place.

Chromoionophore molecules are more sophisticated versions of simple crown ether rings [119]. They are constructed from two functionally different chemical groups: an ionophore, recognising specific ions, and a chromophore, transducing the chemical information produced by the ionophore-ion interaction into an optical signal. Such compounds may be useful for optical sensors since they can change colour on selective complexation with metal cations. Fig. 26 shows the molecular structure of an amphiphilic benzothiazolium styryl dye containing a 1,10-dithia-18-crown-6 ether group that has been synthesised for monolayer studies [120].

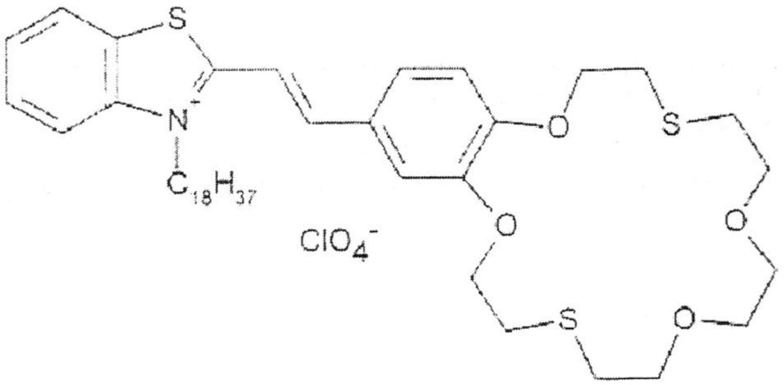

Fig. 26. Molecular structure of an amphiphilic chromoionophore [120].

The pure material forms a stable layer at the air/water interface and the area per molecule can be influenced by the addition of silver perchloroate to the subphase [121]. At low surface pressure, there is a monotonic expansion of the isotherm with increasing silver cation concentration. However, at 36 mN m^{-1}, the dependence is more complex: the isotherm first contracts with $AgClO_4$ concentration in the region 10^{-7} – 10^{-6} M and then expands at higher concentrations. This effect was attributed to the two-step complexation of the chromoionophore with Ag^+ in a floating monolayer. Langmuir-Blodgett layers of this compound can also be built-up on solid supports and the complexation with metal cations studied in the solid state. Fig. 27 shows how the absorption spectrum of one bilayer, deposited on either side of a glass slide, changes with the concentration of $AgClO_4$ in solution [120]. Each spectrum was recorded a few minutes after the concentrated salt solution had been added to the solution in the measurement cell and thoroughly mixed. An isosbestic point is evident at 449 nm.

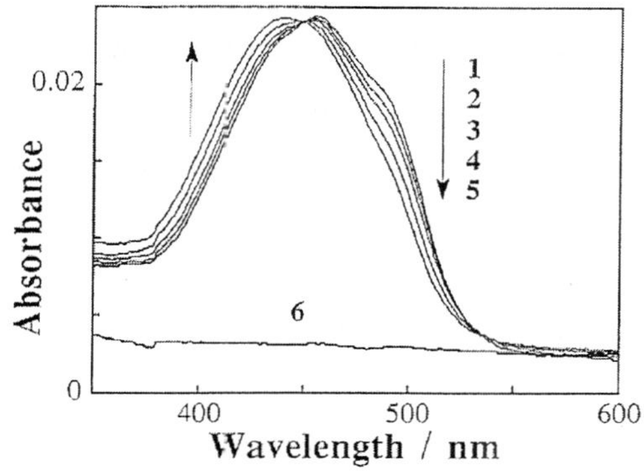

Fig. 27 Absorption spectrum of two bilayers of a chromoionophore deposited onto a hydrophobic glass slide and measured in acidic solution (1.8×10^{-4} M $HClO_4$) in various concentration $AgClO_4$ solutions: (1) 0; (2) 5.3×10^{-6} M; (3) 3.7×10^{-5} M; (4) 1.1×10^{-3} M; and (5) 8.7×10^{-3} M. The LB film was deposited from an acidic subphase (2×10^{-4} M $HClO_4$) and held in acidic water (1.8×10^{-2} M $HClO_4$) for 20 h until the aggregation of chromoionophores was completed. Curve (6) is the absorption spectrum of the solution after the experiment has been completed. After Lednev and Petty [117].

The response to metal cations was found to depend on whether the films were used as-deposited, or kept in acidic water for a few hours prior to the complexation study. The aggregation of the chromoionophore was evident from the absorption spectra in the latter case. Two stages of the complexation process can be seen for a partly aggregated film (as-deposited) and a single stage for a completely aggregated film in Fig. 28 [122]. Non-aggregated molecules are responsible for complexation at low Ag^+ concentrations (10^{-7} – 10^{-6} M) and the aggregated molecules at higher concentration (10^{-6} – 10^{-2} M).

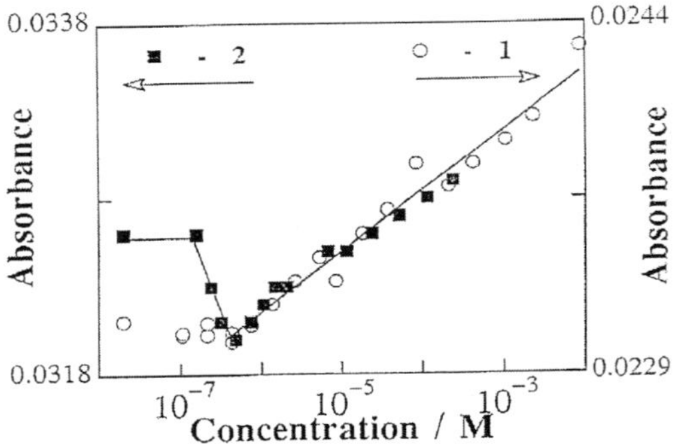

Fig. 28. Change in absorbance of LB films of a chromoionophore at 440 nm as a function of $AgClO_4$ concentration in acidic water (1.8×10^{-4} M $HClO_3$): (1) two bilayers deposited from an acidic subphase (2×10^{-4} M $HClO_4$) and held in acidic water (1.8×10^{-2} M $HClO_4$) for 20 h; (2) four bilayers on two hydrophobic glass sides deposited from an acidic subphase (2×10^{-4} M $HClO_4$). After Lednev and Petty [117].

Complexation with Ag^+ cations results in the break-up of the aggregates. This reaction has been shown to be fully reversible if the complexed film is replaced in acidic water for 20 h.

The complexation of LB films with other ions, such as Hg^{2+}, Pb^{2+} and Ba^{2+} in aqueous solutions, has also been studied [123]. The films were sensitive to quite small amounts of mercury but showed no sensitivity to lead, suggesting that the LB layers have high sensitivity and high specificity to 'soft' (polarisable) metal cations. The ion-selective behaviour of a chromoionophore containing the 18-crown-6 ether group has also been reported [124]. Langmuir-Blodgett layers of this compound exhibited a reversible complexation with K^+, Ag^+, Pb^{2+} and Ba^{2+} ions in solution. As expected, the sensitivity of the films was much less for Ag^+ cations but higher for hard metal cations (Pb^{2+}, Ba^{2+}) than found for the previous compound. A study of the variation in the position of the alkyl chain in amphilphilic chromoionophres has also been reported [125].

Electrochemistry provides a further transduction mechanism for the measurement of ions in solution [126]. Voltammetric and amperometric measurements are possible. Electrochemical transducers offer the advantage of simplicity and low cost; a basic working device can be constructed by immersing two electrodes in an analyte solution and connecting them to a potentiostat. Sensitivity can be high and quantitative data are readily obtained using simulation programmes. The recent incorporation of the electroactive tetrathiafulvalene (TTF) unit into metal-binding structures has demonstrated that they can function as metal-cation sensors in organic media [127]. The presence of the metal cation imposes an inductive effect on the polarisable TTF system, resulting in an anodic shift of the first oxidation potential as indicated by cyclic voltammetry experiments. The second oxidation potential is essentially unchanged, consistent with the expulsion of the metal cation after the first oxidation. Fig. 29 shows the structure of a self-assembled monolayer of a crown-TTF derivative onto a gold surface [128].

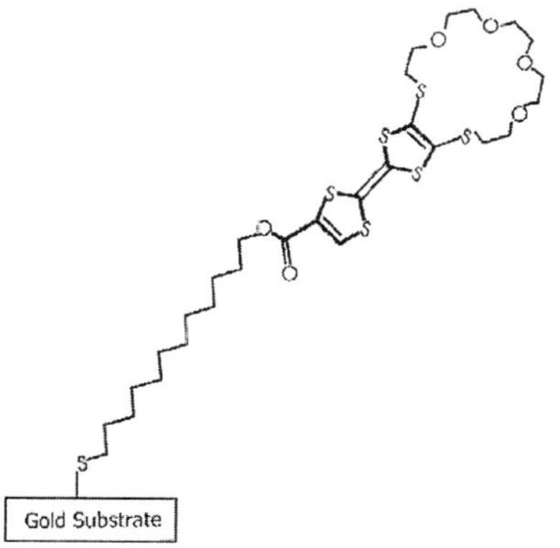

Fig. 29. Proposed orientation of a self-assembled monolayer of crown-TTF derivative. After Moore et al. [128].

The orientation of the molecule on the substrate was suggested by infrared spectroscopy. For electrochemical studies, it was found that adsorption onto platinum gave a slightly more reproducible electrochemical response than adsorption onto gold. Electrochemical recognition was studied in solution for the compound shown in Fig. 27 and for derivatives with different thiol chain lengths. In acetonitrile, there was a small, but reproducible, anodic response to Li^+ and K^+ (10-20 mV in solution), a significant response to Na^+ and Ba^{2+} (45-55 mV in solution) and a greater response to Ag^+. The response to Ag^+ was not dependent on the length of the thiol chain. Such results pave the way to the assembly of analogous self-assembled monolayers that will electrochemically sense heavy-metal cations in organic and aqueous media.

3.5. *Heat Sensors – Pyroelectric Devices*

It is widely recognised that pyroelectric materials have considerable advantages over narrow bandgap semiconductors, such as mercury cadmium telluride (CMT), as detectors of infrared radiation. In addition to their broader spectral sensitivity, pyroelectric detectors possess the advantage of efficient operation at ambient temperatures, obviating the need for expensive cooling systems that are required for their CMT counterparts [13].

In order for a material to be pyroelectric, it must possess a noncentrosymmetric crystal structure with a unique polar axis [129]. Furthermore, if the material is to form the basis of an efficient pyroelectric detector, it must be fabricated in thin-film form. Conventionally, two broad approaches have been taken to producing pyroelectric thin films: the first is to grow a single crystal of a material such as triglycine sulphate (TGS), which belongs to one of the polar space groups, and to thin the crystal by slicing, mechanical grinding or etching; the second, and more satisfactory method is to grow a thin polycrystalline, nonpolar film and to render it pyroelectric by applying a large electric field ('poling'). This is particularly applicable to ceramics, which can be deposited by radio frequency (RF) sputtering, and to polymers, which can be spin-coated into a suitable substrate.

There are several advantages that make LB films particularly attractive candidates for pyroelectric devices. The most important of these is that the sequential deposition of single monolayers enables the symmetry of the film to be precisely defined; in particular, layers of different materials can be built-up to produce a highly polar structure. Secondly, the polarisation of an LB film is 'frozen-in' during deposition, and it is therefore not necessary to subject the film to a poling process. The third advantage is that the LB technique uses amphiphilic organic materials which possess low permittivities ε and the figure of merit for voltage responsively p/ε (where p is the pyroelectric coefficient) is expected to be large. Finally, the LB method enables the preparation of much thinner films than are usually attainable by more conventional techniques.

Fig. 30 illustrates the principle of a superlattice comprising acid and amine molecules whose dipole moments are in opposite senses, but when deposited in Y-type LB film form are aligned in the same direction. Infrared studies have indicated that the deposition results in a proton transfer from the acid head group to that of the amine, giving rise to an overall polarisation component perpendicular to the multilayer plane [17]. Many LB materials deposited as alternate-layer films or as X- or Z-type layers exhibit such behaviour, including polymeric compounds and phospholipid materials [17].

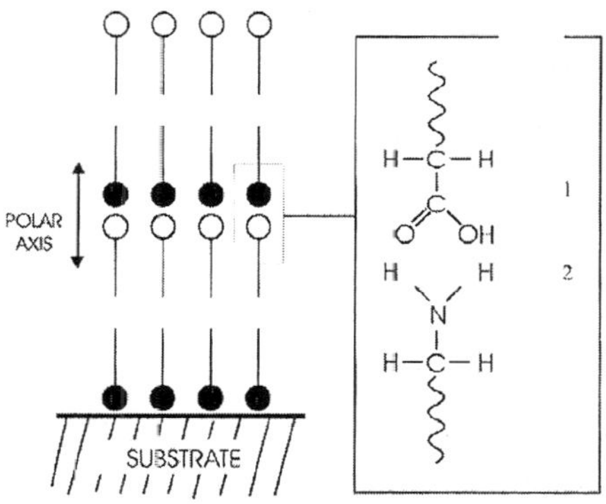

Fig. 30. Schematic diagram of an organic superlattice structure formed by the alternate-layer LB deposition of an acid and an amine. The dipole moments associated with the polar head groups are shown on the right. After Jones et al [13].

The pyroelectric coefficients of such multilayer acid/amine LB films can be about 10 $\mu C\ m^{-2}\ K^{-1}$ and depend on the thermal expansion coefficient of the substrate, indicating that there is a significant secondary contribution to the measured pyroelectric response [17, 130, 131]. Although these pyroelectric coefficients are still less (by about a factor of three) than those measured for polyvinylidene fluoride, a well-known pyroelectric polymer, the dielectric constants are also less, providing comparable figures of merit for infrared detection devices. Alternate-layer LB structures can also exhibit a significant second-order nonlinear optical response (e.g. second-harmonic generation) [17].

4. CONCLUSIONS

Organic compounds possess a range of interesting features that make them attractive candidates for electronic device applications. However, it is not anticipated that these materials will displace silicon in the foreseeable future as the dominant material for fast signal processing. It is much more likely that organic materials will find use in other areas of electronics, where silicon and other inorganic semiconductors cannot compete. Examples already exist, such as liquid crystal displays, pyroelectric plastics and certain chemical sensors. Organic light emitting structures are likely to make a major impact in the market place over the next ten years.

For commercialisation, the organic materials are usually required in the form of thin films. Simple processing methods, such as spin-coating and thermal evaporation are already available. A number of thin film technologies offer the means to manipulate arrays of organic molecules, i.e. molecular engineering. Techniques such as Langmuir-Blodgett film deposition, self-assembly and layer-by-layer electrostatic deposition allow intriguing molecular architectures to be built-up on solid surfaces. These methods allow elegant experiments to be undertaken in the research laboratory that can provide a valuable insight into the physical processes that underpin the device operation. This work will also pave the way for the development of molecular scale electronic (and ionic) devices, which emulate natural processes.

5. ACKNOWLEDGEMENTS

Many staff and students of, and visitors to, the Durham Centre for Molecular and Nanoscale Electronics have contributed to the work described in this paper. We would like to thank particularly: Napolean Agbor, Paul Barker, Martin Bryce, Riccardo Casalini, Da Fu Cui, Catia Granito, Lin Hua, Vaughan Howarth, Carole Jones, Mary Kilitziraki, Igor Lednev, Andy Monkman, Jürgen Nagel, Chris Pearson, Gareth Roberts, Ludovico Valli, Jason Wilde, Changsheng Wang, David Wood, Jack Yarwood and Gun Young-Jung. Particular thanks for sponsorship go the School of Engineering, the University of Durham, the Engineering and Physical Sciences Research Council, the European Commission, the Royal Society, the British Council, the German Academic Exchange Service (DAAD), British Gas (Mary Harris and Russ Pride) and Neotronics (Julian Iredale and John Saffell).

6. REFERENCES

1. M.C. Petty, M.R. Bryce and D. Bloor (eds.), *An Introduction to Molecular Electronics*, Edward Arnold, London (1995).
2. T.H. Richardson (ed.), *Functional Organic and Polymeric Materials*, Wiley, Chichester (2000).
3. J.R. Ferraro and J.M. Williams, *Introduction to Synthetic Electrical Conductors*, Academic Press, Orlando (1987).

4. S. Roth, *One-Dimensional Metals*, VCH, Weinheim (1995).

5. G. Horowitz, Adv. Mater., 2 (1990) 286.

6. H.E. Katz, J. Mater. Chem., 7 (1997) 369.

7. J.H. Burroughes, D.D.C. Bradley, A.R. Brown, R.N. Marks, K. Mackay, R.H. Friend, P.L. Burns and A.B. Holmes, Nature, 347 (1990) 359.

8. J. Kido, Phys. World, 12 (1999) 414.

9. Semiconductor Industry Association Roadmap, see website at: http://www.sematech.org/public/roadmap/.

10. J. Janata, *Principles of Chemical Sensors*, Plenum Press, New York (1989).

11. J.W. Gardner, *Microsensors*, Wiley, Chichester (1994).

12. P.T. Moseley and A.J. Crocker, *Sensor Materials*, IOP Publishing, Bristol (1996).

13. C.A. Jones, M.C. Petty and G.G. Roberts, IEEE Trans. Ultrasonics, Ferroelec. and Freq. Control, 35 (1988) 736.

14. M.C. Petty, *Langmuir-Blodgett Films*, Cambridge University Press, Cambridge (1996).

15. A. Ulman, *Ultrathin Organic Films*, Academic Press, San Diego (1991).

16. M.C. Petty, in *Functional Organic and Polymeric Materials*, ed. T.H. Richardson, Wiley, Chichester (2000) 7.

17. M.C. Petty, in *Langmuir-Blodgett Films* (G.G. Roberts, ed.) Plenum Press, New York (1990)133.

18. R. Casalini, L.M. Goldenberg, C. Pearson, B.K. Tanner and M.C. Petty, J. Phys. D: Appl. Phys., 31 (1998) 1504.

19. M.R. Bryce and M.C. Petty, Nature, 374 (1995) 771.

20. C. Pearson, A.S. Dhindsa, L.M. Goldenberg, R.A. Singh, R. Deing, A.J. Moore, M.R. Bryce and M.C. Petty, J. Mater. Chem., 5 (1995) 1601.

21. S.K. Gupta, D.M. Taylor, P. Dynarowicz, E. Barlow, C.E.A. Wainwright and A.E. Underhill, Langmuir, 8 (1992) 3057.

22. R.B. Rosner and M.F. Rubner, Chem. Mater., 6 (1994) 581.

23. N.F. Mott and E.A. Davis, *Electronic Processes in Non-Crytalline Materials*, Clarendon Press, Oxford (1979).

24. R. Sing, R.P. Tandon, G.S. Sing and S. Chandra, Phil. Mag. B, 66 (1992) 285.

25. P. Sheng, E.K. Sichel and J.I. Gittleman, Phys. Rev. Lett., 40 (1978) 1197.

26. A.R. Blythe, *Electrical Properties of Polymers*, Cambridge University Press, Cambridge (1979).

27. J.R. MacDonald, *Impedance Spectroscopy*, Wiley, New York (1987).

28. M.C. Petty, C. Pearson, A.P. Monkman, R. Casalini, S.Capaccioli and J. Nagel, Coll. Surf. A, 171 (2000) 159.

29. A.K. Jonscher, *Universal Relaxation Law*, Chelsea Dielectrics Press, London (1996).

30. P. May, Phys. World, 8 (1995) 52.

31. E.H. Rhoderick, *Metal-Semiconductor Contacts*, Clarendon Press, Oxford (1978).

32. Y.L. Hua, M.C. Petty, G.G. Roberts, M.M. Ahmad, M. Hanack and M. Rein, Thin Solid Films, 149 (1987) 161.

33. A. Aviram and M.A. Ratner, Chem. Phys. Lett., 29 (1974) 277.

34. A.S. Martin, J.R. Sambles and G.J .Ashwell, Phys. Rev. Lett., 70 (1993) 218.

35. G.J. Ashwell and D.S. Gandolfo, J. Mater. Chem., 11 (2001) 246.

36. C.D. Dimitrakopoulos, B.K. Furman, T. Graham, S. Hegde and S. Purushothaman, Synth. Met., 92 (1998) 47.

37. J.H. Schön, S. Berg, Ch. Kloc and B. Batlogg, Science, 287 (2000) 1022.

38. C.D. Dimitrakopoulos, S. Purushothaman, J. Kymissis, A. Callegari and J.M. Shaw, Science, 283 (1999) 822.

39. J.H. Schön, Ch. Kloc and B. Batlogg, Organic Electronics, 1 (2000) 57.

40. A.R. Brown, A. Pomp, C.M. Hart and D.M. de Leeuw, Science, 270 (1995) 972.

41. Z. Bao, A.J. Lovinger and A. Dodabalapur, Adv. Mat., 9 (1997) 42.

42. A.J. Hudson and M.S. Weaver, in *Functional Organic and Polymeric Materials*, ed. T.H. Richardson, Wiley, Chichester (2000) 365.

43. M.A. Baldo, S. Lamansky, P.E. Burrows, M.E. Thompson and S.R. Forrest, Appl. Phys. Lett., 75 (1999) 4.

44. M.-S. Jang, S.-Y. Song, H.-K. Shim, T. Zyung, S.-D. Jung and L.-M. Do, Synth. Met., 91 (1997) 317.

45. K.H. Choi, D.H. Hwang, H.M. Lee, L.M. Do and T.H. Zyung, Synth. Met., 96 (1998) 123.

46. S. Karg, M. Meier and W. Reiss, J. Appl. Phys., 82 (1997) 1951.

47. X.Z. Jiang, Y.Q. Liu, S.G. Liu, W.F. Qiu, X.Q. Song and D.B. Zhu, Synth .Met., 91 (1997) 253.

48. M. Meier, M.Cölle, S. Karg, E. Buchwald, J. Gmeiner, W. Riess and M. Schwoerer, Mol. Cryst. Liq. Cryst., 283 (1996) 197.

49. I.D. Parker and H.H. Kim, Appl. Phys. Lett., 64 (1994) 1774.

50. L.S. Hung, C.W. Tang and M.G. Mason, Appl. Phys. Lett., 70 (1997) 152.

51. J.S. Kim, M. Granströn, R.H. Friend, N. Johansson, W.R. Salaneck, R. Daik, W.J. Feast and F. Cacialli, J. Appl. Phys., 84 (1998) 6859.

52. S. Karg, J.C. Scott, J.R. Salem and M. Angelopoulos, Synth.Met., 80 (1996) 111.

53. T.M. Brown, J.S. Kim, R.H. Friend, F. Cacialli, R Daik and W J. Feast, Appl. Phys. Lett., 75 (1999) 1679.

54. M.A. Baldo, D.F. O'Brien, Y. You, A. Shoustikov, S. Sibley, M.E. Thompson and S.R. Forrest, Nature, 395 (1998) 151.

55. G.G. Roberts, T.M. McGinnity, W.A. Barlow and P.S. Vincett, Thin Solid Films, 68 (1980) 223.

56. I.D. Parker, J.Appl.Phys., 75 (1994) 1656.

57. M.I. Sluch, C. Pearson, M.C. Petty, M. Halim and I.D.W. Samuel, Synth. Met., 94 (1998) 285.

58. G.Y. Jung, C. Pearson, M. Kilitziraki, L.E. Horsburgh, A.P. Monkman, I.D.W. Samuel and M.C. Petty, J. Mater. Chem., 1 (2000) 163.

59. M. Era, S. Hayashi, T. Tsutsui and S. Saito, J. Chem. Soc., Chem. Commun., (1985) 557.

60. R. Österbacka, G.Juška, K. Arlauskas, A.J. Pal, K.-M. Källman and H. Stubb, J. Appl. Phys., 84 (1998) 3359.

61. A. Bolognesi, G. Bajo, J. Paloheimo, T. Östergård and H. Stubb, Adv. Mater., 9 (1997) 121.

62. L.M. Goldenberg, M. Leclerc, A. Donat-Bouillud, C. Pearson and M.C. Petty, Thin Solid Films, 327-329 (1998) 715.

63. Y. Hua, J. Peng, D. Cui, L. Li, Z. Xu, X. Xu, Z. Yao and X. Chen, Thin Solid Films, 210-211 (1992) 219.

64. V. Cimrov, M. Remmers, D. Neher and G. Wegner, Adv. Mater., 8 (1996) 146.

65. S.E. Burns, N. Pfeffer, J. Grüner, D. Neher and R.H. Friend, Synth. Met., 84 (1997) 887.

66. G.Y. Jung, C. Pearson, L.E. Horsburgh, I.D.W. Samuel, A.P. Monkman and M.C. Petty, J. Phys. D: Appl. Phys., 33 (2000) 1029.

67. Y.-E. Kim, H. Park and J.–J. Kim, Appl. Phys. Lett., 69 (1996) 599.

68. M.C. Petty, J. Batey and G.G. Roberts, IEE Proc., 132 (1985) 133.

69. G.Y. Jung, C. Pearson and M.C. Petty, in *Novel Methods to Study Interfacial Layers*, eds. D. Möbius and R. Miller, Studies in Interface Science, Vol. 11, Elsevier, Amsterdam etc., 2001, pp. 175

70. E.M. Han, L.M. Do, N. Yamamoto and M. Fujihira, Synth. Metals, 273 (1996) 202.

71. L.M. Do, M. Oyamada, A. Koike, E.M. Han, N. Yamamoto and M. Fujihira, Thin Solid Films, 273 (1996) 209.

72. J.R. Sheats, H. Antoniadis, M. Hueschen, W. Leonard, J. Miller, R. Moon, D. Roitman and A. Stocking, Science, 273 (1996) 884.

73. J.C. Scott, J.H. Kaufman, P.J. Brock, R. DiPietro, J. Salem and J.A. Goitia, J. Appl. Phys., 79 (1996) 2745.

74. J. McElvain, H. Antoniadis, M.R. Hueschen, J.N. Miller, D.M. Roitman, J.R. Sheats and R.L. Moon, J. Appl. Phys., 80 (1996) 6002.

75. P.E. Burrows, V. Bulovic, S.R. Forrest, L.S. Sapochak, D.M. McCarty and M.E. Thompson, Appl. Phys. Lett., 65 (1994) 2922.

76. S. Tokito, H. Tanaka, K. Noda, A. Okada and Y. Taga, Appl. Phys. Lett., 70 (1997) 1929.

77. S.A. Carter, M. Angelopoulos, S. Karg, P.J. Brock and J.C. Scott, Appl. Phys. Lett., 70 (1997) 2067.

78. J. Shi and C.W. Tang, Appl. Phys. Lett., 70 (1997) 1665.

79. G.Y. Jung, A. Yates, I.D.W. Samuel and M.C. Petty, Mat. Sci. Eng. C., 14 (2001) 1-10.

80. H.S. Kim, H.S. Noh and T. Tsutsui, Korea Polym. J., 7 (1999) 18.

81. S. Dailey, A.P. Monkman and I.D. W.Samuel, Synth. Met., 102 (1999) 945.

82. C. Wang, G.Y. Jung, Y. Hua, C. Pearson, M.R. Bryce, M.C. Petty, A.S. Batsanov, A.E. Goeta and J.A.K. Howard, Chem. Mater., 13 (2001) 1167.

83. P.T. Walsh, Measurement and Control, 29 (1996) 5.

84. A.S. Snow and W.R. Barger, in *Phthalocyanines: Properties and Applications*, eds. C.C. Leznoff and A.B.P. Lever, VCH Publishers, New York (1989) 342.

85. M. Haug, K.D. Schierbaum, G. Gaulitz and W.Göpel, Sens. Act. B, 11 (1993) 383.

86. R. Casalini, M. Kilitziraki, D. Wood and M.C. Petty, Sens. Act. B, 56 (1999) 37.

87. H. Wohltjen, W.R. Barger, A.W. Snow and N.L. Jarvis, IEEE Trans. Elect. Dev., 32 (1985) 1170.

88. P.S. Barker, C.Di Bartolomeo, A.P. Monkman, M.C. Petty and R. Pride, Sensors and Actuators B, 25 (1995) 451.

89. P.S. Barker, A.P. Monkman, M.C. Petty and R. Pride, IEE Proc. Circuits Devices Syst., 144 (1997) 111.

90. P.S. Barker, M.C. Petty, A.P. Monkman, J. McMurdo, M.J. Cook.and R. Pride, Thin Solid Films, 284-285 (1996) 94.

91. J.W. Gardner and P.N. Bartlett, Sensors and Actuators B, 18 (1994) 211.

92. R. Casalini, J.N. Wilde, J. Nagel, U. Oertel, and M.C. Petty, Sensors and Actuators B, 57 (1999) 28.

93. R. Casalini, J. Nagel, U. Oertel and M.C. Petty, J. Phys. D: Appl. Phys., 31 (1998) 3146.

94. H. Raether, Phys. Thin Films, 9 (1977) 145.

95. J.N. Wilde, M.C. Petty, J. Saffell, A. Tempore, and L. Valli, Measurement and Control, 30 (1997) 269.

96. D.G. Zhu, M.C. Petty and M. Harris, Sensors and Actuators B, 2 (1990) 265.

97. A.W. Czarnik, *Fluorescent Chemosensors for Ion and Molecular Recognition*, American Chemical Society Ser., American Chemical Society, Washington DC, (1993) Vol.538.

98. P.D. Beer, in *Transition Metals in Supramolecular Chemistry*, eds. L. Fabbrizzi and A. Poggi, Kluwer Academic Publishers, Dordrecht (1994) 33.

99. R.K. Iler, J. Colloid Interface Sci., 21 (1966) 569.

100. G. Decher, J.D. Hong and J. Schmitt, Thin Solid Films 210/211 (1992) 831.

101. G. Decher, Y. Lvov and J. Schmitt, Thin Solid Films 244 (1994) 772.

102. A.C. Fou, O. Onitsuka, M. Ferreira, M.F. Rubner and B.R. Hsieh, J. Appl. Phys., 79 (1996) 7501.

103. F. Caruso, D.N. Furlong, K. Ariga, I. Ichinose and T .Kunitake, Langmuir, 14 (1998) 4559.

104. K. Ariga, Y. Lvov and T. Kunitake, J. Am. Chem. Soc., 119 (1997) 2224.

105. Y. Lvov, K. Ariga, M. Onda, I. Ichinose and T. Kunitake, Langmuir, 23 (1997) 6195.

106. C. Pearson, J. Nagel and M.C. Petty, J. Phys. D: Appl. Phys., 34 (2001) 285.

107. J.M. Lehn, *Supramolecular Chemistry: Concepts and Perspectives*, VCH Publishers, Weinheim (1995).

108. G.W. Gokel and A. Nakano, in *Crown Compounds: Towards Future Applications*, ed. S.R. Cooper, VCH Publishers, New York (1992).

109. R.M. Izatt and J.J. Christensen (eds.) *Progress in Macrocyclic Chemistry*, Wiley, New York (1979) Vol.1; (1981) Vol.2.

110. M.N. Hughes, *The Inorganic Chemistry of Biological Processes*, 2nd ed., Wiley, Chichester (1981).

111. G.W. Gokel and J.E. Trafton, in *Cation Binding by Macrocycles: Complexation of Cationic Species by Crown Ethers*, eds. Y. Inoue and G.W. Gokel, Marcel Dekker, New York (1990).

112. J.B. Peng, B.M. Abraham, P. Dutta, J.B. Ketterson and H.F. Gibbard, Langmuir, 3 (1987) 104.

113. V.A. Howarth, M.C. Petty, G.H. Davies and J. Yarwood, Thin Solid Films, 160 (1988) 483.

114. S.Y. Zaitsev, V.P. Zubov and D. Möbius, Biochim. Biophys. Acta., 1148 (1993) 191.

115. V.A. Howarth, M.C. Petty, G.H. Davies and J. Yarwood, Langmuir, 5 (1989) 330.

116. M. Sugawara, H. Sazawa and Y. Umezawa, Langmuir, 8 (1992) 609.

117. V. Vodyanoy, S. Pathirana and W.C. Neely, Langmuir, 10 (1994) 1354.

118. V.A. Howarth, D.F. Cui, M.C. Petty, H. Ancelin and J. Yarwood, Thin Solid Films, 180 (1989) 111.

119. I.K. Lednev and M.C. Petty, Adv. Mater., 8 (1996) 615.

120. I.K. Lednev and M.C. Petty, Langmuir, 10 (1994) 4185.

121. I.K. Lednev and M.C. Petty, J. Phys. Chem., 99 (1995) 4176.

122. I.K. Lednev and M.C. Petty, J. Phys. Chem., 98 (1994) 9601.

123. I.K. Lednev and M.C. Petty, Adv. Mater. Optics & Elec., 5 (1995) 137.

124. I.K. Lednev and M.C. Petty, Thin Solid Films, 284-285 (1996) 683.

125. S.Y. Zaitsev, T.I. Sergeeva, E.A. Baryshnikova, W. Zeiß, D. Möbius, O.V. Yescheulova, S.P. Gromov, O.A. Fedorova and M.V. Alfimov, Mat. Sci. Eng. C, 8-9 (1999) 469.

126. L.M. Goldenberg, J. Mater. Chem., 9 (1999) 1957.

127. T.K. Hansen, T. Jørgensen, P.C. Stein and J. Becher, J. Org. Chem., 57 (1992) 6403.

128. A.J. Moore, L.M. Goldenberg, M.R. Bryce, M.C. Petty, C. Marenco, J. Yarwood, M.J. Joyce and S.N. Port, Adv. Mater., 10 (1999) 395.

129. J.C. Burfoot and G.W. Taylor, *Polar Dielectrics and their Applications*, Macmillan Press, London (1979).

130. D. Lacey, T.H. Richardson, F. Davis and R. Caplan, Mat. Sci. Eng. C, 8-9 (1999) 377.

131. T.H. Richardson, in *Functional Organic and Polymeric Materials*, ed. T.H. Richardson, Wiley, Chichester (2000) 181.

SUBJECT INDEX